Chemical and Biological Sensors for Environmental Monitoring

ACS SYMPOSIUM SERIES **762**

Chemical and Biological Sensors for Environmental Monitoring

Ashok Mulchandani, EDITOR
University of California, Riverside

Omowunmi A. Sadik, EDITOR
State University of New York, Binghamton

American Chemical Society, Washington, DC

Library of Congress Cataloging-in-Publication Data

Chemical and biological sensors for environmental monitoring / Ashok Mulchandani, Omowunmi A. Sadik, editors.

p.   cm.—(ACS symposium series, ISSN 0097-6156 ; 762)

"Developed from a symposium sponsored by the Division of Environmental Chemistry at the 217th National Meeting of the American Chemical Society, Anaheim, California, March 21–25, 1999."

Includes bibliographical references and index.

ISBN 0–8412–3687–9

1. Chemical detectors—Congresses. 2. Environmental monitoring—Equipment and supplies—Congresses. I. Mulchandani, Ashok, 1956– II. Sadik, Omowunmi A., 1964– III. American Chemical Society. Division of Environmental Chemistry. IV. American Chemical Society. Meeting (217th : 1999 : Anaheim, Calif.) V. Series.

TP159.C46 C423   2000
628.5′028′7—dc21   00–25618

The paper used in this publication meets the minimum requirements of American National Standard for Information Sciences—Permanence of Paper for Printed Library Materials, ANSI Z39.48–1984.

PRINTED IN THE UNITED STATES OF AMERICA

Foreword

THE ACS SYMPOSIUM SERIES was first published in 1974 to provide a mechanism for publishing symposia quickly in book form. The purpose of the series is to publish timely, comprehensive books developed from ACS sponsored symposia based on current scientific research. Occasionally, books are developed from symposia sponsored by other organizations when the topic is of keen interest to the chemistry audience.

Before agreeing to publish a book, the proposed table of contents is reviewed for appropriate and comprehensive coverage and for interest to the audience. Some papers may be excluded in order to better focus the book; others may be added to provide comprehensiveness. When appropriate, overview or introductory chapters are added. Drafts of chapters are peer-reviewed prior to final acceptance or rejection, and manuscripts are prepared in camera-ready format.

As a rule, only original research papers and original review papers are included in the volumes. Verbatim reproductions of previously published papers are not accepted.

ACS BOOKS DEPARTMENT

Contents

vii

MICROBIAL-BASED SENSORS

AFFINITY-BASED SENSORS

NUCLEIC ACID-BASED SENSORS

Preface

Environmental monitoring for the detection of pollutants is becoming increasingly important to regulatory agencies, the regulated community, and the general public. This is especially true for compounds that pose a potential human health risk or risk to the environment. The high cost and slow turnaround times typically associated with the measurements of regulated pollutants clearly indicate a need for analytical technologies that are fast, portable, and cost effective. To meet this need, a variety of analytical methods have been introduced. Although a very small number of these methods are commercially available, many are under research and development.

This volume, a result of a one-day symposium held during the 217th American Chemical Society (ACS) National Meeting, presents a cross-section of recent advances in the research and development of chemical and biochemical sensors for environmental monitoring. These chapters demonstrate how many of the key challenges for environmental monitoring are being addressed. This text is critical to keep abreast of the many new environmental sensing technologies that are being researched at the laboratories of prominent scientists in the United States and around the globe.

We gratefully acknowledge the enthusiastic participation and cooperation of all the authors, the reviewers for providing excellent comments and suggestions in time, and of Anne Wilson and Kelly Dennis of the ACS Books Department for guidance and support in making this volume a reality. We thank Dr. Ruth Hathaway and the ACS Division of Environmental Chemistry, Inc. for coordinating and supporting the symposium that resulted in bringing together world-class researchers in the field of environmental sensors. Last but not the least, we warmly acknowledge the gracious support of our families.

ASHOK MULCHANDANI
Department of Chemical and Environmental Engineering
University of California
Riverside, CA 92521

OMOWUNMI A. SADIK
Department of Chemistry
State University of New York at Binghamton
Binghamton, NY 13902–6016

Chapter 1

Chemical and Biological Sensors: Meeting the Challenges of Environmental Monitoring

Omowunmi A. Sadik[1] and Ashok Mulchandani[2]

[1]Department of Chemistry, State University of New York at Binghamton, Binghamton, NY 13902–6016
[2]Department of Chemical Engineering, University of California, Riverside, CA 92521

The focus of this chapter is to address the challenges of developing practical chemical and biological sensors suitable for measuring toxic pollutants, and to highlight how recent works reported in this volume and other literature are helping to address these challenges. It provides an overview of how chemical and biological sensors are meeting the challenges of environmental monitoring including enhanced specificity, fast response times, and the ability to determine multiple analytes with little or no need for sample preparation steps in complex samples.

Chemical sensor and biosensor technologies have emerged as dynamic approaches for identifying and quantitating specific analytes of environmental and human levels of concerns. Due to a growing need for rapid, continuous and multi-component analysis, as well as the necessity for shorter sample preparation methods, new sensing techniques are emerging that make environmental monitoring and surveillance studies much simpler with decreasing costs per sample throughputs. Several sensors utilizing large number of transduction principles have been used for environmental monitoring. These include chemically- and biologically-modified metal or semi-conductor electrodes, ion-selective or gas-sensitive electrodes, thermistors, piezo-electric crystals, field-effect transistors and opto-electronic devices (e.g. fiber-optics and surface plasmon resonance) (1-4). Basically, these transducers convert the input signals into processable electrical signals that can be measured. In general, a chemical sensor consists of a chemically-selective, sensing layer that can respond to certain properties of the substance being analyzed, and is usually in contact with or integrated within a suitable transducer. However, if the sensing layer of the sensor incorporates a biological molecule, such a sensor is generally refereed to as a biosensor. Typically, a biological molecule incorporated in the sensor can be an enzyme, antibody, DNA, receptor proteins or microorganism. The transducer is required to detect the interaction of the bio-specific, sensing element with the analyte. The electrical or optical signal generated by the transducer is then amplified and processed. Hence a signal processor is employed to convert this signal into a processable form.

The monitoring of residue or contamination in soil, water and air can be classified into two main categories. These are: (i) screening or diagnostic techniques in which only a yes-or-no (qualitative) answer is required, and (ii) semi-quantitative or quantitative techniques in which the detection of unwanted chemicals, and the testing of whether or not the residues of the contaminants are within permissible levels are required. It is possible for the former methods to generate false positive or negative results if the sensitivities are insufficient for the detection of the threshold levels. Chemical and biosensors can be employed in both screening and quantitative applications depending on the specific monitoring need.

There is an increasing awareness at research and development laboratories concerning the use of chemical/biological sensors for environmental analysis. These sensors can provide rapid information on the presence of electrolytes, organic pollutants, pesticides, heavy metals and pH levels. Major air pollutants such as sulfur dioxide, nitrogen oxide, carbon monoxide and hydrocarbons (mainly from automobiles) can also be monitored using these sensing techniques. However, in spite of the various types of chemical/biological sensors reported, only a few (primarily pH sensors) have been used successfully for environmental monitoring. This may be attributed to the fact that this area of application usually demands enhanced specificity, fast response times, highest available sensitivity and the ability to determine multiple analytes in complex samples with little or no sample preparation steps. Although a number of pollutant sensing techniques have been reported, very few address those specific requirements.

There are excellent reviews in literature covering the applications of chemical sensors and biosensors for environmental monitoring (2-8). However, the focus of this chapter is to address the challenges of developing a practical chemical and biological sensors suitable for measuring toxic pollutants, and to highlight how recent works reported in this volume and literature are helping to address these challenges.

Challenges in using chemical and biological sensors for environmental monitoring

In spite of the numerous literatures available on environmental sensors, some problems are yet to be solved. These problems concern some of the technical constraints relating to sensor sensitivity, selectivity, multianalyte detection, miniaturization/portability for on-site/field application, time of analysis, fabrication, robustness, instability of biological reagents, and drift in sensor signals. Some solutions to the above problems have been addressed in chapters in this volume and recent literature.

The use of integrated optic chemical sensors for environmental monitoring and remediation applications offers many advantages; the most prominent is that they can easily accommodate multiple sensing elements and can be miniaturized (9-12). Sol-gel technology has been used to fabricate fiber optic and integrated optic chemical sensors for environmental monitoring (13). A complete integrated optic sensor system was demonstrated for the simultaneous identification and quantitation of contaminated trace metal ions in water (13). Signal generation was accomplished by a wavelength-division multiplexed optoelectronic unit, which can be remotely located and/or connected via optical fibers. The use of charge-coupled array devices for simultaneous determination of multiple pesticides using antibody against pesticides has been presented in Chapter 15.

Recently, instrumental odor analysis using a combination of headspace sampling, non-specific chemical sensor arrays and pattern recognition techniques were reported (14-18). These systems are commonly refereed to as "electronic nose" (EN) or electronic olfactometers. EN generally employs different transduction principles. Metal oxides, quartz crystal arrays, surface acoustic wave devices, electrodes, or a combination of these sensors are used to mimic human sense of smell. The transducer is in close contact with arrays of chemically selective polymer layers formed through heterogeneous and quasi-selective, thin films particularly conducting electroactive polymers (CEPs). As analogs to natural olfaction, these films act as sensing receptor units. Commercial EN is gaining wide acceptance for routine applications such as medical diagnosis, smart atmospheric monitoring, environmental analysis, food quality control, packaging materials, cosmetics and perfumery industries (14,18-21). The integration of EN technologies with existing instrumental techniques such as gas chromatography has also been reported for environmental monitoring. Chapter 4 describes a brief overview of electronic nose technologies for chemical sensing and provides the results of using EN for environmentally related analytes.

Of the 19 known chlorinated phenols, the most important congeners include the 2,4-Dinitrophenol (2,4-D), 2,4,5-trichlorophenol, (2,4,5-TCP) and pentachlorophenol. While these compounds can be determined using mass spectrometry and gas chromatographic techniques, the structural similarities of substituted phenols and their derivative posses a significant challenge and thus require the development of rapid, multianalyte techniques. An approach that uses pattern recognition technique to identify and predict environmental compounds was recently demonstrated for a range of phenols and halogenated derivatives (22). Chapter 15 in this volume, describes multiarray sensors for phenols and polyaromatic hydrocarbons. In this method, a 32-array conducting polymer sensor was used and the sensor arrays were found to recognize the structurally similar halogenated derivatives based on the nature and position of their functional groups. Each sensor responded in varying degrees to chlorinated organic molecules with standard deviation of less than 0.05.

Portability, time of analysis and automation are important issues for environmental sensors. Collection and transportation of samples to the laboratory adds to the cost of analysis. Solutions to these issues have been addressed in this volume in Chapters 9, 17 and 18, that report developments of a screen printed disposable electrode for organophosphate pesticides, a compact self-standing immunosensor for bacteria, and spot assay for glucose, respectively. In a recent report development of an enzyme electrode for the remote monitoring, with a very fast response time, of organophosphate pesticides was reported (23). An automated prototype immunosensor called FOBIA (fiber optic biospecific interaction analysis) suitable for the detection of atrazine has been developed by CIBA-Geigy. A commercial, surface plasmon resonance apparatus (i.e. BIAcore from Pharmacia) has been used for the detection of atrazine (24,25). Fiber optic sensors for hydrocarbons and $ClCH:CCl_2$ in water was recently reported (26). These sensor platforms will make on-field, with a possibility of remote, monitoring, away from the laboratory, feasible.

Since most environmental pollutants usually have small molecules, sensing techniques cannot adequately exploit conventional interactions between enzymes or antibodies on natural substrates. Therefore, a general approach to biosensors for small environmental molecules must entail the synthesis of an analogue molecule containing a reactive moiety that can be further chemically manipulated. Other challenges relate to the complex interactions of biomolecules and/or cells with the transducers. Development of environmental sensors, therefore, requires the availability of biological reagents such as antibodies, antigens, and receptor proteins. Moreover, the biocomponents have limited stability and are not likely to be exactly reproducible during measurements in extreme or harsh chemical environments; thus posing problems about quality control and compliance monitoring applications.

In addressing some of these technical challenges, there is a growing interest in rationally assembling dynamic macromolecules capable of providing analogous properties to biological units, with unique abilities to recognize specific analytes and transmit this information into a measurable form. Consequently, many analogs of biological molecules have been synthesized, including cryptands, calixarenes, crown ethers, other inorganic analogs, and genetically engineered proteins. Some of these challenges have been addressed from recent works reported in this volume. These works are used to illustrate how the custom designs of recognition elements are being used to obtain analytically useful signals.

In Chapter 2 for example, the custom design of molecular recognition elements to achieve sensor selectivity suitable for environmental sensing of anions was described. Chapter 3 describes the synthesis of polystyrene-based dithiozone analogues for heavy metal detection. Samples of polystyrene-supported diarylthiocarbazones were shown to undergo distinctive color changes in 0.005M aqueous lead (II) and mercury (II) ions. Chapter 15 describes the synthesis of a new class of protein conjugates to create non-antibody based sensing. This work utilizes metal binding principle to modify 2-puriylazo chelate protein. Practical application was demonstrated using gallium detection and is also applicable to other metals such as cadmium and lead. This illustrates the importance and practicality of using synthetic chemistries to recognize key elements of sensor structures for their respective target analytes. The rational approach to the design of molecular recognition elements may reduce current reliance on biological units to achieve selectivity in environmental applications.

Research and development of chalcogenide glass chemical sensors are gaining more attention. Analytical applications of this type of sensors for environmental monitoring and process control have been reported for the detection of microgram levels of copper (II), iron (III), chromium (VI), lead, cadmium and mercury in natural and waste waters (27,28). In addition, the applications of chalcogenide glass sensors for laboratory analysis, industrial control and environmental monitoring have been reported for heavy metals ions in solution (28). Online chemical sensors in which selective chemical reagents were immobilized have also been demonstrated for low ppb detection of aromatic hydrocarbons, hydrazines and ethylene (29). These sensors are suitable for groundwater monitoring under pH range of 4, with additional optodes for lower pH range. Also reported is the improvement of water quality surveillance using integrated physicochemical and biological sensor control responses.

Other exciting developments include the use of DNA-based biosensors, molecular beacons, and genetically-engineered organisms and proteins as tools for environmental analysis. Chapter 19 provides a critical review of the concept of genetic testing as well as the state-of-the-art analysis of nucleic acid biosensor and chip scale oligonucleotide array technologies. In Chapter 21, the use of DNA as tools for environmental monitoring was described. Possible applications in environmental and healthcare applications are also described. Chapter 20 describes the use of molecular beacons while Chapters 6 and 7 highlight the use of genetically engineered and conformation-induced protein changes for environmentally-related phosphates and heavy metals respectively. Chapters 12 and 14 describe a two-stage, continuous, toxicity monitoring system using recombinant bioluminescent bacteria and bioluminescence-based integrated circuit devices. Chapter 17 reports on environmental toxicity monitoring using a panel of cell-based biosensors containing selected, stress-responsive *Escherichia coli* promoters fused to bioluminescent reporter. Genetic testing methodologies may provide useful strategies for generating recognition diversity for arrays of large volume testing.

Conclusions and Future Trend

The contributions in this volume demonstrate the challenges facing the researchers and developers in their quest for a successful commercial sensor suitable for environmental monitoring and how these challenges are being addressed. However, it provides no definitive answers to the inadequate understanding of the underlying mechanism(s) of sensor-analyte interactions. The greater depth of knowledge of the fundamental interfacial processes at sensor-analyte surfaces, which depends on the morphology of selective sensing layer, the bioactivity of the immobilized molecules, will be required to accurately predict sensor parameters such as sensitivity, selectivity and limit of detection. Understanding the mechanism of sensor-analyte interactions requires the following (i) defining the quantitative structure-activity relationships, (ii) investigating the influence of physical properties e.g. diffusion, hydrophobicity/hydrophilicity, molecular size and shape, and (iii) predicting the effects of (i) and (ii) on selectivity and sensitivity (30). Hence areas of future research in sensors should include greater understanding of interfacial sensor-analyte mechanisms. The future of chemical and environmental sensors looks promising. Current work and continued awareness of the existing problems will set the stage for commercial developments.

References

1. Lowe C. R., *Current Opinion Chem. Biol.* **1999**, 3(1) 106-111.
2. Rogers K. R., Mascini M., *Field Anal. Chem. Tech.* **1998**, 2, 317-331

6

3. Barcelo D., *Anal. Chim Acta* **1999**, 387(3) 225-225.
4. Janata J., Josowicz, M., Vanysek, P., DeVaney, *Anal. Chem.*, **1998**, 70(12), 179R-208R.
5. Janata J, Crit. Rev. *Anal. Chem.*, **1998**, 28(2), 27-34
6. Sadik O. A., Van Emon J. M., *ChemTech*, **1997**, 27 (6) 38-46
7. Hart, J. P., Wring S. A., TrAC, *Trends Anal. Chem*, **1997**, 16(2), 89-103
8. Wang J., *Anal. Chem.*, **1999**, 71-328R-332R
9. Shriver-Lake L., Kusterbeck A., F. S. Ligler, ACS Symposium Series, 646, **1996**, 46-55.
10. Vo-Dinh, Spellicy R., (Editors) *Proceedings of SPIE*, **1998**, 3534
11. Taylor R. F.,, Schultz J. S.(Editors) Handbook of Chemical and Biological Sensors, Inst. Phys., Bristol, UK, **1996**, 550 pp.
12. Kimura J., Kawana Y., Kuriyama T., *Biosensors*, **1998**, 4, 41-52
13. Mendoza E. A., Robinson D., L Lieberman, R. A., Proceedings of SPIE-Int. Soc. Opt. Eng. **1996**, 2836, 76-86.
14. Gardner, J. W.; Bartlett, P. N. *Sensors and Actuators B* **1994**, 18-19, 211-220.
15. Handbook of Biosensors and Electronic Noses; Kress-Rogers, E., Ed.; CRC Press: Boca Raton, FL,**1997**.
16. Stetter J. R., Findlay M., *Anal. Chim. Acta.* **1993**, 284, 1.
17. Masila M., Breimer M., Sadik O. A, Strategies for Improving the Analysis of Volatile Organic Compounds Using GC-Based Electronic Nose, in *Electronic Noses & Sensor Array based Systems*, Hurst, W. J. Ed.; Technomic Publishing Co., Inc., Lancaster, PA, **1999**, 27-42.
18. Goepel, W. and Reinhardt, G. Metal Oxide Sensors: New Devices through Tailoring Interfaces at the Atomic Scale, in *Sensor Update,* Baltes, Goepel, and Hesse, Eds. VCH, Weinheim, Germany **1996**.
19. Misselbrook, T.H.; Hobbs, P. J.; Persaud, K. C. *J. Agric. Engng. Res.* **1997**, 66, 213-220.
20. Hobbs, P. J.; Misselbrook, T. H.; Pain, B. F. *J. Agric. Engng. Res.* **1995**, 60, 137-144.
21. Sensors and Sensory Systems for Electronic Nose, Gardner, J. W., and Bartlett, P. N. Eds.; Kluwer Publishers, London **1992**, 273-301.
22. Masila M., Sargent A., Sadik O. A., "Pattern Recognition Studies of Halogenated Organic Compounds Using Conducting Polymer Array," *Electroanalysis,* **1998**, 10(5), 312-320.
23. Wang J., Chen L., Mulchandani A., Mulchandani P., Chen W. *Electroanalysis*, **1999**, 11 (12), 866-869.
24. Minunni M., Mascini M., *Anal. Lett.*, **1993**, 26, 1441.
25. Mayo C. S., Hallock R. B., *J. Immunological Methods*, **1989**, 120,105.
26. Klainer S., Thomas J., Francis, J. C., *Sensors & Actuators, B* **1993**, B11(1-3), 81-86.
27. Agayn V., Venetka I., Diss. Abstr., Int., B, **1995**, 56(3), 1415.
28. Vlasov., Yu G., Bychkov E. A., Legin A. V., *Talanta*, **1994**, 41(6), 1059-1063.
29. Tabacco, M., B., Zhou, Q., Nelson, B., Shahriari, M. R., *Ceram, Trans.* **1992**, 28 (Solid State Optical Material), 657-672.
30. Sadik O. A., *Electroanalysis,***1999**, 11(No. 12), 839-844.

CHEMICAL SENSORS

Chapter 2

Design of Molecular Recognition Elements for Environmental Potentiometric Sensors

Maria J. Berrocal, R. Daniel Johnson, and Leonidas G. Bachas[1]

Department of Chemistry and Center of Membrane Science, University of Kentucky, Lexington KY 40506-0055

The selectivity of a potentiometric sensor is generally derived from the nature of the molecular recognition element on which the sensor is based. Complex stability, orientation/preorganization, chemical composition, and steric effects all play important roles in governing the selectivity of recognition elements. Thus, a logical approach to custom design of recognition elements is imperative to achieving high sensor selectivity. Various strategies for attaining high sensor selectivity toward environmentally important species, especially anions, via rational design of molecular recognition elements are discussed.

From the analytical chemist's point of view, analyzing environmental samples constitutes a real challenge. The major issue facing analysts is the complexity of environmental matrices. Since usually the main goal is the determination of one particular species in the presence of several interferences, the need for a highly selective method is evident. In addition, studying processes in the environment and in the industrial/environmental interplay necessitate a method capable of continuous monitoring. Finally, as much of environmental investigation requires field work, techniques for environmental application should permit *in situ* analysis. Sensor technology meets all of these requirements. Sensors can be used for routine analysis, as they tend to be easy to operate and maintain. Due to their size and durability, sensors are also ideal for performing continuous monitoring and field work.

Executing *in situ* and other ecological work with minimal sample preparation relies heavily on the ability of a sensor to discriminate for species of interest in highly complex matrices (*1*). A sensor's "discriminatory power" or selectivity is derived almost solely from the molecular (or ionic) recognition element on which the sensor is based (*2*). Thus, it is desirable to construct sensors from recognition elements that have been specially designed to provide maximal selectivity for species of interest.

[1]Corresponding author (e-mail: bachas@pop.uky.edu)

Ion-selective electrodes (ISEs) constitute an example of potentiometric sensors that offer several advantages over other analytical techniques for the analysis of environmentally important ions. Specifically, the sensing platform of a membrane-based ISE consists of an ion carrier (ionophore) entrapped within a liquid polymeric membrane. The membrane does offer some interaction with numerous species, but the main interaction governing the selectivity of the sensor is between the analyte/interferences and the ionophore. Once an ionophore that offers the preferred selectivity has been developed and the polymer components that are ionophore-compatible have been optimized, the production of a functional ISE is rather facile and rapid. Presently, ISEs have been reported for several species including metal ions, anions, surfactants, and gases (3).

EPA requires the concentrations of certain ions to be below a predetermined value (4). For other ions there are recommended concentration limits. In both cases, ISEs can be useful in the analysis of these ions. Table I summarizes the required/recommended limits for some ions of the sulfur, nitrogen and chloride biogeochemical cycles.

Table I. EPA Required/Recommended
Limits for Some Anions

Contaminant	Maximum Level
Cyanide	0.2 mg/L
Nitrate	10 mg/L
Nitrite	1 mg/L
Chloride	250 mg/L
Sulfate	250 mg/L

The selectivity of membrane-based ISEs is somewhat controlled by the relative lipophilicity of the ions. In particular, the response of ISEs based on anion-exchangers, such as tridodecylmethylammonium chloride (TDMAC), that simply form ion pairs with anions follows the Hofmeister series or "classical response": lipophilic organic anions $>$ ClO_4^- $>$ SCN^- $>$ I^- $>$ NO_3^- $>$ Br^- $>$ NO_2^- $>$ Cl^- $>$ HCO_3^- $>$ HPO_4^{2-}, SO_4^{2-}. Analogously, ISEs based on lipophilic borates, such as potassium tetrakis(p-chlorophenyl)borate (KTpClPB), which forms ion pairs with cations, respond to cations in the following order: Ag^+ $>$ Hg^{2+} $>$ Cs^+ $>$ Rb^+, NH_4^+ $>$ K^+ $>$ Li^+ $>$ Na^+ $>$ Fe^{3+}, Pb^{2+}, Cu^{2+}. In order to design an ionophore that has a selectivity pattern different from those predicted by the classical series, molecular recognition principles have to be taken into account. For instance, higher complex stability, and thus increased selectivity, can be achieved by constructing ionophores with proper Lewis acid/base chemistry, structural rigidity, and/or binding cavity size. Selective interaction can also be induced via hydrogen bonding, charge, or π-π interactions.

The selectivity coefficient, K^{pot}, defines the ability of an ISE to distinguish a particular ion from others (5). According to IUPAC, K^{pot} can be evaluated in mixed solutions of primary and interfering ion (Fixed Interference Method), or separate solutions (Separate Solution Method and Matched Potential Method). The smaller the value of K^{pot} the greater the electrode's preference for the principal ion.

Herein, we present several examples of the strategies followed in our group for the rational design of new ionophores that can be used in the development of ISEs for environmentally important anions. These strategies are based on molecular recognition principles of biomimetic receptors and preorganized Lewis acid cavity ionophores.

Sulfur-Containing Species

Sulfur can be found on Earth with oxidation states from -2 to +6. This element is present in the crust in the form of insoluble compounds such as FeS_2, PbS, or $CaSO_4$. Sulfur can also be found as soluble species in seawater, like SO_4^{2-} and HS^-, as well as volatile compounds in the atmosphere, like SO_2, H_2S and $(CH_3)_2S$ (6). Sulfur is also one of the constituents of several biomolecules.

The origin of these sulfur-containing compounds can be attributed to several natural sources, including volcanic and biological activities. Other species are generated by chemical reactions in the atmosphere. But by large, the main source of some sulfur compounds comes from pollution generated by human activities. In particular, fossil fuel burning and industrial emissions can dramatically change the natural cycling of sulfur (see Figure 1). Specifically, SO_2 is an example of air pollutant whose origin is mostly anthropogenic. SO_2 has to be controlled, not only because of its negative impact on human health but also for its role in the atmospheric generation of H_2SO_4, one of the main components of acid rain. The following examples illustrate the use of ISEs in the determination of environmentally important sulfur-containing species.

Sulfide-Selective Electrode

Sulfides can be found in several waste effluents, such as oil refinery and viscose rayon wastes. The presence of these compounds gives water a characteristic "rotten eggs" smell, but most importantly, some sulfides are highly toxic. Commercially available sulfide-selective electrodes (7) are based on a Ag_2S membrane. These solid-state electrodes exhibit excellent selectivity for sulfide with respect to other sulfur-containing anions like sulfate, thiocyanate and dithiosulfate. EPA has approved this electrode for the analysis of drinking and surface waters as well as domestic and industrial waste (method 9215).

Hydrogen Sulfite-Selective Electrode

Biological compounds, which exhibit extremely selective chemical recognition properties, are a good source of inspiration for the design of highly selective ionophores (8). Hydrogen bonding and "lock and key" (or more appropriately "hand in glove") type of interactions can be found in biomolecules such as proteins, nucleic acids, antibodies, etc. These compounds have served as model for the design of several synthetic receptors (9). Recently, Copley and Barton (10) reported that the amino acid that was found most frequently in oxoanion-binding sites of proteins is arginine (Figure 2). Arginine's guanidinium moiety (boxed in Figure 2) is thought to

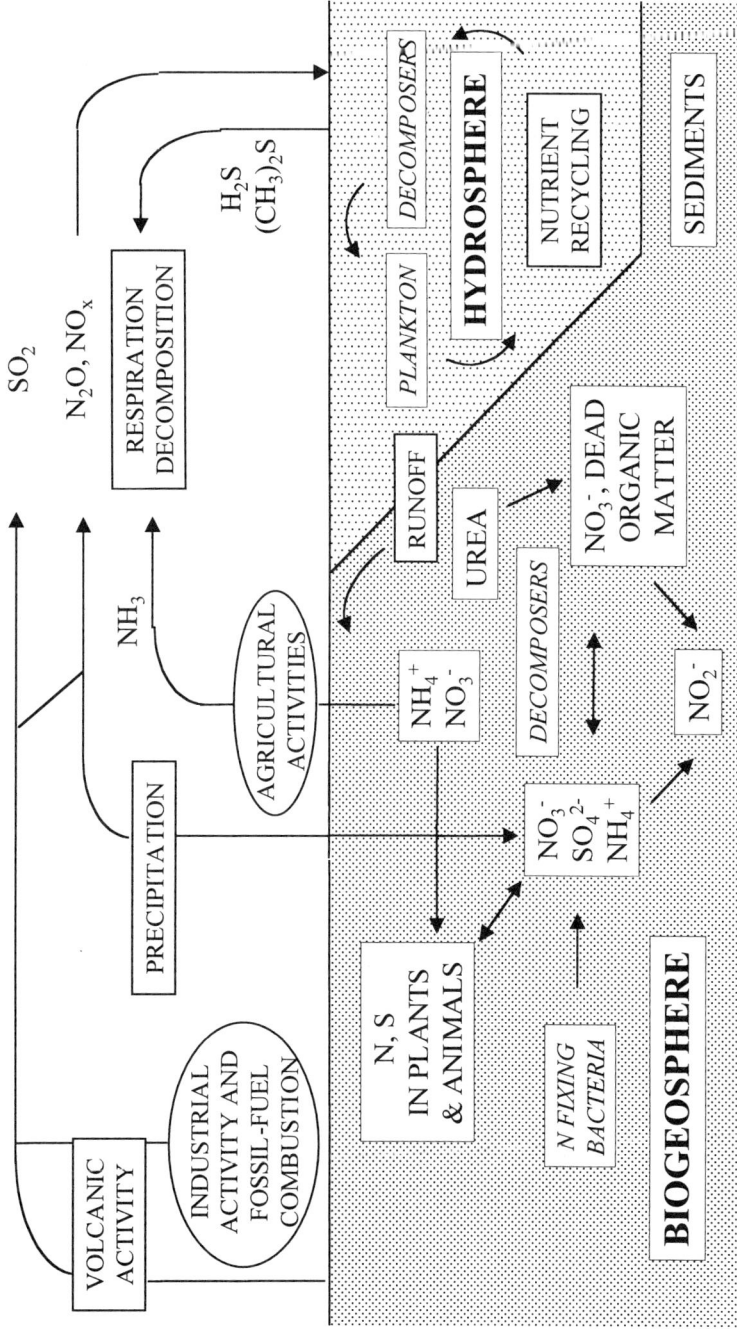

Figure 1. Enviromental cycles involving sulfurous and nitrogenous species.

play an important role in the binding of anions like sulfate and phosphate. Compounds containing the guanidinium functionality have been used successfully as ionophores in anion-selective electrodes for oxoanions and carboxylates (*11, 12*).

The ionophore shown in Figure 3 incorporates the guanidinium functionality found in arginine into a rigid, multi-cyclic framework. The corresponding ISE shows good selectivity towards hydrogen sulfite over the other tested anions (*11*). The major interferences for this ionophore are salicylate and perchlorate with K^{pot} values of 4.6×10^{-3} and 7×10^{-3} relative to hydrogen sulfite, respectively. The working range of the electrode at pH 6.0 is from 5.0×10^{-5} to 0.500 M, with a sensitivity of -47 mV/decade, and a detection limit of 3.9×10^{-5} M for hydrogen sulfite.

Sulfur Dioxide Gas Sensor

As it can be observed in Figure 4A, a Severinghaus-type gas sensor for SO_2 can be obtained by placing a pH electrode behind a SO_2 permeable membrane (*13*). SO_2 diffuses through the membrane and is converted into its corresponding anion, HSO_3^-. Since the solution behind the membrane is not buffered, its pH changes as a result of SO_2 hydrolysis, causing a response at the pH electrode. This electrode suffers from interferences by gases like NO_x, CO_2, and Cl_2, which can also diffuse through the membrane and cause a change in the pH of the internal solution.

The hydrogen sulfite-selective electrode described above can be incorporated into a gas sensor for SO_2 (*14*), as shown in Figure 4B. By placing the hydrogen sulfite electrode behind the gas-permeable membrane, a gas sensor for sulfur dioxide with excellent selectivity is produced (Figure 5). Not only are non-gaseous interferences like perchlorate and iodide avoided, but this arrangement also has advantages over Severinghaus-type gas sensors regarding gaseous interferences. In particular, this setup allows discriminating against gases such as CO_2 and NO_2 since the HSO_3^--selective electrode does not respond appreciably to HCO_3^-, NO_2^-, and NO_3^-. Also, unlike in the Severinghaus-type sensor, the solution behind the gas-permeable membrane is buffered. This buffering generates a buffer-trap effect that shifts the SO_2 hydrolysis equilibrium behind the gas-permeable membrane to the right, which, in turn, improves the detection limit of the sensor.

Sulfate-Selective Electrodes

The presence of high concentrations of sulfate in water has been implicated in the corrosion of pipes (*15*). Further, high levels of sulfate in drinking water may, when consumed, lead to gastrointestinal disorders in humans. For this reason, the EPA considers this anion to be a secondary contaminant in its national drinking water regulations, and thus recommends, but not requires, a maximum concentration of SO_4^{2-} of 250 mg/L (2.6×10^{-3} M) in drinking water (*16*).

The potentiometric detection of sulfate can be performed by Gran's plot titration of the sulfate solution with a commercially available barium-selective electrode (*17*). Sulfate-selective electrodes based on solid state crystals have also been reported (*18*). More recently, polymeric-membrane sulfate-selective electrodes have been developed using imidazole and thiourea derivatives as ionophores (*19, 20*).

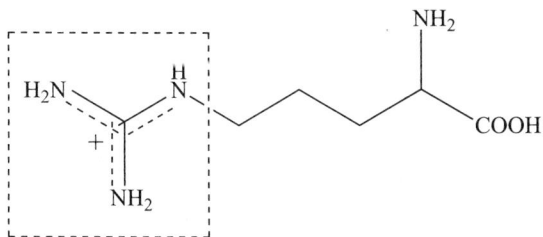

Figure 2. Structure of the amino acid arginine.

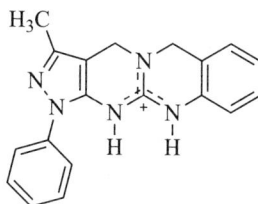

Figure 3. Structure of the guanidinium-based ionophore on which hydrogen sulfite-selective electrodes are based.

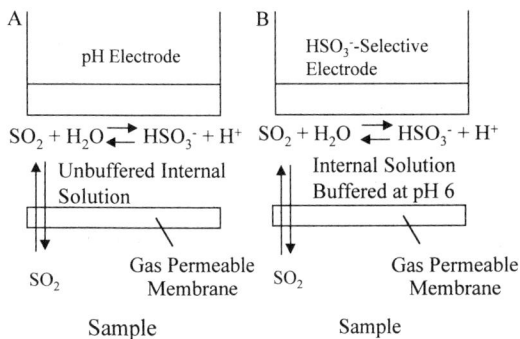

Figure 4. Schematic diagrams of (A) a Severinghaus-type SO_2 gas sensor and (B) a SO_2 gas sensor based on a HSO_3^--selective electrode.

Urea and thiourea compounds substituted with aminochromenone groups show high association constants with oxyanions, as determined by NMR titration (*21*). The ionophore shown in Figure 6 is a chromenone-based receptor (*22*). This compound has the appropriate geometry to form simultaneous hydrogen bonds with oxoanions like phosphate or sulfate. Preliminary results obtained in this group, with an ISE using this ionophore, point to a deviation from the Hofmeister series for sulfate with respect to other more lipophilic anions (Berrocal, M. J.; Cruz, A.; Badr, I. H. A.; Raposo, C.; Morán, J. R.; Bachas, L. G., unpublished data).

Nitrogen-Containing Species

Nitrogen cycling in the environment is one of the most important and, simultaneously, most complex biogeochemical processes (see Figure 1). The atmosphere is composed of 78.1% N_2 (by volume in dry air), but species such as NH_3 and NO_x tend to be the most environmentally important of the atmospheric nitrogen-containing molecules. Ammonia gas from biological excretion reacts with aerosol sulfur and nitrogen containing acids, and, thus, is an important chemical in the neutralization of acid rain-causing molecules. On the other hand, NO_x gases, formed naturally by lightning and anthropogenically by automobile and industrial emissions, play important parts in the reduction of ozone to O_2 and the formation of photochemical smog (*6*). Nitrogen species such as NH_4^+, NO_3^-, and NO_2^- are assimilated by land biota and are a crucial part of growth and development of plant species. For this reason, nitrogenous fertilizer has become widely used, but an imbalance of these ions in the soil and water has adverse affects on pH, photosynthesis, and the overall health of land/water ecosystems. The following examples demonstrate how rationally designed ISEs can be used for monitoring the nitrogen-containing molecules and ions mentioned above.

NH_4^+-Selective Electrodes and NH_3 Gas Sensor

Nonactin is a naturally occurring macrotetrolide antibiotic known to transport alkali cations through biological membranes. In the 1970's it was discovered that this compound demonstrates selectivity for potassium and ammonium, and could be used as an ionophore in cation-selective electrodes. Later studies found there to be a slight preference for ammonium over potassium, and today the nonactin (and its derivative monactin)-based ISE remains one of the best potentiometric sensors for ammonium (*3*). The mechanism of interaction between ammonium and nonactin consists on the formation of hydrogen bonds between the cation and the ether oxygens, as confirmed by IR, ^{13}C NMR, and crystallographic methods. This type of governing interaction, however, leads to problems from interferences from the alkali metals, particularly sodium and potassium with log K^{pot} of ~ -2.9 and -0.8 for Na^+ and K^+, respectively (*3*). Attempts to improve selectivity by employing an ionophore cocktail of nonactin and other similar macrotetrolides have been performed with only a slight improvement being noted.

An ammonia gas sensor has been constructed in the Severinghaus-type manner. As stated before, this type of gas sensor suffers from several interferences, such as those species that can diffuse through the gas-permeable membrane and change the pH

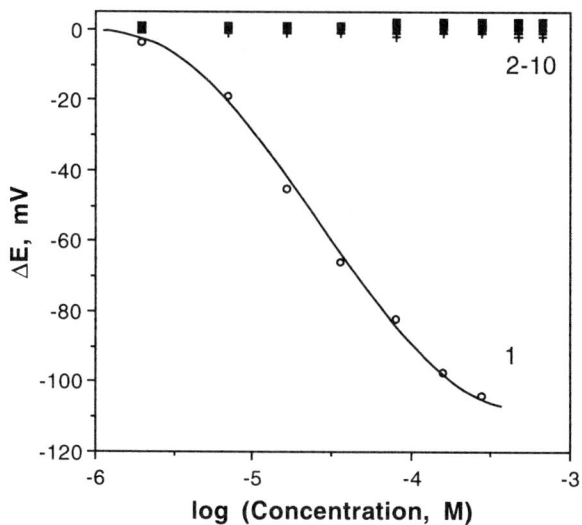

Figure 5. Selectivity of SO₂ gas sensor toward: hydrogen sulfite (1), nitrite (2), bicarbonate (3), fluoride (4), acetate (5), perchlorate (6), salicylate (7), thiocyanate (8), iodide (9), and bromide (10). Adapted from reference 14.

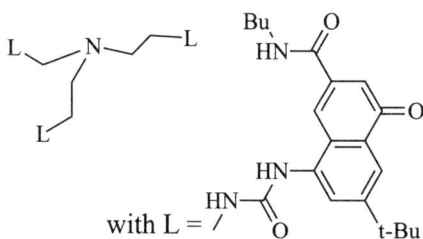

Figure 6. Chromenone-based ionophore for sulfate.

of the solution behind the membrane. In this case, many amines interfere strongly since they have a higher pK_a than ammonia. By using an ammonium-selective electrode instead of a pH electrode to construct the ammonia sensor (in a manner analogous to the one described above for SO_2), the interferences from amines disappear as the nonactin-based ISE responds to ammonium approximately 1000-fold greater than toward organic amines (23). In addition, the major interferences of the nonactin ISE (sodium and potassium) do not form species that can transverse the gas-permeable membrane and, hence, do not affect NH_3 measurements.

Nitrate-Selective Electrode

The majority of ionophores for nitrate that are commercially available are based on quaternary ammonium salts, such as tridodecylmethylammonium and tetraoctadecylammonium salts. Electrodes based on these salts, however, lack selectivity and, most often, the selectivity pattern of the electrodes follows the classical Hofmeister selectivity series. The response of these electrodes, then, is primarily toward ClO_4^- with a log K^{pot} of 3.0 relative to nitrate, determined by the separate solution method (3, 24). In fact, in many cases, it would be necessary to pretreat sample solutions to remove some of the more highly interfering anions (EPA method 9210).

In our laboratory, we have taken a molecular imprinting approach to constructing a nitrate-selective electrode. Molecular imprinting is based on the same principles of size, shape, and/or charge complementarity as the recognition of ligands by ionophores. The process of molecular imprinting involves polymerization in the presence of a template, during which the template induces the formation of cavities that are complimentary to the analyte of interest. To develop a nitrate-sensing element, we formed polypyrrole via constant current deposition of pyrrole in the presence of $NaNO_3$ (25).

The selectivity of the molecularly-imprinted NO_3^- ISE is much improved versus those based on the quaternary ammonium salts. For example, the log K^{pot} for perchlorate in this case is -1.24 relative to nitrate (fixed interference method). Thiocyanate, which was also a major interferent for the quaternary ammonium salt-based ISE, with log K^{pot} of 1.7 (24), is better discriminated against by the molecularly-imprinted ISE. For this nitrate-selective ISE the most important interference is chloride.

Nitrite-Selective Electrode

Vitamin B_{12} is a naturally occurring biochemical molecule containing complexed cobalt with a coordination number of 6. In this structure, four coordination sites are occupied by nitrogens in the molecule's corrin ring. Cobalt's remaining two coordination sites are filled by a cyano group and the ribonucleotide portion of the molecule. Vitamin B_{12} derivatives, where the ribonucleotide portion has been removed and where the peripheral amide side chains have been replaced by groups that increase the lipophilicity of the molecule, have been used by us and Simon's group to produce ISEs that are selective toward nitrite and thiocyanate (26, 27). These vitamin B_{12} derivatives are called cobyrinates, and two such ionophores are available commercially from Fluka (aquo-cyano-cobyrinic acid-heptakis-(2-phenylethylester) and dicyano-cobyrinic acid-heptapropylester) (28).

Our group has studied the development of ISEs from the cobyrinate shown in Figure 7, which differs from vitamin B_{12} in that the propionamide and acetamide perimeter groups of the vitamin have been replaced with carboxylic esters. ISEs constructed using this cobyrinate ionophore yielded near-Nernstian responses to salicylate, thiocyanate, and nitrite (26). The ionophores employed by Simon and co-workers are very similar to that depicted in Figure 7, also yielding ISEs with selectivity for nitrite and thiocyanate. All NO_2^--selective ISEs based on cobyrinates demonstrate only small differences in selectivity patterns that can be attributed to the nature of the side chains (27).

Potentiometric NO_x Gas Sensor

A NO_x sensor based on the Severinghaus design suffers from interferences similar to those mentioned for the SO_2 sensor. Thus, constructing a NO_x sensor from a nitrite or nitrate-selective ISE would offer several advantages. In our laboratory, we have constructed a NO_x sensor based on a vitamin B_{12} derivative-based ISE (29). Here, the nitrite selective-electrode is placed behind a gas-permeable membrane and in contact with a solution buffered at pH 5.5. Once behind the gas-permeable membrane, the NO_x species are hydrolyzed to generate NO_2^- and NO_3^-. The ISE, then, responds in a proportionate manner to the nitrite species generated in the buffer. This sensor was able to achieve a detection limit of 4×10^{-7} M, while a commercially available Severinghaus-type sensor demonstrates detection limits as low as 2×10^{-6} M. In addition, K^{pot} values for the ISE-based sensor were at least 1 to 2 orders of magnitude lower than the values reported for Severinghaus-type sensors. The selectivity coefficient for our sensor toward thiocyanic acid was 6.6×10^{-4}, a marked improvement over the 0.21 value for the commercial Severinghaus sensor (26).

A NO_x gas sensor constructed from a molecularly-imprinted nitrate-selective electrode has also been developed in our laboratory (30). This gas sensor is produced by placing the nitrate-selective electrode behind a gas-permeable membrane. Also, there is a buffer compartment present between the gas-permeable membrane and the ISE. As before, NO_x species that pass through the gas-permeable membrane are trapped in the buffer compartment as NO_2^- and NO_3^-. In this case, the electrode responds proportionately to the amount of nitrate in the buffer, rather than nitrite as in the NO_x sensor described above. The gas sensor based on the NO_3^--selective electrode offers advantages over the Severinghaus arrangement similarly to the other aforementioned gas sensors (SO_2 and NO_x). Detection limits for this gas sensor were on the order of 1×10^{-6} M, with response characteristics being retained for over 80 days. This lifetime is consistent with lifetimes of the molecularly-imprinted nitrate-selective electrodes.

Chloride-Selective Electrode

The primary interest in monitoring chloride in the environment lies in the analysis of the hydrosphere. Measuring chloride content in waters is important in

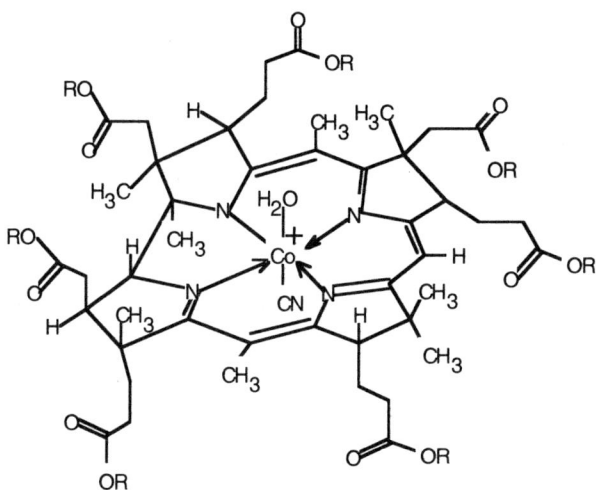

Figure 7. Cobyrinate ionophore for NO$_2^-$, (R = -CH$_2$CH$_2$CH$_3$).

understanding mechanisms controlling the global water cycle and monitoring waters for chemicals that lend to corrosion (e.g., of ships coatings) (*31*).

Preorganized macrocyclic Lewis base hosts, such as crown ethers and calixarenes, have been widely used as ionophores for cation-selective electrodes. One advantage of these types of ionophores is the ease with which the macrocyclic host can be synthetically altered by adding side arm chains; for example, to produce desired complex stability and selectivity. Also, the "macrocyclic effect", which describes the effect of solvation, preorganization, and the number of conformations on the free energies and enthalpies of ion-ionophore binding, is an added advantage of the preorganized host.

Macrocyclic anion hosts, however, are much rarer and less widely used. A class of metallo-macrocyclic compounds known as mercuracarborands (Figure 8) has recently been developed (*32*). The mercuracarborands consist of carborane cages and mercury centers, which provide Lewis acid character to the compound, thus forming a host for anions. Conceptually, these compounds can be described as charge-reversed analogs of crown ethers in terms of architecture and Lewis acid/base chemistry. Mercuracarborands have a larger preorganized cavity than observed in crown ethers facilitating the accommodation of anions. These compounds also take advantage of the previously mentioned macrocyclic effect to provide stability in anion binding. Finally, in addition to varying cavity sizes to alter selectivity toward differently sized anions, the substituents on the carborane cages can be changed. Such rational design can lead to a variance in the inductive effect of the carboranes on the mercury centers. Altering the effect of carborane cages on the mercury centers, in turn, influences the Lewis acidity of the host and, ultimately, selectivity.

An ISE prepared from [9]mercuracarborand-3 (MC3) yielded an electrode selective for chloride (*33*). Indeed, the MC3-based electrode demonstrated improvements in selectivity over all anions in the Hofmeister series, with only iodide and bromide being interferences (log $K^{pot} > 0$). Iodide and bromide are less interfering, however, for this electrode than common Ag/AgCl-based ISEs for chloride. Preliminary results from electrodes constructed from the hexamethyl form of MC3 (-R = -CH$_3$) show that carborane substitution has a profound effect on the acid/base properties of the mercury centers of the ionophore, yielding different selectivity patterns than for MC3-based ISEs. Aside from providing a marked improvement in selectivity, the MC3-based ISEs demonstrate retention of response characteristics for approximately two months (Figure 9).

Conclusion

Ion-selective electrodes are well suited as tools for environmental analyses. Finding ionophores from which to create electrodes with desired or appropriate selectivites, however, has been challenging. Biomimetic design is one method of meeting the challenges of ionophore design by taking advantage of the accomplished optimization of natural systems. Employing concepts of rational design can also facilitate the task of ionophore development. Through both of these approaches, it has become possible to create sensors for traditionally elusive species, particularly anions.

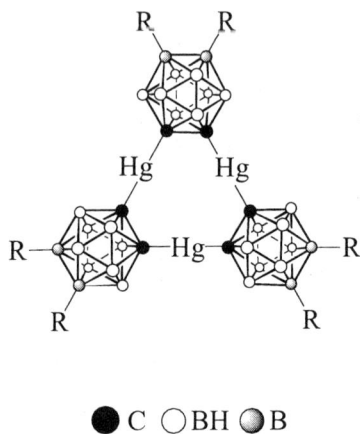

●C ○BH ◐B

Figure 8. Structure of [9]mercuracarborand-3 (R = -H) and derivatives.

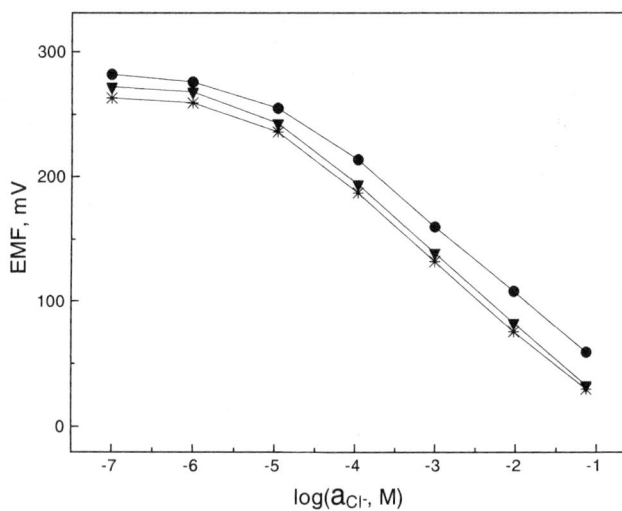

Figure 9. Lifetime of MC3-based ISE. Times: 1 day (▼), 25 days (✳), and 53 days (●). Adapted from reference 33.

Acknowledgments

The authors would like to thank the National Science Foundation and the National Aeronautics and Space Administration for their financial support of this work. R. D. J. would like to thank the NSF-IGERT fellowship program and M. J. B. would like to thank the Kentucky Research Challenge Trust Fund for the continued support of their graduate work. The authors would also like to thank members of our group and our collaborators whose contributions are described in references 11, 12, 14, 25, 26, 29, 30, and 33.

Literature Cited

1. Christie, I.M.; Rigby, G. P.; Treloar, P.; Warriner, K.; Maines, A.; Ashworth, D.; Eddy, S.; Reddy, S. M.; Vadgama, P. In *Biosensors for Direct Monitoring of Environmental Pollutants in Field;* Nikolelis, D. P.; Krull, U. J.; Wang, J.; Mascini, M., Eds.; Kluwer Academic: Netherlands, 1998; pp 41-42.
2. Spichiger-Keller, U. E. In *Chemical Sensors and Biosensors for Medical and Biological Applications;* Wiley-VCH: New York, 1998; pp 38-57.
3. Bühlmann, P.; Pretsch, E.; Bakker, E. *Chem. Rev.* **1998,** *98,* 1593-1687.
4. *National Primary and Secondary Drinking Water Regulations;* Office of Ground Water and Drinking Water; US EPA, 1998.
5. Umezawa, Y.; Umezawa, K.; Sato, H. *Pure Appl. Chem.* **1995,** *67,* 507-518.
6. Cox, P. A. In *The Elements on Earth: Inorganic Chemistry in the Environment;* Oxford University Press: New York, 1995.
7. Schmitt, S. E.; Pungor, E. *Anal. Lett.* **1971,** *4,* 641-652.
8. Hernández, E.; Bachas, L. G. In *Biosensors for Direct Monitoring of Environmental Pollutants in Field;* Nikolelis, D. P.; Krull, U. J.; Wang, J.; Mascini, M., Eds.; Kluwer Academic Publishers: Netherlands, 1998; pp 97-106.
9. Wotring, V. J.; Johnson, D. M.; Daunert, S.; Bachas, L. G. In *Immunochemichal Assays and Biosensor Technology for the 1990's;* Nakamura, R. M; Kasahara, Y.; Rechnitz, G. A., Eds.; American Society of Microbiologists: Washington, DC, 1992; pp 355-376.
10. Copley, R. R.; Barton, G. J. *J. Mol. Biol.* **1994,** *242,* 321-329.
11. Hutchins, R. S.; Molina, P.; Alajarín, M.; Vidal, A.; Bachas, L. G. *Anal. Chem.* **1994,** *66,* 3188-3192.
12. Hutchins, R. S.; Bansal, P.; Molina, P.; Alajarín, M.; Vidal, A.; Bachas, L. G. *Anal. Chem.* **1997,** *69,* 1273-1278.
13. Ross, J. W.; Riseman, J. H.; Krueger, J. A. *Pure Appl. Chem.* **1973,** *36,* 473-487.
14. Mowery, M. D.; Hutchins, R. S.; Molina, P.; Alajarín, M.; Vidal, A.; Bachas, L. G. *Anal. Chem.* **1999,** *71,* 201-204.
15. Mancy, K. H.; Weber, W. J. *Analysis of Industrial Wastewaters;* Wiley-Interscience: New York, 1971; pp 525-527.
16. *Health Effects from Exposure to High Levels of Sulfate in Drinking Water;* US EPA, 1999.
17. Lawbli, M. W.; Dinten, O.; Pretsch, E.; Simon, W.; Votgle, F.; Bongardt, F.; Kleiner, T. *Anal. Chem.* **1985,** *57,* 2756-2758.

18. Umezawa, Y. In *Handbook of Ion-Selective Electrodes: Selectivity Coefficients;* CRC Press: Boca Raton, FL, 1990.
19. Li, Z.; Liu, G.; Duan, L.; Shen, G.; Yu, R. *Anal. Chim. Acta* **1999**, *382*, 165-170.
20. Nishizawa, S.; Bühlmann, P.; Xiao, K. P.; Umezawa, Y. *Anal. Chim. Acta* **1998**, *358*, 35-44.
21. Raposo, C.; Crego, M.; Mussons, M. L.; Caballero, M. C.; Morán, J. R. *Tetrahedron Lett.* **1994**, *35*, 3409-3410.
22. Raposo, C.; Almaraz, M.; Martín, M.; Weinrich, V.; Mussons, M. L.; Alcázar, V.; Caballero, M. C.; Morán, J. R. *Chem. Lett.* **1995**, 759-760.
23. Lee, H. L.; Meyerhoff, M. E. *Analyst* **1985**, *110*, 371-376.
24. Wegman, D.;Weiss, H.; Ammann, D.; Morf, W. E.; Pretsch, E.; Sugahara, S.; Simon, W.*; Microchim. Acta III* **1984**, 1.
25. Hutchins, R. S.; Bachas, L. G. *Anal. Chem.* **1995**, *67*, 1654-1660.
26. Daunert, S.; Witkowski, A.; Bachas, L. G. *Prog. Clin. Biol. Res.* **1989**, *292*, 215-225.
27. Stepánek, R.; Kräutler, B.; Schulthess, P.; Lindemann, B.; Amman, D.; Simon, W. *Anal. Chim. Acta* **1986**, *182*, 83-90.
28. Fluka Chemical Corp. *Selectophore: Ionophores, Membranes, Mini-ISE;* Fluka Chemie AG: Buchs, Switzerland, 1996; pp 83-84.
29. O'Reilly, S. A.; Daunert, S.; Bachas, L. G. *Anal. Chem.* **1991**, *63*, 1278-1281.
30. Hernández, E.; Mortensen, C.; Bachas, L. G. *Electroanalysis* **1997**, *9*, 1049-1053.
31. O'Neill, P. In *Environmental Chemistry;* George Allen and Unwin: London, 1985; pp 50-54, 161.
32. Hawthorne, M. F.; Zheng, Z. *Acc. Chem. Res.* **1997**, *30*, 267-276.
33. Badr, I. H. A.; Diaz, M.; Hawthorne, M. F.; Bachas, L. G. *Anal. Chem.* **1999**, *71*, 1371-1377.

Chapter 3

Synthesis of Polystyrene-Supported Dithizone Analogues for Use as Chemical Sensors for Heavy Metals

R. B. King and I. Bresińka

Department of Chemistry, University of Georgia, Athens, Georgia 30602

Fiber optic sensors for detection of heavy metals require chromogenic ligands which change color upon contact with the metal ions of interest and which can be immobilized onto an insoluble matrix incorporated into the fiber optic system. Diarylthiocarbazone ligands such as dithizone (diphenylthiocarbazone) are of particular interest because of the intense and distinct colors of their complexes with toxic heavy metals of environmental concern such as lead and mercury. This paper describes methods for chemical immobilization of a diarylthiocarbazone unit onto a polystyrene matrix. The key step in the preparation of polystyrene-supported diarylthiocarbazones uses the coupling of nitroformaldehyde β-naphthylhydrazone with diazotized aminopolystyrene. The resulting polystyrene-supported diarylnitroformazan can then be converted to the corresponding polystyrene-supported diarylthiocarbazone by treatment with aqueous ammonium sulfide followed by oxidation. Using this method samples of polystyrene-supported diarylthiocarbazones have been obtained which undergo distinctive color changes in 0.005 M aqueous Pb^{2+} and Hg^{2+}. The properties of such polystyrene-supported diarylthiocarbazones obtained from diazotized aminopolystyrene and nitroformaldehyde β-naphthylhydrazone, including their reactions with heavy metal ions, are compared with materials containing an extra azo chromophore obtained by the method of Chwastowska and Kosiarska using the coupling of diazotized aminopolystyrene with di-β-naphthylthiocarbazone.

Fiber optic sensors (1,2,3,4,5,6,7) require chromogenic ligands for monitoring heavy metals through color changes occurring in the presence of small quantities of the metals of interest. Of particular interest are ligands suitable for monitoring the toxic lead and mercury cations, which are of considerable environmental concern. In this connection, lead and mercury are "soft" acids in the Pearson sense thereby having a higher affinity for sulfur ligands than oxygen ligands. Ideal chromogenic ligands for such metals should have the following properties:

   (1) The metal-ligand complex must be highly colored even in low metal concentrations and must have a color distinct from that of the uncomplexed reagent.

For this reason ligands which have the highly chromophoric azo group are of particular interest.

(2) The metal complexation should be chemically reversible so that optical sensors containing such ligands can monitor varying metal ion concentrations.

(3) The chromogenic indicator should be selective for particular metal ions to reduce chemical interference.

(4) Both the ligand and its metal complexes must be stable in the environment where they are used.

This objective thus requires the development of highly colored chromogenic ligands with selective heavy metal complexing abilities as well as methods for incorporating such ligands into optic sensors through suitably chosen immobilization procedures.

Among the diverse immobilization methods, the use of chemical bonding of the chromogenic ligand onto a polystyrene matrix appears to be very attractive since polystyrene can be readily converted to aminopolystyrene (8,9,10) and then diazotized with nitrous acid (Figure 1). The resulting polymer-supported diazonium ion can provide a source of the required polymer-supported azo chromophore through a suitably chosen coupling reaction leading to a ligand of desired type. Most other immobilization methods do not appear to be suitable for the generation of immobilized diazonium ions as synthetic intermediates.

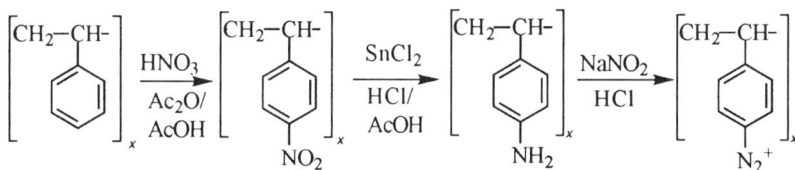

Figure 1. Conversion of polystyrene to aminopolystyrene followed by diazotization.

Chromogenic ligands of particular interest in this work are diarylthiocarbazones such as dithizone (diphenylthiocarbazone) and its analogs, which form highly colored complexes with mercury and lead (11,12). In this connection a number of methods are available for the synthesis of dithizone and related compounds (13,14). The method of greatest interest in connection with this project is the procedure of Hubbard and Scott (15) based on earlier work by Bamberger, Padova, and Omerod (16) since it provides a method for the conversion of an aromatic amine to the corresponding diarylthio-carbazone (Figure 2).

The diarylthiocarbazones can be anchored onto a polystyrene matrix by coupling with a polystyrene-supported diazonium salt using a procedure reported by Chwastowska and Kosiarska (Figure 3)(10). However in this case the coupling process introduces a new azo chromophore which might dilute the effect of the heavy metal to be detected on the visible spectrum. This difficulty can be avoided by chemically anchoring a diarylthiocarbazone functionality onto a polymer so that the polymer matrix become an integral part of the diarylthiocarbazone chromophore. Such an approach requires a sequence of reactions such as those depicted in Figures 4 and 5. In the first step a nitroformaldehyde arylhydrazone is prepared by coupling the corresponding arenediazonium salt with the nitromethane anion in a 1:1 ratio (Figure 4)(16). Subsequent coupling of the nitroformaldehyde arylhydrazone with the diazonium salt of polystyrene (Figure 1) can then give an unsymmetrical diarylnitro-formazan in which one of the aryl groups is part of the polystyrene matrix (Figure 5). Such unsymmetrical diarylnitroformazans can be then converted to the corresponding unsymmetrical diarylthiocarbazones by the same methods as used for the symmetrical dithizones (Figure 5)(14,15). This synthetic approach to the unsymmetrical diaryl-thiocarbazones through nitroformaldehyde arylhydrazones (Figure 4) has been developed by Pupko and Pel'kis (17,18) but had not yet been adapted for the preparation of polymer-supported diarylthiocarbazones when the work described in this paper was initiated.

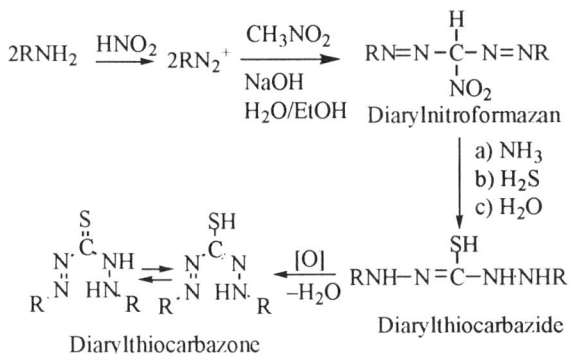

Figure 2: Conversion of a primary aromatic amine to the corresponding diarylthiocarbazone.

Experimental

Materials

Polystyrene SP5009 (1% divinylbenzene crosslinking) and polystyrene SP5018 (2% divinylbenzene crosslinking) were purchased from Advanced ChemTech. More highly crosslinked polystyrene macroreticular beads (copolymer XFS 4022 with 20% divinylbenzene) were obtained from Dow Chemical Company. The remaining reagents were purchased from standard suppliers.

Synthesis of Nitroformaldehyde β-naphthylhydrazone

The β-naphthylamine hydrochloride (0.4 mole) was stirred overnight in a mixture of 100 mL of water and 120 mL of HCl and then diazotized with 28 g of $NaNO_2$ dissolved in 60 mL of water at –5 to –8°C using a cold bath of dry ice and o-xylene. The sodium salt of nitromethane was prepared in a separate vessel by combining 0.4 mole (22 mL) of nitromethane in 100 mL of absolute ethanol with 16 g of NaOH in 100 mL of water at –5 to –8°C. The resulting white precipitate of the sodium salt of nitromethane was dissolved in 50 mL of water to give a yellow solution. This solution was added dropwise to the filtered diazonium salt solution kept at –8°C . The final mixture was stirred for 3 hr. The precipitate was filtered off and washed with water. The crude product was purified by dissolving in 5% aqueous sodium hydroxide.

After separation of the insoluble fraction the nitroformaldehyde arylhydrazone was precipitated with dilute hydrochloric acid (0.25 M). Three such cleaning cycles were required to obtain reasonably pure product. The reaction yield was 42% based on the amine used in the synthesis. Air dried solid nitroformaldehyde β-naphthylhydrazone is orange and melts at 90–94°C. It was characterized by infrared and ultraviolet-visible spectroscopy.

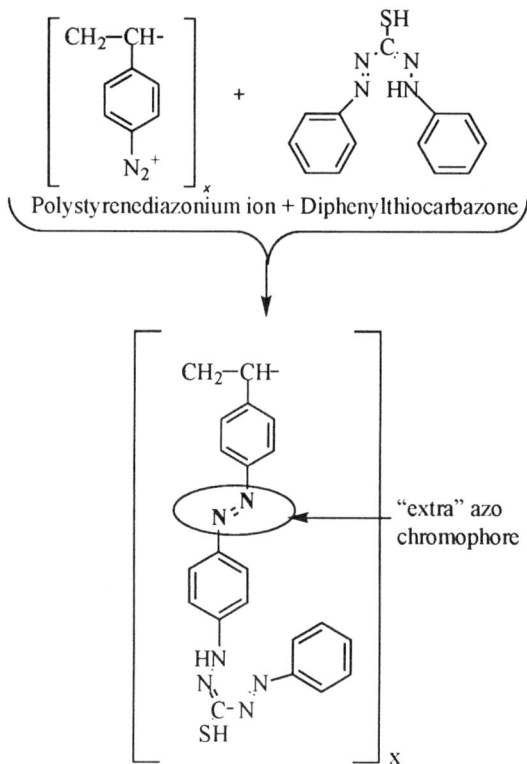

Figure 3: The procedure of Chwastowska and Kosiarska (10) for coupling diazotized aminopolystyrene with diphenylthiocarbazone (dithizone). This procedure leads to a supported diarylthiocarbazone with the indicated "extra" azo chromophore. An analogous procedure is used in this work to couple diazotized aminopolystyrene with di-β-naphthylthiocarbazone.

Preparation of Nitro- and Aminopolystyrenes

A 30 g-portion of the polystyrene was suspended in a mixture of acetic anhydride and acetic acid and cooled to –4°C. Nitric acid was slowly introduced and the mixture was stirred at low temperature (~ –4°C) . After the reaction was completed the solids were filtered off, washed, and air dried. The nitration of polystyrene was also carried out in which the acetic acid-acetic anhydride mixture was substituted by sulfuric acid (samples **3-18**, **3-20**, **3-22**, **3-24**). The reaction conditions are listed in Table I. The nitropolystyrenes were reduced to the corresponding aminopolystyrenes with stannous chloride in hydrochloric acid at room temperature or 90°C (see Table I).

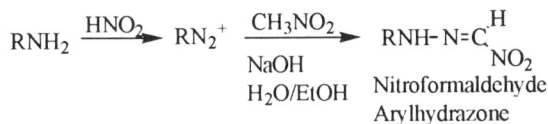

Figure 4: Conversion of a primary aromatic amine to the corresponding nitroformaldehyde arylhydrazone

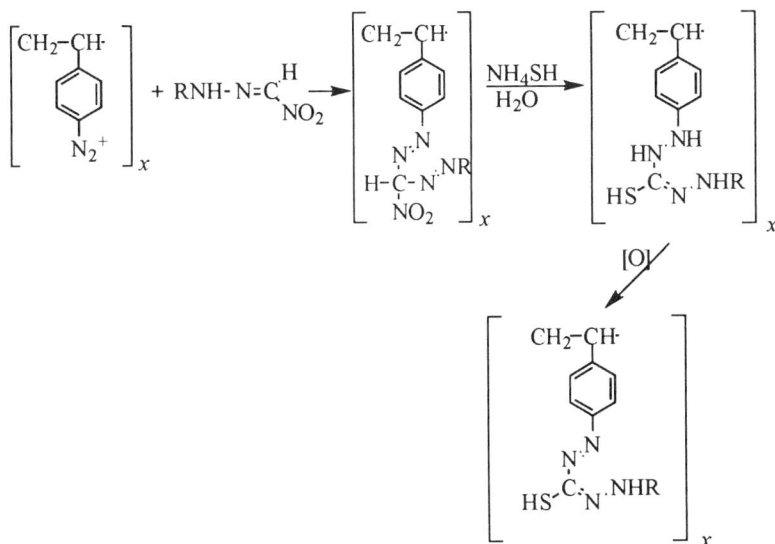

Figure 5: The procedure used in this work to anchor a diarylthiocarbazone chromophore onto a polystyrene matrix.

Table I. Conditions for preparation of nitro- and amino- polystyrenes

	Nitration of Polystyrenes						Reduction of Nitropolystyrenes				
Sample number	Polystyrene (g)	mL of HNO_3	mL acetic anhydride	mL acetic acid	Time (hr)	Sample number	g of sample	g of $SnCl_2$	mL of aq. HCl	Temp. °C	Time (hr)
2-112	60	15	150	50	2	2-113	57	125	250	room	24
3-2	59	20	150	45	4	3-7	57	125	250	room	48
3-24	30	15	50 H_2SO_4		4	3-25	28	125	250	room	72
1-16A	10	3.2	41	9	5	1-16B	5	5	5	room	48
2-60	32	50	120	30	2.5	2-61	34	125	250	reflux	30
2-88	50	15	200	45	1	2-94	45	125	250	reflux	28
3-20	30	15	50 H_2SO_4		3	3-21	28	125	250	room	72
2-87	50	30	200	45	3	2-97	58	150	250	reflux	48
3-22	30	15	50 H_2SO_4		4	3-23	28	125	250	room	72
2-89	50	15	200	45	0.5	2-93	58	125	250	reflux	48
3-18	30	15	50 H_2SO_4		3	3-19	28	125	250	room	72

Functionalization of diazotized aminopolystyrene with nitroform-aldehyde β-naphthylhydrazone and its conversion to thiocarbazone

A 0.5 to 10 g-portion of aminopolystyrene was suspended in a mixture of hydrochloric acid and water cooled to ~ -4°C and then diazotized with 0.1 mole of $NaNO_2$ dissolved in 40 mL of distilled water. The mixture was stirred for 1 hr to allow complete reaction. The resulting solid was separated by filtration, washed with cold water to remove an excess of acid, and then immediately transferred into a cold buffer solution of 36 g of sodium acetate, 40 mL of acetic acid, and 20 mL of water. The nitroformaldehyde arylhydrazone (~0.5 to 1 g) dissolved in 200 mL of ethanol was filtered and cooled to 0–2°C. This ethanolic solution was slowly added to the polymer suspension. The final mixture was stirred at –2 to +4°C for 6 hr. The dark colored solid was separated and washed with water and then with ethanol until the filtrate was clear. In several cases Soxhlet extraction was used to minimize use of ethanol.

In the next step the functionalized polystyrene was treated with an aqueous solution of ammonium sulfide at room temperature for at least 3 hr. using 20 to 100 mL of 20% aqueous ammonium sulfide for a 1 to 10 g-portion of functionalized polymer. A more rapid reaction was obtained with solutions also containing 5 ml of 21 % ammonium hydroxide .

Anchoring of di-β-naphthylthiocarbazone to polystyrene (procedure of Chwastowska and Kosiarska)

A portion of aminopolystyrene (~5 g) suspended in a mixture of 120 mL of concentrated hydrochloric acid and 100 mL of water was cooled to about –4°C and then diazotized with 18.6 g of sodium nitrite dissolved in 50 mL of water for 4 hr. Solids were filtered off, washed with cold water, and transferred into a cold acetate buffer solution prepared by dissolving 36.4 g of sodium acetate and 50 mL of acetic acid in 100 ml of water. An acetone solution (200 mL) containing approximately 0.5 g of di-β-naphthylthiocarbazone was filtered and slowly added and the mixture was stirred for 4 to 18 hr . After the reaction was complete, the crosslinked polystyrene was washed with water, then with acetone to remove unreacted di-β-naphthylthiocarbazone, and air-dried. Attempts to separate unreacted di-β-naphthylthiocarbazone from the uncrosslinked polymer met with difficulty owing to significant acetone solubility of the polymer itself. Therefore a water-acetone mixture (3:1) rather than pure acetone was used to wash these samples. Air-dried samples obtained from the uncrosslinked polymer did not appear to be homogeneous.

Spectroscopic methods

The infrared spectra were recorded on a Perkin Elmer 1600 Series FT-IR infrared spectrometer using KBr disks. The electronic spectra of chloroform solutions were recorded on a Hewlett Packard Model 8452A Diode Array Spectrophotometer. The electronic spectra of solid samples were taken in Nujol mulls placed between two methacrylate ultraviolet grade plates. Proton nuclear magnetic resonance spectra of nitroformaldehyde arylhydrazones, diarylnitroformazans and the corresponding diaryl-thiocarbazones were recorded in $CDCl_3$ solutions on a Bruker AC-300 spectrometer operating at 300 MHz for protons. However since most of polymers used in the preparation of the functionalized polystyrenes are insoluble in organic solvents, solution NMR spectra of these polymers could not be obtained.

Results and Discussion

Synthesis of nitroformaldehyde β–naphthylhydrazone

The synthetic method reported by Pupko and Pel'kis (16) (Figure 4) was investigated in which 1 mole of aryldiazonium salt was coupled with 1 mole of the sodium salt of nitromethane to form the corresponding nitroformaldehyde arylhydrazone. By using this method nitroformaldehyde arylhydrazones were synthesized in which the aryl group was phenyl, 2,4,6-trimethylphenyl and β-naphthyl. However, the phenyl and 2,4,6-trimethylphenyl derivatives were quite unstable at room temperature and decomposed over a period of several days. Therefore we concentrated our efforts on the more stable nitroformaldehyde β-naphthylhydrazone .

Since the paper by Pupko and Pel'kis (16) did not describe adequately the synthesis of nitroformaldehyde β-naphthylhydrazone, the details of the preparative procedure needed to be worked out. In our initial experiments based on the Pupko and Pel'kis procedure (16) 0.1 mole of β-naphthylamine hydrochloride was suspended in water and then diazotized with the stoichiometric amount of $NaNO_2$ with cooling to –5 to –8°C. The sodium salt of nitromethane (0.1 mole) was prepared separately. The aryldiazonium salt solution was added dropwise to the nitromethane solution at –5 to –8°C and the mixture was stirred for 4 hr. Several cleaning cycles were required to obtain reasonably pure product and the final yield was relatively poor not exceeding 20%. We found that this coupling reaction is strongly pH-dependent and is not a smooth reaction since considerable amounts of tarry byproducts are produced, especially when aniline and 2,4,6-trimethylaniline are used as starting materials. The main side product of the reaction was the corresponding diarylnitroformazan. In a number of experiments we found that if the reaction conditions suggested by Pupko and Pel'kis (16)were used, the final pH of the mixture was about 2.5 and considerable diarylnitroformazan side product was produced at the beginning of the coupling reaction when the mixture still was very basic. Our modification of the Pupko and Pel'kis procedure (16) using an excess of acid for the preparation of the aryldiazonium salt and reversing the order of mixing so that the nitromethane solution was added to the aryldiazonium solution rather than vice versa was found to suppress significantly the formation of the diarylnitroformazan side product by keeping the pH low during the coupling reaction. Such modified conditions were used for synthesis of nitroformaldehyde β-naphthylhydrazone on a large scale.

The nitroformaldehyde β-naphthylhydrazone was characterized by its electronic and infrared spectra. Thus the maximum in the electronic spectrum of nitroformaldehyde β-naphthylhydrazone occurs at 412 nm as compared with 488 nm for di-β-naphthylnitroformazan. The infrared spectrum of nitroformaldehyde β-naphthylhydrazone consists of sets of bands in the 600-1700 cm^{-1} region. The most intense absorption occurs between 1050 and 1400 cm^{-1} where several bands with shoulders are observed. The di-β-naphthylnitroformazan exhibits a very poor infrared spectrum dominated by bands at 1271, 1343 and 1550 cm.

Preparation of aminopolystyrene

We studied the functionalization of the following three samples of commercially available polystyrenes of different particle sizes and crosslinked with different amounts of divinylbenzene (DVB).

1. Polystyrene SP5009 with a 100-200 mesh particle size and containing 1% DVB.

2. Polystyrene SP5018 with a 200-400 mesh particle size and containing 2% DVB.

3. Copolymer XFS 4022 consisting of beads and containing 20% DVB. We also studied a sample of commercially available uncrosslinked polystyrene (Polystyrene MW 45,000) in form of crushed plates.

The samples of crosslinked polystyrene with different divinylbenzene (DVB) contents and different particle sizes as well as the sample of uncrosslinked polystyrene as suspensions were nitrated either with nitric acid in a mixture of acetic anhydride and acetic acid according to the published procedures (8,9) or in the standard sulfuric acid medium frequently used for nitrations of benzenoid derivatives (samples **3-18**, **3-20**, **3-22**, **3-24** as listed in Table I).   The samples of the resulting nitropolystyrene were then reduced to the corresponding aminopolystyrenes using stannous chloride in hydrochloric acid solution (Table I).

In order to follow changes in the polymers after nitration and reduction, infrared spectra of starting materials were recorded. The infrared spectra of starting polystyrenes were found to be similar. Some changes in the relative intensity of bands were noticeable presumably owing to the different DVB content. The uncrosslinked polymer was found to exhibit a poor spectrum. A set of bands between 1400 and 1700 cm^{-1} was found to be more intense in the crosslinked polymers.

Several batches of nitropolystyrene were prepared (Table I). An initial batch (sample **1-16A**) was synthesized using the method of King and Sweet (9) including purification with different mixtures of organic solvents . The materials were then washed with pure water to remove acids following by ethanol. The degree of nitration was estimated from the infrared spectra of the solid nitropolystyrenes, which were yellow solids exhibiting infrared bands around 1520 and 1345 cm^{-1} not present in the original polystyrenes indicative of aromatic nitro groups. The 1345 cm^{-1} band overlaps with the ~1320 and 1370 cm^{-1} bands of polystyrenes. The product of the nitration in THF solution (sample **3-2**) exhibited very low intensity bands in this region. The intensity of these bands increased for samples synthesized in a sulfuric acid medium. The degree of nitration was found to depend on the particle size of solid rather than the divinylbenzene content with smaller particles leading to lower degrees of nitration.

The nitropolystyrenes were reduced to the corresponding aminopolystyrenes using stannous chloride in hydrochloric acid at room temperature or 90°C (Table I).   The uncrosslinked nitropolystyrene can be completely reduced to the corresponding aminopolystyrene by increasing the reduction time at room temperature.   In the infrared spectra of the aminopolystyrenes the bands at 1345 and 1520 cm^{-1} assigned to nitro groups are no longer observable. The crosslinked nitropolystyrenes are more resistant towards reduction so that higher temperatures are recommended for complete reduction to aminopolystyrenes. However the rate of reduction depends on the particle size of the solid rather than the DVB content as we have previously observed for the nitration reaction discussed above. For the partially reduced samples the infrared bands at ~1520 cm^{-1} and ~1345 cm^{-1} are still observed but in lower intensities.

Functionalization of diazotized aminopolystyrene with nitroformaldehyde β-naphthylhydrazone

Several batches of each type of aminopolystyrene functionalized with nitroformaldehyde β-naphthylhydrazone were prepared and characterized by infrared and ultraviolet spectroscopy. In a typical experiment aminopolystyrene was diazotized with an excess of sodium nitrite in hydrochloric acid and then coupled with nitroformaldehyde β-naphthylhydrazone to give a polymer-supported diarylnitroformazan (Figure 5) following the reaction pathway described earlier for synthesis of unsymmetrical diarylnitroformazans (17,18). The best results in the synthesis of unsupported diaryl-nitroformazans were obtained when the reaction mixture was buffered to pH about 4 to 5 as noted above. However the diazotization reaction required a very low pH of ~0-1. Therefore in order to avoid neutralization of excess acid with alkali the coupling procedure was modified. The aminopolystyrene diazotized at low pH was quickly filtered off from the reaction mixture, washed generously with cold water, and immediately transferred into a cold buffer solution of sodium acetate and acetic acid followed by coupling with the nitroformaldehyde arylhydrazone dissolved in ethanol. The dark colored solids were separated and washed with water to remove acetates followed by ethanol to remove excess nitroformaldehyde arylhydrazone. In several cases Soxhlet extraction was used to minimize the use of ethanol.

The electronic spectra of these products showed a very broad absorption in the 400-700 nm range without distinct maxima similar to the spectra of solid diarylnitroformazans. The infrared spectra of functionalized polymers in general exhibit changes in the relative intensities of bands in the 1100–1700 cm^{-1} range as well as new bands at 1273, 1320, 1520, 1541, 1628 and 1653 cm^{-1} compared to the starting aminopolystyrene. Those absorptions can be observed as distinct bands or as shoulders depending on the degree of functionalization which would depend of course on the number of active sites on the polymer generated during diazotization. The positions of new bands or shoulders in general are comparable with the those of most intense bands for di-β-naphthylnitroformazan and the corresponding unsymmetrical phenyl-β-naphthylnitroformazan.

In the next step the functionalized polystyrenes were treated with an aqueous solution of ammonium sulfide at room temperature for at least 3 hr. In some cases this reaction was carried out for 18 hr in order to observe visually a color change to beige in accord with expectations based on the color of the corresponding di-β-naphthylthiocarbazide. However in many cases the products after treatment with aqueous ammonium sulfide displayed dark green or gray colors similar to solid diarylthiocarbazones.

After treatment with aqueous ammonium sulfide the infrared spectra of the functionalized polystyrenes exhibited bands characteristic of polystyrene and some shifts in the band positions compared to the β-naphthylnitroformazan modified polymers. The most significant difference is that the band at ~1273 cm^{-1} originally present in the infrared spectra of polystyrenes functionalized with nitroformaldehyde β-naphthylhydrazone vanished after treatment with ammonium sulfide similar to observations with unsupported di-β-naphthylnitroformazan indicating conversion of the supported diarylnitroformazan functionalities to the corresponding diarylthio-carbazide functionalities.

The green or olive color of the solid products suggests that the supported diarylthiocarbazides were at least partially oxidized to the corresponding diarylthiocarbazones. Therefore we decided to test the products at this stage of the reaction sequence with heavy metal cations. In this connection these solids were shaken vigorously with 0.005 M aqueous solutions of Pb^{+2} and Hg^{+2}. The green or olive colors of these solids were replaced with reddish or pinkish colors indicating metal complexation. These metal-containing polymers were separated and air dried and their infrared spectra were recorded giving the following results:

Lead-containing polymer: 672, 752, 834, 905, 1016, 1067, 1160, 1178, low intensity 1230, 1341, 1438, 1498, 1578 with several shoulders, 1654 cm^{-1}.

Mercury-containing polymer: 675, 752, 829, 903, 1015, 1054, 1154, 1268, 1326, 1340, 1432, 1465, 1471, 1508, 1560, 1576, 1654 cm^{-1}.

It was not surprising that in these experiments only low intensity bands corresponding to the metal complexes were observed in addition to those of the diarylthio-carbazide since the degree of the oxidation of the functionalized polymer is presumably very low. However some changes in the infrared spectra are visible. The infrared band around 1382 cm⁻¹ in the metal-free polymers vanished from spectra of samples treated with the heavy metals. Instead a set of bands at 1320, 1342, and 1360 cm^{-1} were observed. The bands at ~1022, 1067, 1179, 1490, and 1597 cm^{-1} for supported β–naphthyl species are shifted to 1016 and 1015, 1059 and 1054, 1160 and 1154, 1500 and 1474, 1578 and 1576 cm^{-1} for samples treated with Pb^{2+} and Hg^{2+}, respectively. Such changes in the infrared spectra are consistent with literature data (19,20,21,22,23,24) concerning formation of heavy metal complexes of diphenyl-thiocarbazone and di-β-naphthylthiocarbazone.

Since there are no data in the literature concerning oxidation of polymer-supported diarylthiocarbazides to the corresponding diarylthiocarbazones, it was necessary to find suitable conditions for these oxidations. In this connection the polymer-supported diarylthiocarbazides were oxidized using methods described above including treatment with ethanolic KOH, air oxidation in a chlorinated solvent (1,2-dichloroethane), and treatment with dilute hydrogen peroxide.

Polymer-supported diarylthiocarbazides were expected to be more difficult to oxidize than unsupported diarylthiocarbazides. Therefore the polymer-supported diarylthiocarbazides were allowed to react with potassium hydroxide solution for 1 or 2 hr instead of the 15 min used for di-β-naphthylthiocarbazide. The infrared spectra of these products indicated no significant differences between those two samples, which is not surprising since diarylthiocarbazides and diarylthiocarbazones absorb in similar regions. However some changes in the relative intensities and band positions are observed, especially around 1152, 1179, 1342 and 1654 cm^{-1} indicative of thiocarbazone formation. Air oxidation in 1,2-dichloroethane appeared to give the same product. Oxidation with dilute hydrogen peroxide produced an almost black solid. The infrared spectra exhibited changes in the 1000–1340 cm^{-1} region with domination of a band around 1152 cm.$^{-1}$

Presently available methods do not allow a quantitative comparison of supported diarylthiocarbazones obtained by different oxidation methods . Therefore these products were tested in reactions with heavy metals (Pb^{2+} and Hg^{2+}). In this connection all samples were found to give a positive reaction with 0.005 M Pb^{2+} and Hg^{2+}. There were changes in the infrared spectra of such samples treated with heavy metals similar to those discussed above.

Both KOH oxidation/disproportionation and air oxidation were found to give similar products based on infrared spectroscopy and reactions with heavy metal cations. However owing to the properties of uncrosslinked polystyrene, the KOH oxidation/disproportionation was chosen for further experiments.

All types of functionalized polystyrenes were converted into the corresponding polystyrene-supported diarylthiocarbazides. The longest time required to observe color changes upon treatment of the polystyrene-supported diarylnitroformazan with aqueous ammonium sulfide occurred with samples prepared from polystyrene containing 2% DVB. Furthermore, treatment of these polystyrene-supported diarylthiocarbazides with potassium hydroxide gave no observable color changes; however the resulting products appeared to react with heavy metal cations giving pinkish to violet reflections in the solutions. However, such reactions were not spontaneous requiring some time (up to 18 hr). to develop these color changes. For polystyrene-supported dithizones made from polystyrene containing 2% of DVB, these color changes with heavy metals could not be observed visually. In experiments using different polystyrenes it was found that the reactions of polystyrene-supported diarylnitro-formazans with aqueous ammonium sulfide in basic ethanolic suspension were more rapid than those in neutral solution giving observable color changes after 3 hr of stirring. All samples appeared to react with heavy metal cations (Pb^{2+} and Hg^{2+}) but a long time up to 3 days was required to observe the color changes.

Anchoring of di-β-naphthylthiocarbazone to polystyrene using the procedure of Chwastowska and Kosiarska

The second approach to the synthesis of spectroscopic indicators for heavy metal cations is based on work of Chwastowska and Kosiarska (10) in which previously synthesized diarylthiocarbazones are anchored to the diazotized aminopolystyrene through an additional coupling reaction (Figure 3). Several batches of polymers functionalized in this way were synthesized from the corresponding aminopolystyrene derived from uncrosslinked as well as crosslinked polystyrene. The products of this reaction are brown or black solids, which were characterized by infrared and ultraviolet spectroscopy.

The electronic spectra of polystyrenes functionalized by the method of Chwastowska and Kosiarska (10) display a maximum around 268 nm and a very flat, broad absorption between 400–700 nm. The solid di-β-naphthylthiocarbazone itself in a Nujol mull exhibits absorptions at 270, 340, and 600–750 nm but the maxima cannot be estimated precisely. In the infrared spectra of polystyrenes functionalized by the method of Chwastowska and Kosiarska (10) new bands noticeable as shoulders around 1200, 1383, 1475, and 1583 cm^{-1} indicate anchored di-β-naphthylthiocarbazone.

The di-β-naphthylthiocarbazone-supported polystyrenes functionalized by the method of Chwastowska and Kosiarska (10) were tested in reactions with Pb^{2+} and Hg^{2+} cations in aqueous solution. In this connection several mg of modified polymer were shaken with 0.005 M aqueous solutions of Pb^{2+} and Hg^{2+} cations for several minutes. The resulting suspensions exhibited pink and violet-red reflections in contrast to the corresponding metal-free polymers. The solids were filtered off, washed with water, and air dried. The amounts of solids were not sufficient to obtain reasonable electronic spectra in the solid state but allowed us to obtain their infrared spectra.

The most significant differences in the infrared spectra of di-β-naphthylthio-carbazone-supported polystyrenes functionalized by the method of Chwastowska and Kosiarska (10) upon treatment with heavy metals were observed in the 1100–1600 cm^{-1} region. Thus the infrared band around 1382 cm^{-1} vanished from spectra of samples treated with Pb^{2+} and Hg^{2+}. Instead a set of bands around 1310, 1341, and 1360 cm^{-1} was observed. The bands at ~1152, 1474 and 1583 cm^{-1} for anchored di-β-naphthylthiocarbazone were shifted to 1176 and 1173 cm^{-1}, 1491 and 1486 cm^{-1}, and 1594 and 1597 cm^{-1} for Pb^{2+} and Hg^{2+} treated samples, respectively. Such changes in the infrared spectra are consistent with literature data concerning formation of metal dithizonates (19,20,21,22,23,24).

We were uncertain if chemically anchored or unwashed physically adsorbed di-β-naphthylthiocarbazone was responsible for the color reactions with heavy metal cations in di-β-naphthylthiocarbazone-supported polystyrenes functionalized by the method of Chwastowska and Kosiarska (10). Therefore we prepared samples containing 1% by weight of the diarylthiocarbazone by making a tetrahydrofuran solution containing both the pure uncrosslinked polymer and di-β-naphthylthio-carbazone and allowing this solution to evaporate at room temperature in a big Petri dish. After evaporating overnight a polystyrene thick film was obtained. This film had the characteristic dark green color of the di-β-naphthylthiocarbazone ligand but did not undergo color changes with 0.005 M Pb^{2+} and Hg^{2+} even after 3 days. The infrared spectrum shows only changes in the relative intensities of bands owing to low loading. Similar results were obtained by using polyvinylchloride (PVC) rather than polystyrene films containing di-β-naphthylthiocarbazone. A very thin PVC film prepared by coating a very dilute solution on a large surface exhibited an infrared spectrum similar to a pure PVC matrix with some shifts in band positions. Thus the band at ~ 1097 cm^{-1} in PVC is shifted to 1064 cm^{-1} for the film and a shoulder at ~1131 cm^{-1} can be observed . The electronic spectra of the modified PVC film exhibited absorptions at 328 and 650 nm characteristic of pure di-β-naphthylthio-carbazone. Samples prepared in this way appeared to react with 0.005M aqueous solutions of heavy metal cations (Pb^{2+} and Hg^{2+}) after 3 days.

Summary

This work demonstrates a method for chemical immobilization of a diarylthio-carbazone unit onto a polystyrene matrix using the coupling of nitroformaldehyde β-naphthylhydrazone with diazotized aminopolystyrene followed by treatment with aqueous ammonium sulfide and oxidation. This method has been used to prepare samples of polystyrene-supported diarylthiocarbazones which undergo distinctive color changes in 0.005 M aqueous Pb^{2+} and Hg^{2+}. Materials of this type are potentially useful as chromogenic ligands in fiber optic sensors for detection of toxic heavy metals of environmental concern.

Acknowledgment

We are indebted to the U. S. Department of Energy for support of this work through the Westinghouse Savannah River Corporation and the Education Research, and Development Association of Georgia Universities. We would like to acknowledge helpful discussions with Dr. Patrick O'Rourke, currently of EquaTech International, New Ellenton, South Carolina.

References

1. J. F. Adler, D. C. Ashworth, R. Narayanaswamy, R. E. Moss, I. O. Sutherland, An Optical Potassium Ion Sensor, *Analyst*, **1978**, *112*, 1191–1192.
2. W. A. de Oliveira, R. Narayanaswamy, A Flow-Cell Optosensor for Lead Based on Immobilized Dithizone, *Talanta*, **1992**, *39*, 1499–1503.
3. D. J. Edlund, D. T. Friesen, W. K. Miller, C. A. Thornton, R. L. Wedel, G. W. Rayfield, J. R. Lowell, Thin-film Polymeric Sensors for Detection and Quantification of Multivalent Metal Ions, *Sensors and Actuators B*, **1993**, *10*, 185–190.
4. R. Czolk, J. Reichert, H. J. Ache, An Optical Sensor for the Detection of Heavy Metal Ions, *Sensors and Actuators B*, **1992**, *7*, 540–543.
5. P. V. Lambeck, Integrated Opto-chemical Sensors, *Sensors and Actuators B*, **1992**, *8*, 103–116.
6. A. Morales-Bahnik, R. Czolk, J. Reichert, H. J. Ache, An Optochemical Sensor for Cd(II) and Hg(II) Based on a Porphyrin Immobilized on Nafion® Membranes, *Sensors and Actuators B*, **1993**, *13-14*, 424–426.
7. G. Frishman, G. Gabor, Surface Characteristics of Optical Chemical Sensors, *Sensors and Actuators B*, **1994**, *17*, 227–232.
8. R. E. Buckles, M. P. Bellis, o-Nitrocinnamaldehyde, *Organic Syntheses*, Coll. vol. IV, Wiley, New York, **1963**, p.722.
9. R. B. King, E. M. Sweet, Polymer-Anchored Cobalt (II) Tetraarylporphyrin Catalysts for the Conversion of Quadricyclane to Norbornadiene in a Solar Energy Storage System, *J. Org. Chem.*, **1979**, *44*, 385–391.
10. J. Chwastkowska, E. Kosiarska, Synthesis and Analytical Characterization of a Chelating Resin Loaded with Dithizone, *Talanta*, **1988**, *35*, 439–442.

11. E. B. Sandell, Diphenylthiocarbazone, in *Photometric Determination of Traces of Metals*, Wiley-Interscience, New York, **1978**, Chapter 6G, pp. 597–647.
12. M. M. Harding, The Crystal Structure of the Mercury Dithizone Complex, *J. Chem. Soc.*, **1958**, 4136–4143.
13. I. B. Suprumovich, Preparation of Dinaphthylthiocarbazone and its Metal Salts, *J. Gen. Chem. U. S. S. R.*, **1938**, *8*, 839–843,
14. E. Bamberger, R. Padova, E. Omerod, Über Nitro- and Amlno-formazyl, *Liebigs Ann.*, **1926**, 260–307.
15. D. M. Hubbard, E. W. Scott, Synthesis of Di-beta-naphthylthiocarbazone and Some of its Analogs, *J. Am. Chem. Soc.*, **1943**, *65*, 2390–2393.
16. L. S. Pupko, N. Berzina, P. S. Pel'kis, The Synthesis of Substituted Phenylhydrazones of Nitroformaldehyde, *Zhur. Obshch. Khim.*, **1963**, *33*, 2217–2220.
17. L. S. Pupko, P. S. Pel'kis, Synthesis and Investigation of the Properties of Unsymmetrical Derivatives of Thiocarbazone, *Zhur. Obshch. Khim.*, **1954**, *24*, 1640–1645.
18. L. S. Pupko, P. S. Pel'kis, Synthesis and Investigation of Unsymmetrical Diarylthiocarbazones with Halo and Carboxy Substituents, *Zhur. Org. Khim.*, **1965**, *1*, 735–738.
19. H. M. N. H. Irving, *Dithizone*, Analytical Science Monograph The Chemical Society, London, **1977**.
20. A. M. Kiwan, A. Y. Kassim, 1,5-Di(2-Fluorophenyl)-3-Mercaptoformazan, A New Metal Extractant, *Anal. Chim. Acta*, **1977**, *88*, 177.
21. S. Liu, J. Zubieta, Chemical and X-ray Structural Characterization of Diphenylcarbazone and Diphenylthiocarbazone Complexes of Molybdenum(VI), $(n\text{-}Bu_4N)[MoO_2(C_6H_5NNC(X)NNC_6H_5)(C_6H_5NNC(X)NN(H)C_6H_5)]$ (X = O or S) and $(nBu_4N)MoOCl_3(C_6H_5NNC(O)NNC_6H_5)]$; Isolation of 2,3-Diphenyl-tetrazolium-5-thiolate, $C_6H_5NNC(S)NNC_6H_5$, from the Reaction of Diphenyl-thiocarbazone with Molybdate, *Polyhedron*, **1989**, *8*, 677–688.
22. Y. Singh, S. B. Tyagi, R. N. Kapoor, Synthesis of Some New Organo-zirconium(IV) Dithizonates, *Acta Chim. Hungarica*, **1989**, *126*, 665–672.
23. S. B. Tyagi, B. Singh, R. N. Kapoor, Reactions of Hafnium(IV) Isopropoxide Isopropanolate with 1,5-Diarylthiocarbazones, *Acta Chim. Hungarica*, **1986**, *122*, 229–234.
24. Y. Singh, R. Sharan, R. N. Kapoor, Synthesis and Spectra Characteristics of Bis(indenyl)titanium(IV) Dithizonate Complexes, *Synth. React. Inorg. Met.-Org. Chem.*, **1987**, *17*, 759–771.

Chapter 4

Electronic Nose for the Detection of Organochlorines and Polyaromatic Hydrocarbons

Miriam M. Masila and Omowunmi A. Sadik[1]

Department of Chemistry, State University of New York at Binghamton, Binghamton, NY 13902–6016

Halogenated organic compounds and polyaromatic hydrocarbons (PAHs) are traditionally analyzed using gas chromatography with mass spectrometry (GC/MS) and high performance liquid chromatography (HPLC) techniques. However, due to the costs and time involved between sample collection and analysis, *in-situ*, real-time monitoring techniques are required for on-site characterization. This paper describes the development of multiarray sensors coupled to integrated pattern recognition techniques for the detection of organochlorines and PAHs. The sensor elements consist of an array of thin films having unique fingerprints for each analyte. The output sensor signals were integrated into software-based artificial neural network for the identification of the analytes. The sensor array showed good reproducibility to the different volatile and was classified with respect to the nature of analyte functional groups and physico-chemical properties. Classification rates of greater than 98% for achieved for the organochlorines and PAHs.

[1]Corresponding author.

37

Introduction

Organochlorine compounds and aromatic hydrocarbons are the most prevalent contaminants found in groundwater, underground storage tanks and hazardous waste sites in the United States. These contaminants require constant monitoring due to their toxicity, persistence and ability to bioaccumulate in the environment. The high costs and time required to analyze these compounds using conventional GC/MS have resulted in an urgent need for *in-situ*, remote-controlled, monitoring techniques capable of providing rapid identification, quantitation and long-term mapping of polluted sites.

In recent years, *in-situ* techniques which are commonly referred to as Electronic Nose (EN) capable of mimicking the mammalian olfactory systems have been developed for odors, perfumes, beverages, flavorings and alcohlos [1-5]. Using EN, it is possible to retrieve a desired, multi-component information with some degree of accuracy. For example, when arrays are presented with pulses of organic vapors, they generate the characteristic temporal response for each vapor. These response patterns can then be used to train computational neural networks capable of recognizing unknown samples. EN has been used in a wide range of applications, including clinical diagnosis, malodor measurements in agricultural environments, and quality control issues in the food industry [6-11].

However, odor and taste analyses with the EN are only suitable for the analysis of non-hazardous compounds. In addition, limited instrumental sensitivity of the EN sensing elements precludes the use of hazardous compounds with low vapor pressure. Since certain organic thresholds can be extremely low, compounds possessing vapor pressure too low to register in many of the EN profiles could still be important odorants. Therefore, to achieve total in-situ analysis using these systems, there is a need for alternative techniques that can define sensor-odor activity relationships in terms of the whole analytical process and not just the sensor arrays alone (12,13). In this paper, we present the use of quantitative sensor-solvent activity relationship (QSSR) to enhance the performance of *in-situ* volatile analysis in EN systems. Using polyaromatic hydrocarbons (PAHs), industrial solvents and chlorinated phenols as model compounds, the concept of QSSR was tested by establishing the relationship between the sensor's structural parameter and the chemical compound, volatile detectability and analyte intensity. QSSR was tested using chlorinated phenols, low molecular weight organics and PAHs. These analytes were chosen due to their toxicity, persistence in the environment and bioaccumulation.

Mimicking Natural Olfaction Using Chemical Sensor Arrays

The term electronic nose commonly refers to multiarray sensors coupled to headspace sampling and pattern recognition modules. The system employs metal oxides, quartz crystal arrays, surface acoustic wave devices, electrochemical cells and conducting polymers, or a combination of these sensors to mimic human sense of smell. In analogy to natural olfaction, array-based sensors use certain processes to mimic the mammalian olfactory system. These processes are used to generate signals for different odors or volatiles.

As outlined in Table 1, there are several basic, parallel steps involved in odor recognition in both the EN and natural olfaction. These include sampling, reception, signal generation, transmission, detection/recognition and cleansing. Sampling in natural olfaction occurs during sniffing, which draws the odorants from the outside world into the nasal cavity. In EN the volatile molecules within the headspace generated inside a sample bag are drawn into the sensor surface. In natural olfactory, reception occurs in the olfactory epithelium where the odorants bind to the olfactory receptors, thus generating an electrical signal. Adsorption/desorption kinetics on the sensor-odorant interface in EN result in changes in the sensor resistance of conducting polymer, which correspond to the generated signal. Olfactory axons transmit the signals from the olfactory epithelium to the olfactory bulb for initial processing, detection and recognition and finally to the cerebral cortex for further signal processing as well as discrimination (14,15). In EN, the database generated from sensor responses is saved for further manipulations that are used to obtain normalized sensor responses as well as for pattern recognition.

Electronic Nose Technology

Metal Oxides
Metal oxide semiconductors, (MOS), e.g. tin oxide and zinc oxide, exhibit a change in electrical resistance/conductivity when exposed to volatile compounds due to the adsorption or catalytic oxidation of gaseous molecules on the sensor surface [2,3,16]. These changes are then registered as signals corresponding to the sensor response. Metal oxide sensors are operated at high temperatures (100-600°C). The sensitivity and specificity to a particular analyte can be improved by selecting the appropriate operating temperature and type of metal oxide. However, a broad, overlapping selectivity can be encountered which makes metal oxides ideal for sensors in EN systems. Metal oxides have sensitivity range of 5-500 ppm [2]. The limitations of metal oxide sensors include high operating temperatures, high sensitivity to humidity, poisoning by sulfur compounds and baseline shift. They are however readily available and relatively cheap to make [16].

Conducting Polymers
Conducting eletroactive polymer (CEP) sensors represents another type of EN sensing element in which signals are generated as a result of changes in resistance due to adsorption/desorption of analyte molecules at the polymer surface. The synthesis, characterization and applications of CEPs based sensors such as polypyrrole, polyaniline and polythiophene have been extensively studied [17-21]. The advantages of CEP sensors over other sensor types include operation at ambient

Table I: Odor recognition in natural olfaction versus electronic nose

	Natural Olfaction	Electronic Nose
Sampling	Sniffing: Mixing of molecules, dissolution in mucus and transportation to olfactory receptors	Headspace sampling: Filling bag with samples and reference air, sample conditioning and exposure to sensor surface
Reception	Olfactory receptor: Interaction of odorants with receptors (locating binding site). Binding of odorant molecules to receptor.	Sensor array: Diffusion and interaction of odorants with sensor surface. Adsorption/desorption processes (non-covalent binding) of molecules.
Signal Generation	Chemical stimulation generates electrical signal.	Sensor interface: Changes in electrical resistance of sensor elements, generating responses used for database
Transmission	Olfactory axon: Signal is transmitted to olfactory bulb.	Computer: sensor responses stored in the computer for further manipulation.
Signal Recognition and processing	Olfactory bulb: Subconscious association of odor with known smell	Software program: Pattern recognition software:
	Cerebral cortex: Conscious sensation of smell and emotional reactions.	Pattern recognition techniques: e.g. CA, PCA, ANN, Unique fingerprints, % recognition
Cleansing	Breathing fresh air.	Purging sensor surface.

temperatures, high sensitivity (0.1-100 ppm) [2], ease of fabrication and reversibility in analytical signals obtained when exposed to volatiles and fast kinetics at the sensor surface.

Piezoelectric Sensors
Piezoelectric sensors measure change in mass resulting from the interaction of analytes and the sensor material. Examples include surface acoustic wave (SAW) and quartz crystal microbalance (QCM) devices [22-23]. QCM consists of a gold or platinum disk modified with a polymer film to serve as the sensing material. Two electrodes on each side of the disk are used as connectors to an oscillator circuit. The disk resonates at a fundamental frequency (10MHz) after excitation with an oscillating signal. When gaseous molecules are adsorbed on the sensor surface or absorbed into the polymer, an increase in mass occurs resulting in reduction of the resonance frequency of the disk. Frequency reduction is related to the mass of the analyte adsorbed from Sauerbrey equation and is given by:

$$\Delta f = \frac{-2 f_0^2 \Delta m}{A(\mu_q \rho_q)^{1/2}}$$

....Equation 1

where Δf is the change in fundamental frequency, f_0 is the frequency of the quartz crystal prior to mass change. Δm is the change in mass, A is the piezoelectrically active area of the electrode, μ_q is the shear modulus, and ρ_q is the density of the quartz [24].

Optical Sensors
An optical sensor utilizes the optical properties of the sensing surface such as fluorescence, chemiluminescence, photobleaching or phosphorescence to monitor the interaction between the analyte molecules and the active material [25,26]. It may consist of a conventional optical fiber, which guides the radiation from the source through the sensing material and back to the spectrophotometer. The sensor surface is coated with a fluorescence dye whose frequency shifts upon interaction with the analyte. This shift in frequency is registered as a signal, which is then analyzed to give the characteristic properties of the analyte [27,28].

Pattern Recognition

Array of sensors used in EN systems generate a multi-dimensional response pattern (n-dimensional space), where n is the number of sensors in the array. When n is more than 3, human perception of the structural relationships is difficult unlike in 2- and 3-dimensions. A number of approaches are therefore necessary to map such a multi-dimensional pattern to a 2- or even 3- dimension [29]. Certain types of methods are used for processing EN data. These include:

1. Linear or nonlinear mapping algorithms: Linear mapping algorithms are simple and general in their applications. Nonlinear mapping algorithms are much more complicated mathematical formulations used when linear mapping can not preserve complex data structures.

2. Supervised and unsupervised learning methods: In supervised learning, mathematical rules are employed to train the output of the sensor array by relating the output pattern to a set of known descriptors (K classes) held in a knowledge base. The unknown odor is tested against the knowledge base and a class membership is given as unique discriminant. In unsupervised learning the program discriminates between the response patterns without a separate training step. Unsupervised methods are closer to how the brain works.

3. Parametric and non-parametric methods: Parametric methods depend on known probability distribution of the variables. They are based on an assumption that the response pattern generated can be described by a probability density function. A knowledge base, which contains the appropriate probability density function, is created using the response pattern.

Some of the common linear, parametric and supervised pattern recognition techniques include principal component regression (PCR), discriminant function analysis (DFA), and partial least squares (PLS). Principal component analysis (PCA) is a linear, supervised but non-parametric technique; while cluster analysis (CA) is a linear, non-parametric and unsupervised pattern recognition technique. Other cluster methods such as Sammon mapping and fuzzy logic (e.g. kohonen network) are nonlinear, non-parametric and unsupervised pattern recognition methods. Back-propagation artificial neural network (ANN) is nonlinear, non-parametric and supervised.

In ANN, a three-layer back-propagation method, which mimics the olfactory system, used for the system training. The three layers are input, hidden and output layers. The input layer is equivalent to the olfactory receptor cells, the hidden layer to glomeruli nodes and the output layer to the mitral cells. The sample database is fed into the input layer and the hidden layer sums the data in every other database to form a set of temporary sensory models. The third layer tries to reconstruct the original data. One pass through the three layers is called a training epoch. If the result of the third layer matches the original response pattern, and no error is reported, then the training is considered complete. In addition, the software is able to recognize such pattern when it encounters it again. If the third layer results do not match the original pattern, another pass through the layers is repeated [3,31].

Problems of Conventional EN Sensors

In EN instruments, sensor signals are significantly dependent on the nature of the sensing materials. However, other parameters likely to influence the performance of the resulting system include flow profile and nature of carrier gas, the reaction (adsorption/desorption) kinetics of the odor and the active material, the extent of diffusion of the odor within the active material, and the nature of the sensing material [13]. Existing EN instruments use either equilibrium and/or dynamic stripping sampling techniques. In each case, the air from the headspace is drawn into contact with the sensor surface for analysis. This introduces an unknown quantity of the vapor samples into the analysis. Consequently, E-nose technique is good for qualitative analysis, but not particularly suitable for quantitative analysis. Therefore, the integration of GC and EN reported previously should ensure that a finite (known) quantity of the sample is introduced into the system for analysis. This approach not only allows a better sample handling, but also provides a way of separating samples containing more than one component. Another way of enhancing analyte signals using these systems involves the use of QSSR discussed below.

Quantitative Sensor-Solvent Activity Relationship (QSSR)

In analogy to biological systems, we describe the concept of QSSR as a rational approach that enhances the EN signals. QSSR is an empirical way of connecting the odor potency of a series of chemically related compounds to sensor structures. As a simple example, an odor–structure relationship should exists between the adsorption-desorption energies of a molecule and air–sensor interfaces. Other important measurable parameters may include polarity, partition coefficients between volatile and sensor interface, electron donor-acceptor interactions, molecular size and shape. The application of QSSR can be evaluated to predict the properties of an odorous sample prior to experimental measurements. This can be achieved using the most important concept in odor analysis such as detectability or intensity. Detectability relates to the number of times an odor sample must be diluted with an equal volume of free-air and yet is detected 50% of the time. Intensity is a function of the odor concentration.

Ideally, if a mathematical function can be established between the structural parameter of the sensor and the chemical or odorous molecule, the odor detectability or intensity properties can be estimated. When coupled with gas chromatography, the quantitative relationship between a sensor response and partition coefficient (K) can then be related to odor polarity, chain length or molecular weight. If the value of K of an organic molecule having chain lengths C_{18}, C_{19}, and C_{21} is known, it would be possible to estimate the strength of the odor by interpolation of partition coefficient versus the number of carbon atoms. In addition, if the perceived odor of a series of phenolics can be related to their binding strength, it would be feasible to predict the odor of different phenols from the measurement of the sensor intensity [30]. By assigning empirical values to specific structural features of a molecule, multiple linear regression therefore produces a quantitative structure-activity relationship of the following form:

Odor Intensity = a x chain length + b x vapor pressure + c X partition coefficient +
d x solubility etc ....*Equation (2)*

The magnitudes of the coefficients *a*, *b*, plus their standard deviation will reveal which structural features are important in determining the intensity of the odor. Large coefficient implies important factors and small coefficient or wide interval signifies relatively unimportant factor.

Experimental Section

Reagents and Stock Solutions

2,4-Dichlorophenol (2,4-DCP), 2,4,6-trichlorophenol (2,4,6-TCP) 98%, and 2-chlorophenol (2-CP) were purchased from Sigma. 2,4,5-trichlorophenol (2,4,5-TCP) 99.9%, 4-bromo-2,5-dichlorophenol (4-B-2,5-DCP) 99.3%, 1,1-bis(4-chlorophenyl)-2,2,2-trichloroethane (DDT) 99.3%, and 3-chlorophenol (3-CP) 99% were EPA reference standards (EPA-Research Triangle Park, NC). I-butanol and methanol (both certified ACS reagents) were purchased from Fisher chemicals, pyrene (98%) from Aldrich and toluene (99.9%) was from J. T. Baker. All reagents were used as supplied by the manufacturers and 2% butanol in nanopure water was used as the wash solution for the sensor arrays. Pyrene was dissolved in toluene/methanol mixture (10:1, v/v).

Instrumentation

AromaScanner model A32S (AromaScan, Inc. NH) was used for all electronic nose experiments. The block diagram of this system is shown in Figure 1. The instrument consists of a sample station, sample analyzer and A32S software. The sample station includes an air inlet system, the sample port, the wash, and sample conditioning station where the temperature and humidity of the sample are preconditioned for a specified period of time before analysis. The analyzer consists of an array of 32 sensors, based on conducting polypyrrole films prepared from a wide range of counterions. The electrical resistance of these polymers changes upon interaction with the sample vapors, thus generating the sensor signal. For the analysis of PAHs, the sample port was connected to a gas chromatographic system using a homemade sampling port. The A32S software was used for data acquisition, manipulation and mapping, moreover, ANN training was done using ANN software which is incorporated in the A32S software.

Sampling

Equilibrium sampling was used for the chlorinated phenols. In equilibrium sampling, the sample plus the filtered reference air are introduced into a sample

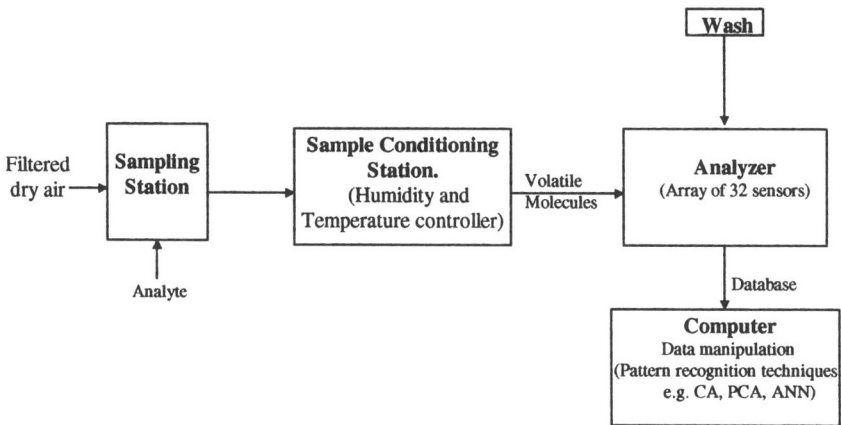

Figure 1. Block diagram of AromaScan A32S electronic nose.

pouch at a defined humidity. The sample vapors are allowed to accumulate in the headspace while the sample pouch is kept at constant temperature (30 minutes) so as to attain an equilibrium between the sample and the sample vapors. The air from the headspace is then drawn across the sensors during analysis. The data acquisition method involves several steps. First, a stream of reference air is drawn across the sensors for 30 sec to establish a stable baseline. Secondly, air from the headspace is passed over the sensor array for 80 sec and the analyte signals are recorded. Next, vapors from the wash are then passed over the sensors for 60 sec to wash off sample molecules from the sensor surface. Finally the reference air is again drawn across the sensor surface for 30 sec to confirm that the sensors are effectively clean.

0.01g of the solid samples was weighed into the sample pouch and the equivalent volumes of the liquid phenols calculated from their densities were used. The sampling conditions used include a relative humidity of 50% and a temperature of 25°C. In order to maintain a constant headspace volume for all the analytes, the filling time of the reference air was approximately 110s for each bag. The samples were directly analyzed after 30mins of incubation time in the sample conditioning station. For pyrene analysis, 1μL samples were injected at a temperature of 150°C.

Data Processing

The raw data obtained during data acquisition was fed into the A32S pattern recognition software to generate a database. The A32S data mapping software converge the 32-dimensional space into a 2- or 3-dimensional space using a nonlinear mapping algorithm that preserves the internal structures as much as possible. It was therefore possible to display the results in a 2- or 3-D plot called Sammon map [29]. Differences were observed and quantified by the Euclidean distance (ED). This method of treating sensor data has been used for many years [32]. A small ED is indicative of similarity in odor [30]. An array of n sensors generates n response patterns when exposed to a single analyte corresponding to a high n-dimensional space. The Euclidean distance is calculated as follows

$$D_{ij} = [\sum_{k=1}^{n} (Y_{ik} - Y_{ij})^p]^{\frac{1}{p}}$$

.............Equation 3

where d_{ij} is the difference between two patterns (i and j) in the new d-dimensional space. The initial values of the d-dimensional space are chosen arbitrarily from small random values. A gradient method (or steepest descent method) is employed such that at each iteration, the error E is minimized. E is zero when $d_{ij}* = d_{ij}$. P is normally set to 2 (the Euclidean metric). Using similarity value S_{ij} is given by:

$$S_{ij} = 1 - \frac{D_{ij}}{Max[D_{ij}]}$$

.....................Equation 4

Results and Discussion

We have previously shown that odor recognition at CEP sensors could be based on the functional groups, size, shape as well as the physical properties of the odorants [30]. For compounds with the same structural shape, different positioning of the same functional group and the presence of different functional groups result in different electronic distribution, electronegativity, polarity and other physical constants. These physico-chemical parameters influence the desorption/adsorption kinetics responsible for the non-covalent interactions at the sensor-odorant interface. Subsequently the sensor response gives rise to a unique pattern for compounds whose parameters differ significantly. In this work, we tested the validity of this assumption by examining some structurally similar organochlorine compounds including; 2-CP, 3-CP, 2,4-DCP, 2,4,5-TCP and 4-B-2,5-DCP and PAHs (Scheme1).

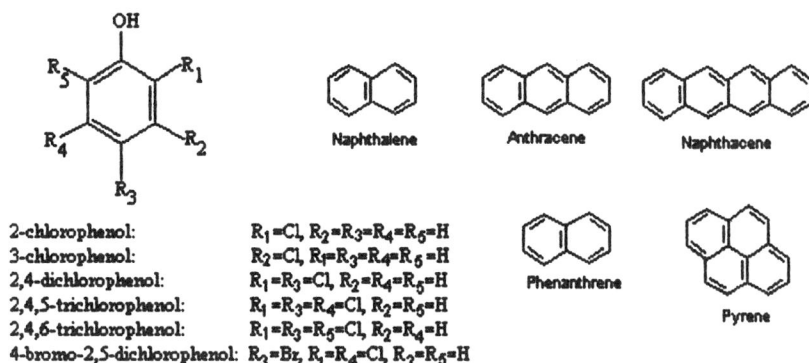

2-chlorophenol:	R_1=Cl, R_2=R_3=R_4=R_5=H
3-chlorophenol:	R_2=Cl, R_1=R_3=R_4=R_5=H
2,4-dichlorophenol:	R_1=R_3=Cl, R_2=R_4=R_5=H
2,4,5-trichlorophenol:	R_1=R_3=R_4=Cl, R_2=R_5=H
2,4,6-trichlorophenol:	R_1=R_3=R_5=Cl, R_2=R_4=H
4-bromo-2,5-dichlorophenol:	R_3=Br, R_1=R_4=Cl, R_2=R_5=H

Scheme 1: Structures of important chlorinated phenols and PAH

The most important chlorinated phenols selected include 2, 4-dinitrophenol, 2, 4, 5-trichlorophenol and pentachlorophenol. PAHs are distinguished by fused, chain-linked cyclic ring structure with a wide range of molecular sizes, structural types as well as isomers. The major sources of PAHs are crude oil, coal and oil shale. Coal tars and petroleum residues from refinery plants contain high levels of PAHs. PAHs are found at significant levels in soil, water, plants, and animals. PAHs in the atmosphere largely result from fossil fuel combustion, refuse burning, coke ovens, vehicle emissions, forest fires, and volcanic activities. PAHs are known to be bioaccumulative as well as carcinogenic. Examples include benzo(a)pyrene and dibenz(a,h)anthracene and nitro-substituted PAHs.

Discrimination Based on Physical Constants and Polarity:

The effects of certain physical constants on analyte recognition, such as melting point, solubility and vapor pressure were examined. Figure 2 shows a plot of sensor response for sensor number 24 versus the melting points of chlorinated phenols. As evidenced from the plot, the sensor response decreases with increase in the melting point. Low melting point results from weak intermolecular interactions between analyte molecules, and inherently high volatility. High volatility increases the number of molecules subsequently results in lower sensor response. The less volatile molecules should absorb more completely and produce higher sensor response (slope of the mg/unit polymer per concentration curve. This observation is consistent with other work where higher sensitivity was observed for lower vapor pressure analytes (33).

A very high, positive, sensor response was observed for 2-CP compared with other chlorinated phenols. This unique pattern for 2-CP can be attributed to its relatively low melting point (8°C) compared to the others having melting points ranging between 34°C for 3-CP to 65°C for 2,4,6-TCP. The low melting point of 2-CP can be associated with high volatility at the sensing temperature (25°C). At this temperature, 2-CP exists in the liquid state while the other phenols investigated are solids at room temperature. In addition, a direct correlation was observed between the sensor signals and melting points for the other organochlorines. For example, distinct discrimination was recorded for 2-CP compared to the other chlorinated phenols. This is evidenced from the Euclidean distance measurements for all 32 sensors investigated.

Similar response patterns were recorded for 2,4-DCP, 2,4,5-TCP and 2,4,6-TCP respectively (Figure 3). The difference in sensor responses for these three analytes on one hand, and the two-monosubtituted phenols, 2-CP and 3-CP, on the other hand is very large with the latter giving the largest difference. This difference, and hence the discrimination could be in terms of their ring substitution. 2-CP and 3-CP have only one chlorine atom substitution on the phenolic ring, a factor that could result in well-defined discrimination from chlorinated phenols with higher number of substitution. The difference in substitution results in different electron distribution within the ring system due to the electron withdrawing effect of chlorine atoms and subsequently variation in polarity of the compounds. The different polarities imply that the adsorption/desorption kinetics at the sensor surface and the sensor intensity will be

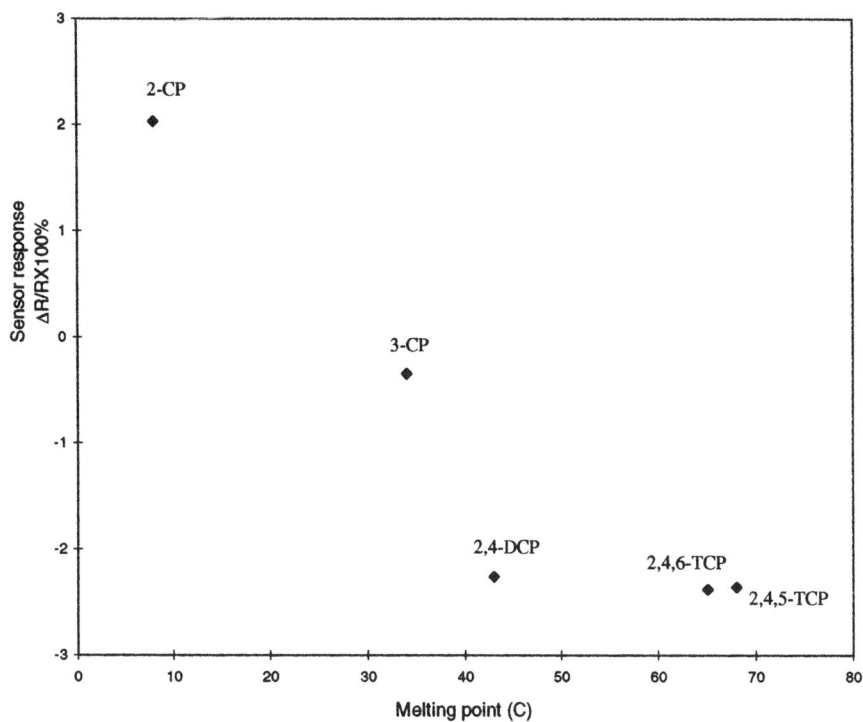

Figure 2. *A plot of sensor response versus the melting points of chlori-nated phenols.*

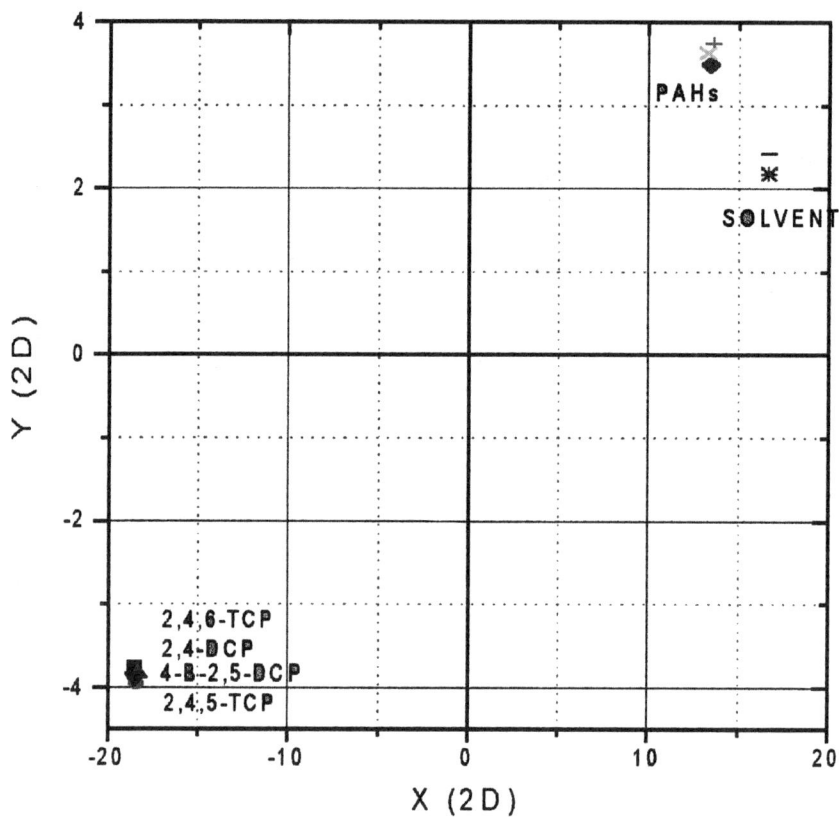

Figure 3. Pattern recognition of different classes of compounds, a 2-dimensional and Sammon map for 2,4-DCP, 2,4,5-TCP, 2,4,6-TCP, 4-B-2,5-DCP, pyrene and toluene/methanol solvent mixtures.

Table 2: Experimental GC-EN data for Industrial Solvents

Industrial Solvents	Retention Time (Rt)	Capacity Factor (k')	Partition Coefficient (K)	Sensor response	Boiling point (Bpt °C)
Acetone	2.9	1.04	260	0.44	56
Toluene	4.95	1.78	445	0.34	110.6
Octane	4.1	1.89	472.5	0.248	125
m-Xylene	4.95	2.48	620	0.298	138
Decane	7.48	4.25	1062.5	0.205	174

significantly different. This could be used in differentiating other chlorinated organic compounds. Moreover, monosubstituted 2-CP and 3-CP are relatively smaller in size compared to the di- and tri-substituted phenols. The small molecules are able to diffuse within the sensors more readily, thus resulting in facile interactions.

Gas Chromatography with Multiarray Detection (GC-EN)

The multiarray sensors coupled with gas chromatography were used for the analysis of different classes of chemical compounds. In order to correlate odor intensity to partition coefficient and sensor intensity, we used an array of gas sensor coupled to a GC. This involves (i) Using the GC to identify and separate the odor components, (ii) Increasing the sensor sensitivity by modulating the amount of volatile reaching the sensor arrays, and (iii) Utilizing neural network software for further identification of the odor components. Table 2 is a summary of GC-EN data obtained for the industrial solvents. The sensor response was inversely proportional to the capacity factor. For example, acetone has the lowest value of k', and it yielded the highest sensor response of 0.44. The sensor responses were also inversely related to the retention time, the capacity factor, and the boiling points for the analyte investigated (Table 2).

Using Equation 1, the QSSR concept was tested to determine which specific structural features of the analyte molecules are important for detectability. The QSSR testing data obtained by plotting the sensor response versus the parameters shown in Table 3 for each analyte. Multiple linear regression using partition coefficient, solubility, vapor pressure and molecular weight was obtained in order to estimate the quantitative structure-activity relationship. The slopes correspond to the coefficient a,b, etc., obtained for the industrial solvents and chlorinated phenols. The magnitudes of the coefficients a, b, plus the standard deviation reveal which structural features are important in determining the intensity of the odor.

Table 3: Slopes for different QSSR parameters obtained.

Parameter	Industrial Solvents[a]	Chlorinated phenols[a]
Retention time	-4.30×10^{-2}	NA
No. of carbon atoms	-3.38×10^{-2}	NA
Partition coefficient	-2.00×10^{-4}	NA
Molecular weight	-2.90×10^{-3}	-4.60×10^{-2}
Boiling point	-2.00×10^{-3}	-5.32×10^{-2}
Solubility	2.00×10^{-4}	1.00×10^{-4}
Vapor pressure	8.00×10^{-4}	2.918

(a) = The values represent the magnitudes of the coefficients described from Equation 2

NA = not available or not applicable.

Large coefficients were obtained for solubility and vapor pressure while the other parameters gave negative slopes. This implies that for the organochlorine and the industrial solvents tested, the most important factor is vapor pressure with the highest coefficient. The small coefficient obtained for other parameters signifies relatively unimportant factors. The magnitude of the response was as follows:

Solvents: Vapor pressure > solubility > boiling point > molecular weight > partition coefficient > # of C atoms > Rt.

Chlorinated phenols: Vapor pressure> solubility> log P> Molecular weight> boiling
Point > melting point

These trends confirm the hypothesis that there is a predictable relationship between the aspects of sensor structures with analyte signals. We are currently studying a wider range of different functional groups in order to extend this concept for other important analytes.

Discrimination Based on the Nature of the Functional Groups

GC-EN was estimated for pyrene, four chlorinated phenols and some industrial solvents. Each of the classes of three analytes gave different sensor response pattern as shown in Figure 3. This shows that one sensor can respond to a wide range of chemicals indicating that the sensor is not specific. However, the intensity of the response depends on the interaction between the sensor and the analyte. Chlorinated phenols exhibited similar fingerprints, which are different from those of pyrene or the solvent. Such discrimination can be based on the functional groups, size and shape of the analytes. Pyrene dissolved in toluene/methanol solvent was distinctively discriminated from the pure solvent mixture as shown by the differences in the Euclidean distance indicating that the solvent was recognized as another analyte. This shows that the sensor array can recognize dissolved analytes not just pure compounds. Moreover, the pattern recognition could be based not only on the nature of functional groups, but discrimination due to differences in electronegativity producing distinct discrimination [30]. The normalized sensor response patterns for pyrene and solvent exhibited unique features for all 32 sensor arrays as shown in Figure 4. The normalized sensor response was obtained by dividing each response by the total responses of the 32 sensors. Sensor numbers 22, 23 and 24 exhibit lower normalized responses for pyrene/solvent than the pure solvents. Thus an array sensors can provide both the specificity and resolution that a single sensor can not achieve.

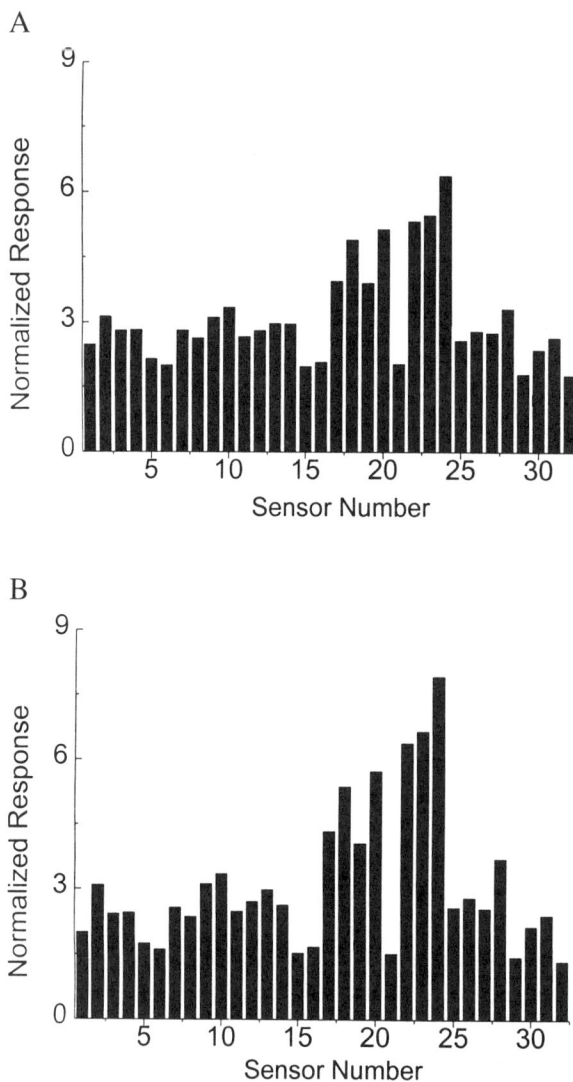

Figure 4. Normalized sensor response for a) 0.0005M pyrene in toluene/ methanol (10:1) and b) Toluene/methanol.

Qualitative Analysis of PAHs and ANN Training

The use of the multiarray sensor for the qualitative identification of component mixtures (concentrations and compositions) was tested using ANN training. The sensor response for pyrenes, chlorinated organics and solvents was used as the source file to train the ANN. Different concentrations of pyrene were recognized by the sensor array as evidenced by the Euclidean distances between the analytes (Figure 5). Cluster analysis of results obtained from different concentrations of pyrenes show good discrimination between the samples. ANN training using 2-CP, 2,4-DCP, 2,4,6-TCP and DDT was carried out using the Fuzzy back-propagation training method. After training, an unknown sample chlorinated phenol was fed into the system, and the network determined its composition. Pattern recognition using cluster method for unknown 2,4-DCP with 100% recognition showed that the database lies within the cluster of the 2,4-DCP response patterns for the standard used for the training (Figure 6). Thus the neural network allowed a quantitative estimation of the composition of mixtures by recognizing the analytes with an average confidence level of 99.7%.

Conclusions

This study has demonstrated the use of electronic nose employing conducting polymer sensor arrays and pattern recognition techniques in the analysis of chlorinated phenols and PAHs. Cluster analysis and ANN methods were employed for the qualitative analysis based on the nature of the analyte, which includes functional groups, size, shape, electronic distribution, polarity as well as physical properties. Pattern recognition of dissolved chemicals was used for quantitative analysis. ANN method was successfully used for the identification of unknown analytes and the recognition rates ranged from 98.8 to 100% for the tested analytes. QSSR approach has the potential for successfully relating sensor-structures with analyte intensity. This could be used in characterizing unknown contaminated sites.

Acknowledgements

The authors would like to acknowledge AromaScan Inc., NH, for the loan of AS32S instrument used in this work and the US-Environmental Protection Agency, Office of Research and Development for funding.

Literature Cited

1. Gardner, J. W.; Bartlett, P. N. *Sensors and Actuators B* **1994,** 18-19, 211-220.
2. Nagle, H. T.; Gutierrez-Osuna, R.; Schiffman, S. S. *IEEE Spectrum,* **September 1998,** 22-34.

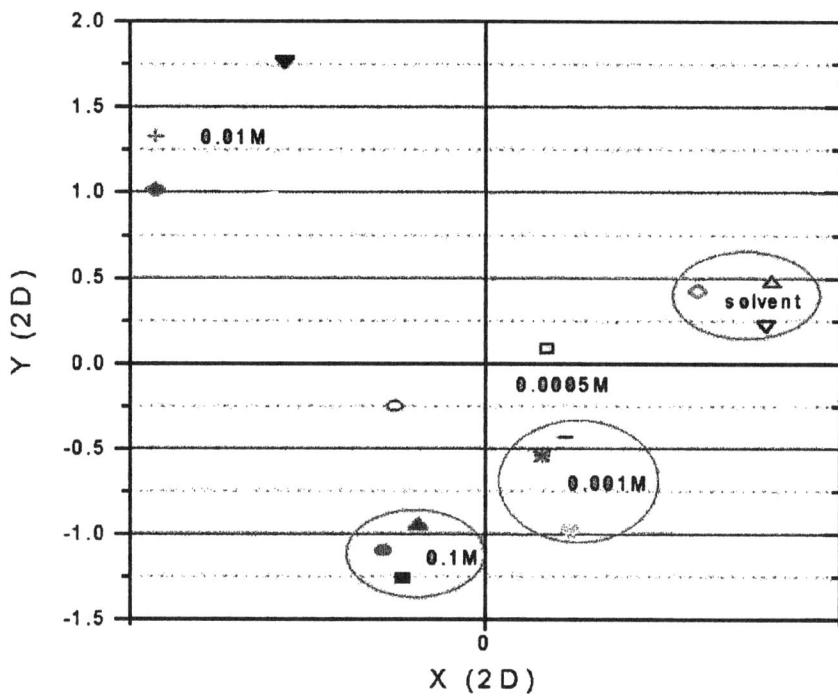

Figure 5. 2-dimensional pattern recognition map for database used in ANN training for the analysis of 2-CP, 2,4-DCP, 2,4,6-TCP and DDT as well as recognition of the unknown 2,4-DCP sample.

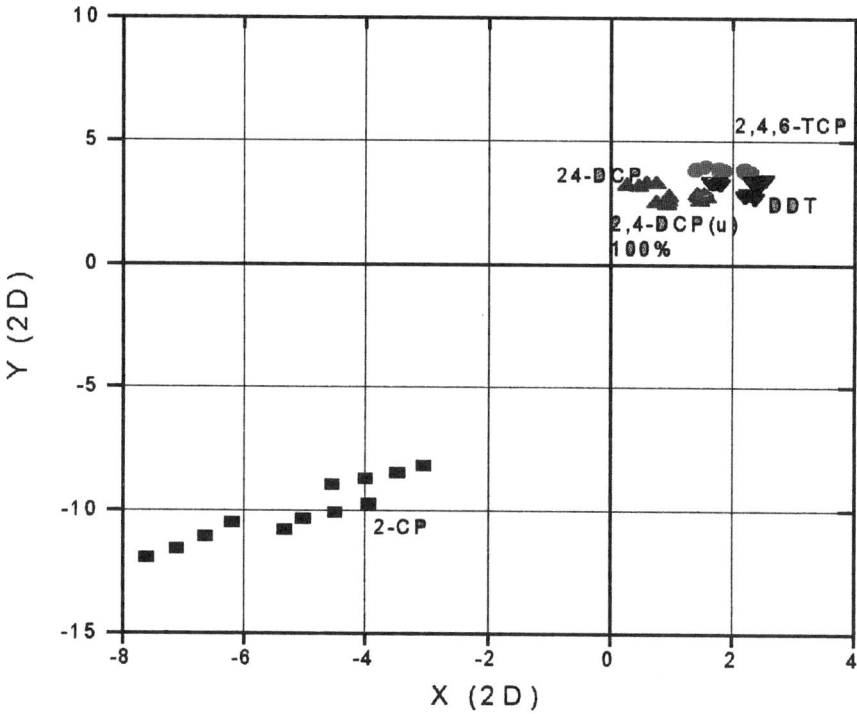

Figure 6. 2-Dimensional pattern recognition map for database used in ANN training for the analysis of 2-CP, 2,4-DCP, 2,4,6-TCP and DDT as well as recognition of the unknown 2,4-DCP sample.

58

3. Handbook of Biosensors and Electronic Noses; Kress-Rogers, E., Ed.; CRC Press: Boca Raton, FL,1997.
4. Pearce, T. C.; Gardner, J. W.; Friel, S.; Bartlett, P. N.; Blair, N. *Analyst* **1993,** 118, 371-377.
5. Gardner, J. W.; Shurmer, H. V.; Tan, T. T. *Sensors and Actuators B* **1992,** 6, 71-75.
6. Hodgins, D. *Sensors and Actuators B* **1995,** 26-27, 255-258.
7. Persaud, K. C.; Khaffaf, S. M.; Hobbs, P. J.; Sneath, R. W. *Chemical Senses* **1996,** 21, 495-505.
8. Misselbrook, T.H.; Hobbs, P. J.; Persaud, K. C. *J. Agric. Engng. Res.* **1997,** 66, 213-220.
9. Hobbs, P. J.; Misselbrook, T. H.; Pain, B. F. *J. Agric. Engng. Res.* **1995,** 60, 137-144.
10. Gibson, T. D.; Prosser, O.; Hulbert, J. N.; Marshall, R. W.; Corcoran, P.; Lowery, P.; Ruck-Keene, E. A.; Heron, S. *Sensors and Actuators B* **1997,** 44, 413-422.
11. Stetter J. R., Findlay M., *Anal. Chim. Acta.* **1993,** 284, 1.
12. Masila M., Breimer M., Sadik O. A, "Electronic Nose as a Detector for Gas Chromatography," Pittsburgh Conference of Analytical Chemistry & Applied Spectroscopy, March 1-6, **1998,** New Orleans.
13. Masila M., Breimer M., Sadik O. A, Strategies for Improving the Analysis of Volatile Organic Compounds Using GC-Based Electronic Nose, in *Electronic Noses & Sensor Array based Systems,* Hurst, W. J. Ed.; Technomic Publishing Co., Inc., Lancaster, PA, **1999,** 27-42.
14. Amoore J. Molecular basis of Odor, Charles Thomas, Springfield, IL, **1970.**
15. Stephen S. Ed. Handbook of Experimental Psychology, John Wiley & Sons, NY, **1951.**
16. Goepel, W. and Reinhardt, G. Metal Oxide Sensors: New Devices through tailoring interfaces at the Atomic Scale, in *Sensor Update,* Baltes, Goepel, and Hesse, Eds. VCH, Weinheim, Germany **1996.**
17. Sadik, O. A.; Wallace, G. G. *Electroanalysis* **1993,** 5, 555-563.
18. Sadik, O. A.; Wallace, G. G. *Electroanalysis* **1994,** 6, 860-864.
19. Sadik, O. A. *Analytical Methods and Instrumentation* **1995,** 2, 293-301.
20. Rocha, S.; Delgadillo, I.; Ferrer Correia, A. J.; Barros, A.; Wells, P. *J. Agric. Food Chem.* **1998,** 46, 145-151.
21. Handbook of Conducting polymers; Skotheim, T. A., Ed.; Marcel Dekker, Inc.: New York, NY, **1986,** Vol. 1.
22. Takakubo, M. *Synthetic Metals* **1986,** 16, 167-172.
23. Buttry, D. A.; Ward, M. D. *Chem. Rev.* **1992,** 92, 1355-1379.
24 Ballantine, D. S., White, R. M., Martin, S. J., Ricco, A. J., Zeller, E.T, Frye, G. C., and Wohltjen, H. Acoustic Wave Sensors, Academic Press, NY, **1997.**
25 Dickinson, T. A., White J., Kauer J., Walt D., *Anal. Chem.* **1996,** 68, 2191.

26 Grate J. W., Abraham M. H., "Solubility Interactions and the Design of Chemically Selective Sorbent Coatings for Chemical Sensors and Arrays," *Sensors & Actuators,* B 3, **1991**, 85-111.

27 Janata J. Principles of Chemical Sensors, Plenum Press, NY, **1989**.

28 Diamond, D. Ed., Principles of chemical and Biochemical Sensors; John Wiley & Sons, Inc., NY, **1998**.

29 Sammon, J. W., Jr. *IEEE Trans. On Computers,* **1969**, 18C, (5), 401-409.

30 Masila, M. M.; Sargent, A.; Sadik, O. A. *Electroanalysis* **1998**, 10, 312-320.

31 Gardner, J. W. and Batlett P. N.; Pattern Recognition in Gas Sensing, in *Techniques and Mechanisms in Gas Sensing,* Moseley, P. T., Norris, J., and William, D. Eds.; Adam-Hilger, Bristol, **1981**.

32 Sensors and Sensory Systems for Electronic Nose, Gardner, J. W., and Bartlett, P. N. Eds.; Kluwer Publishers, London **1992**, 273-301.

33 Patrash S., Zellers E., *Anal. Chem.* **1993**, 65, 2055.

Chapter 5

Selection in System and Sensor

W. Olthuis, S. Böhm, G. R. Langereis and P. Bergveld

**MESA+ Research Institute, University of Twente,
P.O. Box 217, 7500 AE Enschede, The Netherlands**

Since the dawn of chemical sensors, there were high hopes of rapid and widespread utilization of these devices. This predicted success, however, has not yet been achieved, due to several causes: (bio)chemical fouling of the sensor, instability in the sensor signal resulting in drift, and lack of available selector materials, specific for all species to be detected.

In this chapter, a more realistic approach is pursued, i.e., the incorporation of the sensing element in a Total Analysis System (TAS). This system comprises a double-lumen microdialysis probe, providing a coarse first-stage selection of molecules, to be allowed into the system, decreasing the possibility of sensor fouling. Additionally, the system is provided with integrated electrochemically driven pumps to precisely dose nanoliter amounts of calibration liquid, for periodic calibration of the possibly drifting sensing element.

A possible and simple sensing element in these kind of TAS's is a conductivity probe. Electrolyte Conductivity (EC) is not selective for specific ions. It is shown in this chapter, however, that separate ion concentrations can be calculated, when the EC is measured at several temperatures, using the characteristic temperature responses of the ionic conductivities of these ions.

Selection with a Total Analysis System (TAS)

General

An ideal chemical sensor would transfer concentration information of a specific chemical compound with 100 per cent selectivity, i.e., without any cross-sensitivity to the electrical (or optical) domain. Additionally, such a sensor must be able to operate directly in the sample solution. Schematically, such a sensor is shown in Figure 1.

Despite all efforts, such sensors do not exist. Nevertheless, with full scale laboratory equipment, analysts are often able, off-line, to specifically characterize a certain sample after laborious sample pre-treatment. The development of a TAS is an attempt to integrate component-wise the several relevant sub-systems involved in laboratory equipment on a small scale. The functions these components should fulfill are typically sampling, transport, mixing, reaction, separation and detection. If this TAS is very close to the place of the actual measurement, then the system is commonly referred to as micro- or μ-TAS *(1)*. An impression of such a μ-TAS, showing some of the relevant sub-systems, is given in Figure 2. It is beyond the scope of this chapter to describe all the possible advantages of μ-TAS, already partly realized and described in, e.g., *(2)*.

In this section, we restrict ourselves to two problems, related to the direct use of chemical sensors, i.e., (bio)chemical fouling and instability of the sensor signal, which can to a certain extent be solved by using a TAS, as shown in the next sub-sections.

Microdialysis probe

A realistic approach to the problem of fouling of the sensor surface, especially by large, solid particles and proteins is by separating these interfering compounds from the chemical sensor. Precisely this is accomplished by the microdialysis technique *(3)*. A perfusion liquid is pumped through a hollow fiber, in contact with the sample solution to be investigated. A part of this fiber consists of a semi-permeable membrane, such as cellulose acetate. During the passage of the perfusion liquid through the fiber, only substances of small molecular weight can diffuse through the membrane, preventing the fouling substances from entering the system. Subsequent analysis of the dialysate and detection of the species of interest is carried out off-site, but on-line in the TAS on a semi-continuous base. This system is schematically shown in Figure 3.

There are many parameters determining how well the concentration of the analyte in the dialysate agrees with that in the sample solution. This property is expressed by the relative recovery ratio R:

$$R = \frac{[analyte]_{dialysate}}{[analyte]_{sample}} * 100\% \tag{1}$$

As can be imagined, R depends on the diffusion coefficient of the entering analyte, the residence time of the perfusate in the semi-permeable part of the lumen, and thus of the volumetric flow rate, and on dimensional parameters, like the length and radius of the semi-permeable part of the lumen.

Now let us calculate how long the perfusate should minimally stay in the semi-permeable part of the lumen in order to obtain a high R-value of, say, 90%. For this, flow is considered to be absent, and the length of the semi-permeable part of the lumen is considered to be much longer than its diameter. Mass transport is now

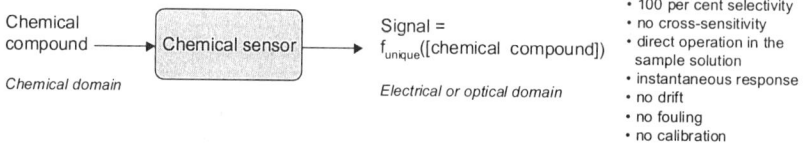

Figure 1. The ideal chemical sensor visualized; its function and its requirements.

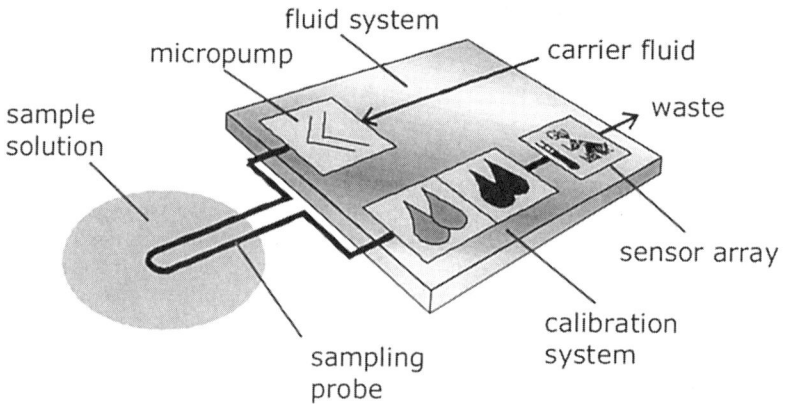

Figure 2. Impression of a Micro Total Analysis System showing several required sub-systems.

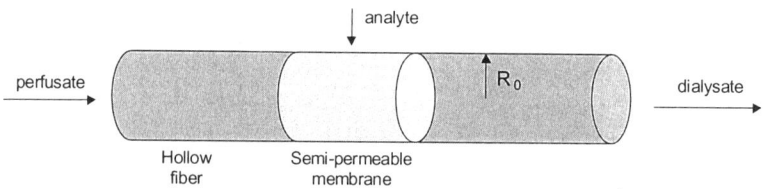

Figure 3. Illustration of the microdialysis technique, showing the hollow lumen of which a part consists of a semi-permeable membrane through which the analyte can enter the system.

governed by diffusion only. Of course, mass transport takes place from the semi-permeable membrane to the inner volume of the lumen. This process is described by the differential equation for diffusive mass transport in cylindrical co-ordinates:

$$\frac{\partial C}{\partial t} = \frac{D}{r}\cdot\frac{\partial}{\partial r}\left(r\frac{\partial C}{\partial r}\right) = D\left(\frac{\partial^2 C}{\partial r^2}+\frac{1}{r}\frac{\partial C}{\partial r}\right), \qquad 0<r<R_0 \qquad (2)$$

C and D being the concentration of the analyte in the lumen, and its diffusion coefficient, respectively. R_0 is the radius of the lumen, as also shown in Figure 3. At the start, we consider no analyte to be present in the lumen, and the semi-permeable membrane itself does not form an extra obstacle for diffusive transport. Then, the initial and boundary conditions become

$$C=0 \qquad 0<r<R_0 \qquad t=0 \qquad (3a)$$
$$C=C_0 \qquad r=R_0 \qquad t\geq0 \qquad (3b)$$

with C_0 the concentration of the analyte in the sample. This concentration is considered to remain constant, due to a commonly relatively large sample volume and the presence of convection outside the lumen. Then the solution of the occurring concentration profile within the lumen, using eqns. 2, 3a and b is (4)

$$C(r,t) = C_0 - 2C_0\sum_{n=1}^{\infty} e^{-Dt\frac{\beta_n^2}{R_0^2}}\cdot\frac{J_0\left(\frac{r}{R_0}\beta_n\right)}{\beta_n J_1(\beta_n)} \qquad (4)$$

with J_0 and J_1 the Bessel function of the first kind and order 0 and 1, respectively, and with β_n the roots of $J_0(\beta)=0$. For a worst case estimation, the longest distance to be traversed for the analyte is considered: from the wall to the center of the lumen, being the lumen radius, R_0. There, at $r=0$, the concentration is at its lowest. The expression for the concentration at $r=0$, relative to the analyte concentration in the sample, C_0, considering that $J_0(0)=1$, becomes

$$\frac{C(r=0,\tau)}{C_0} = 1 - 2\sum_{n=1}^{\infty} e^{-\beta_n^2\tau}\cdot\frac{1}{\beta_n J_1(\beta_n)} \qquad (5)$$

introducing the dimensionless variable

$$\tau = \frac{Dt}{R_0^2} \qquad (6)$$

A plot of $C(r=0, \tau)/C_0$ as a function of τ is shown in Figure 4a. A practical example is shown in Figure 4b, for a lumen radius $R_0=150\ \mu m$ and diffusion coefficient $D=1\cdot10^{-9}\ m^2/s$. It can be seen, that it takes ca. 11 s to reach 90% of the original sample concentration at the axis of the lumen, $r=0$.

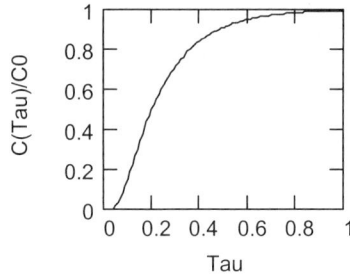

Figure 4a. Graph of the normalized concentration due to cylindrical diffusion as a function of the dimensionless variable, τ.

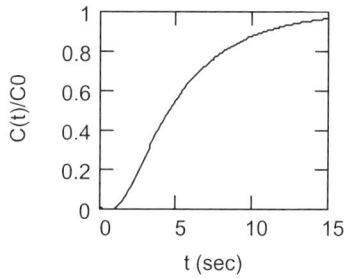

Figure 4b. Graph of the normalized concentration due to cylindrical diffusion as a function of time, t, for $D=1 \cdot 10^{-9}$ m^2/s and $R_0=150$ μm.

A more realistic and milder approach is to calculate the mean concentration of the sample in the lumen, averaged over its cross section *(4)*:

$$\frac{C_{mean}(\tau)}{C_0} = \frac{1}{\pi R_0^2} \int_{\varphi=0}^{2\pi} \int_{r=0}^{R_0} \frac{C(r,\tau)}{C_0} r\,dr\,d\varphi$$

$$= 1 - 4\sum_{n=1}^{\infty} \frac{e^{-\beta_n \tau}}{\beta_n^2} \tag{7}$$

Also for this case, plots of this normalized mean concentration both as a function of the dimensionless variable τ and, as an example, as a function of time for $D = 1 \cdot 10^{-9}$ m^2/s and $R_0 = 150 \ \mu m$ is shown in Figure 5a and b, respectively. Now, already after ca. 7.5 s 90% of the original sample concentration is reached.

With this, the minimum residence time of the perfusate in the semi-permeable part of the lumen is estimated. Of course, due to convection, the non-zero distance between plug-formation of the dialysate and the detector, and the position of the sensing part of the detector itself, this calculated minimum residence time only forms a lower limit of the total sampling time of the probe. Precise calculations of these additional effects are beyond the framework of this chapter.

Microdialysis as a sampling tool for a μ-TAS is even more favourable, when a double-lumen microdialysis probe is used *(5, 6)*. Now, the perfusate is driven through a small inner tube, located within an outer tube. The fluid leaves the inner tube at the probe-tip and subsequently flows past the semi-permeable membrane back via the outer tube as schematically shown in Figure 6.

For an optimized design of the probe, it is important to know, that R is inversely proportional to the volumetric flow rate of the perfusate. In order to obtain a R-value near 100 per cent, which is favourable, because calibration of the probe is then not necessary, the flow rate should be low. However, to maintain a reasonable response time of the TAS, any dead volume between the sampling probe and the detection in the TAS should be minimized. Better still, the probe and the sensors should be integrated as much as possible, resulting in a true μ-TAS. Therefore, a silicon micromachined connector was designed and realized, as schematically shown in Figure 7. The funnel-like structure between the large opening surrounding the outer tube of the lumen and the V-shaped channel for the inner tube act as a guide for the insertion of the inner tube. A dead volume remained of only 180 nl, being only a fraction of that of a conventional probe, which is ca. 1.5 μl. Thus, the lag time of the TAS, equipped with this micromachined probe, is considerably reduced. An array of five finished probes, ready for cutting apart is shown in Figure 8.

Now that the sensor can operate in a relatively clean sample, the possible fouling of the semi-permeable membrane should be addressed. On the one hand, the probe - and operational parameters- can be designed in such a way, that the recovery ratio, R, is very high, near 100%. Such a design could guarantee that R remains high, say > 95%, even after mild fouling of the membrane. On the other hand, there is more

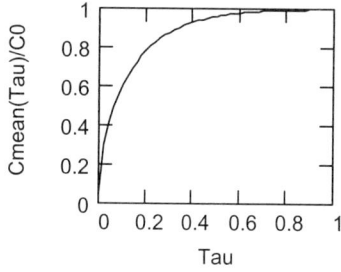

Figure 5a. Graph of the normalized mean concentration due to cylindrical diffusion as a function of the dimensionless variable, τ.

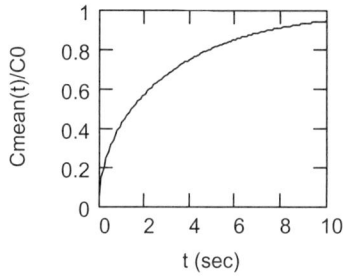

Figure 5b. Graph of the normalized mean concentration due to cylindrical diffusion as a function of time, t, for $D=1 \cdot 10^{-9}$ m^2/s and $R_0=150$ μm.

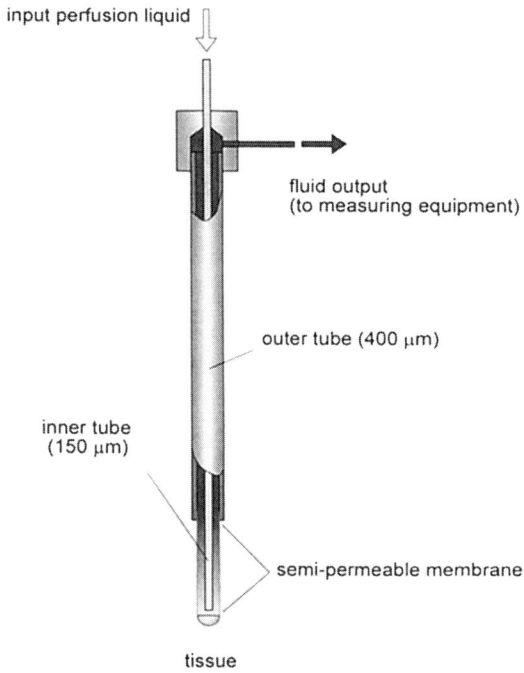

Figure 6. Impression of the double-lumen microdialysis probe. Reproduced with permission from reference [5]. Kluwer Academic Publisher, 1998.

68

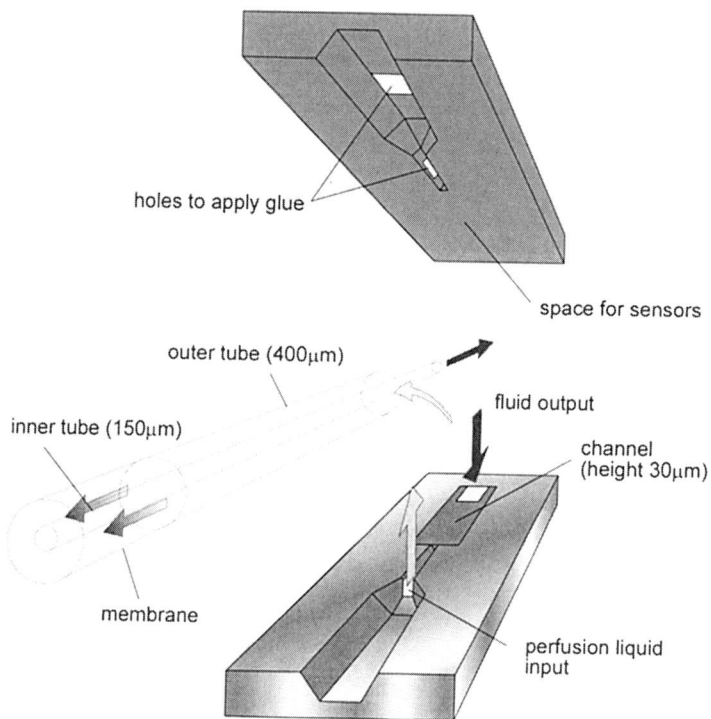

Figure 7. Exploded view of the microdialysis probe with connector. Reproduced with permission from reference [5]. Kluwer Academic Publisher, 1998.

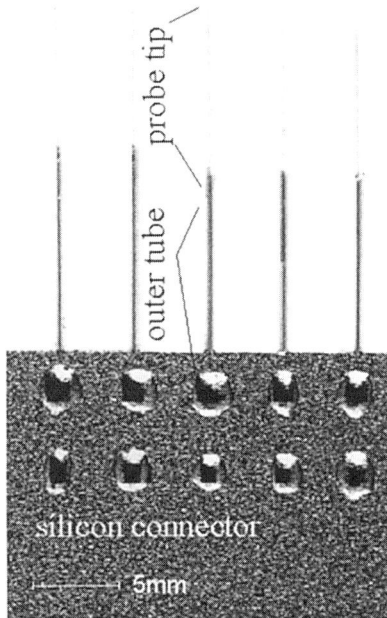

Figure 8. Realized Si-micromachined microdialysis probes. Reproduced with permission from reference [5]. Kluwer Academic Publisher, 1998.

freedom to chose a membrane that is less prone to fouling or better biocompatible than the material of the selector part of the sensor itself. In addition, the probe can be subjected to an overpressure of the perfusate with respect to the analyte, under the control of the TAS, thereby actively cleaning the membrane.

Calibration system

Although the sensor in the TAS as proposed in this chapter can operate in a relatively clean solution, due to the microdialysis probe as sample inlet, regular calibration of the sensor is still required. The system performing this calibration should of course preferably be a part of the TAS itself, as shown in Figure 2. For this purpose a dosing system is required, capable of dispensing calibration liquid in the carrier channel, leading to the sensor. The response to this plug of calibration liquid can be applied to determine the off-set and sensitivity of the sensor (i.e., if two reservoirs of calibration liquid are implemented). The system proposed in this chapter is based on the displacement of a fluid from a reservoir by the electrochemical production of gas bubbles *(6, 7)*.

In the realized structure, the calibration liquid to be dispensed is stored in a meander-shaped channel, starting in a reservoir filled with an electrolyte in which two electrodes are placed, as schematically shown in Figure 9.

By sending an electric current through the electrodes, gas bubbles are produced by the well-known anodic and cathodic electrolysis of water.

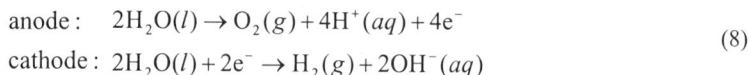

$$\text{anode}: \quad 2H_2O(l) \rightarrow O_2(g) + 4H^+(aq) + 4e^-$$
$$\text{cathode}: \quad 2H_2O(l) + 2e^- \rightarrow H_2(g) + 2OH^-(aq)$$

(8)

These bubbles expand in the reservoir and drive out the calibration liquid, stored in the meander. After saturating the electrolyte with hydrogen and oxygen, the volume, V, of the gas bubbles and thus of the calibration plug can be precisely controlled according to

$$V = \frac{3i\Delta t}{4F} V_m$$

(9)

with i and Δt the amplitude and duration of the applied current pulse, respectively, F Faraday's constant, and V_m the molar gas volume at given temperature and (atmospheric) pressure.

The proposed system has been realized in Si micromachining technology and a photograph of the device, including two integrated electrochemical pumps and meander-shaped reservoirs for calibration liquid is shown in Figure 10 *(8)*. A detail of the pump, where the meander channels of the calibration units connect to the carrier channel, is shown in Figure 11. This figure clearly illustrates the precise design, possible by photolithographic techniques and silicon Deep Reactive Ion Etching. The calibration system is constructed of two layers. A bottom layer of silicon in which a channel structure is etched and a top layer of glass to create closed channels and to

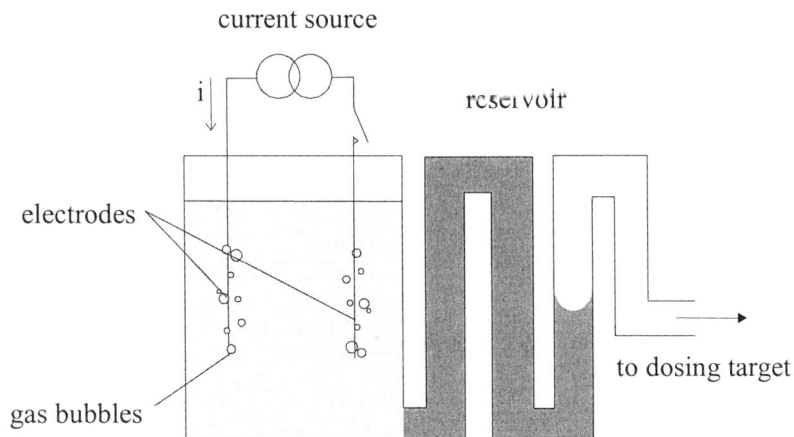

current source

i

reservoir

electrodes

gas bubbles

to dosing target

Figure 9. Principle of the proposed electrochemically actuated pump and dosing system. Reproduced with permission from reference [6]. Kluwer Academic Publisher, 1998.

Figure 10. Photograph of the realized device, showing two complete calibration units.

Figure 11. Photograph of a detail, showing the connection of the two meander-shaped channels to the carrier channel.

visually examine the flow through the channels. The electrodes for the electrolysis are deposited on the glass top layer.

A calibration system, capable of performing 20 calibration of 100 nl each, must contain 2 µl of calibration fluid. This 2 µl of calibration fluid is stored in the meander channel. A meander channel of 100 µm × 200 µm must therefore have a length of 10 cm. To push all the calibration fluid out of the meander channel, there must be at least 2 µl of electrolyte in the reservoirs. To have sufficiently electrolyte left for electrolysis when the 20th calibration is performed a reservoir volume of 5 µl is chosen.

A measurement setup is build to verify the calculated production of gas during electrolysis according to eq 9. A computer controlled current source (CCCS) is used for electrolysis. The current source is equipped with two current ranges, 0-100 µA and 0-2 mA. The connection holes of the reservoirs are sealed with epoxy (Araldit). The calibration system is then immersed in an electrolyte solution. The electrolyte consists of a 200 mM KNO_3 solution to which some detergent (Tween 20) is added. Using a vacuum system, air inside the calibration system is replaced by the electrolyte solution.

Using a CCD camera and a color monitor, as shown in Figure 12, the displacement of the fluid during electrolysis can be visualized. By measuring the displacement in combination with the channel cross-sectional area, the volume of the produced gas can be calculated. Precise dosing is shown to be possible in Figure 13, where the measured volumetric flow for $\Delta t=5$ s at the mentioned current values is shown, including the theoretical curve from eq 9.

To show an actual example of the calibration procedure of a pH-sensitive ISFET, the response of a two-point calibration on two injected calibration samples obtained with a system like the one described here is included in Figure 14.

Selection with an Electrolyte Conductivity (EC) sensor

One of the important sub-systems in any TAS is the detection. In many cases, optical techniques, such as luminescence, fluorescence or spectroscopy are applied, using full-size optical equipment. Only in a few cases, the optics, necessary for the detection, are integrated as well, which is a prerequisite for a real TAS (9, 10). Considering size, electrochemical sensors seem to be better suited as detector in a TAS. Two regularly encountered types of sensors are amperometric and conductometric sensors (11, 12). Due to their simple construction and the fact that selectivity is often already obtained in an earlier stage during separation, these sensors can give satisfying results. Better results, however, can possibly be obtained, when some selectivity is built-in in the detector. Although this seems to be impossible with an EC sensor, due to the inherently non-selective nature of EC, we will in the second part of this chapter, present a method to add selectivity to EC sensing.

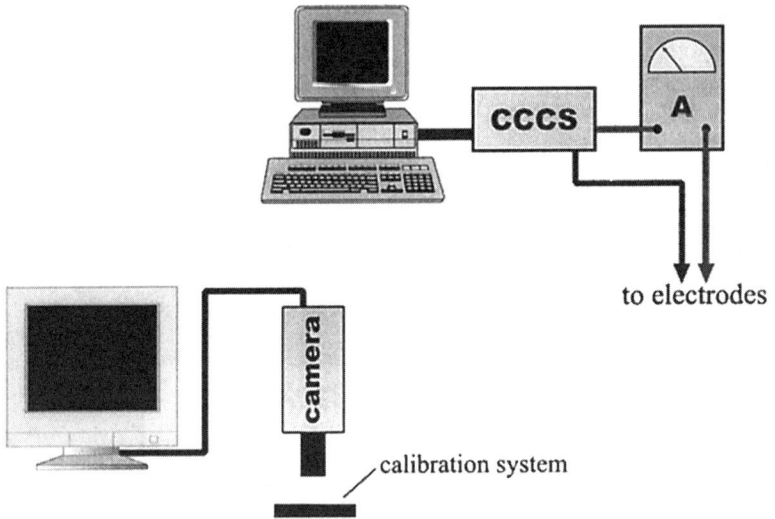

Figure 12. The measurement set-up, consisting of a computer controlled current source (cccs) to evoke gas bubbles by electrolysis and a ccd camera to monitor the fluid displacement.

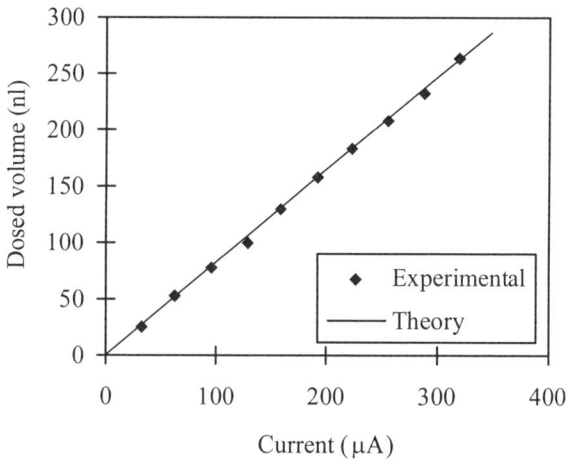

Figure 13. Measured dosed volumes after 5 s current actuation with given current amplitudes.

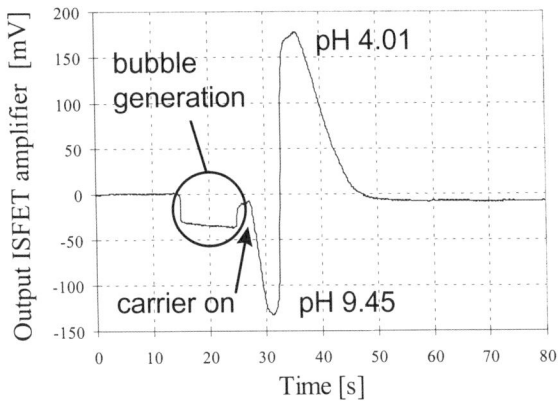

Figure 14. Measurement result of a two-points ISFET calibration downstream in the carrier flow. Reproduced with permission from reference [6]. Kluwer Academic Publisher, 1998.

Operational principle

A single EC measurement does not give any specific information on the ion concentrations that are present in the solution, when more than 2 types of ions are involved. However, a range of measurements at various temperatures of the electrolyte does, since the temperature dependence of the mobility of an ion is unique for that ion (13, 14).

The total conductivity of an electrolyte expressed in terms of the limiting molar conductivities of the separate ions can be expressed by

$$\Lambda = \sum_{i=1}^{I} |z_i| c_i \lambda_i(T) \tag{10}$$

with z_i the charge of ion i, c_i the concentration, I the number of different types of ions and $\lambda_i(T)$ the limiting molar conductivity of ion i. This last factor is dependent on temperature and is specific for every single ion. The generalized polynomial fit of $\lambda_i(T)$ of order J with respect to temperature T_0 can be written as

$$\lambda_i(T) = \lambda_i^0 \sum_{j=0}^{J} k_{i,j} (T - T_0)^j. \tag{11}$$

with λ_i^0 the limiting molar conductivity for ion i at $T=T_0$. Since the coefficients $k_{i,j}$ are unique for ion i, the total conductivity Λ is a unique linear combination of the limiting molar conductivities $\lambda_i(T)$ with the specific ion concentrations as coefficients.

Using equations 10 and 11 the total conductivity of an electrolyte having I types of ions can now be calculated:

$$\Lambda(T) = \sum_{i=1}^{I} \left[|z_i| c_i \lambda_i^0 \sum_{j=0}^{J} k_{i,j} (T - T_0)^j \right] \tag{12}$$

Manipulation yields:

$$\Lambda(T) = \sum_{j=0}^{J} \left[(T - T_0)^j \sum_{i=0}^{I} \left(|z_i| c_i \lambda_i^0 k_{i,j} \right) \right]. \tag{13}$$

For conductivity measurements at N different temperatures, I types of ions and a polynomial fit of order J, a matrix equation can be formed using eq 13:

$$
\begin{bmatrix} \Lambda_1 \\ \Lambda_2 \\ \cdot \\ \cdot \\ \Lambda_N \end{bmatrix}
=
\begin{bmatrix}
1 & (T_1 - T_0) & (T_1 - T_0)^2 & \cdot & (T_1 - T_0)^J \\
1 & (T_2 - T_0) & (T_2 - T_0)^2 & \cdot & \cdot \\
\cdot & \cdot & \cdot & \cdot & \cdot \\
\cdot & \cdot & \cdot & \cdot & \cdot \\
1 & (T_N - T_0) & (T_N - T_0)^2 & \cdot & (T_N - T_0)^J
\end{bmatrix}
\cdot
\begin{bmatrix}
k_{1,0} & k_{2,0} & \cdot & k_{I,0} \\
k_{1,1} & k_{2,1} & \cdot & \cdot \\
\cdot & \cdot & \cdot & \cdot \\
\cdot & \cdot & \cdot & \cdot \\
k_{1,J} & k_{2,J} & \cdot & k_{I,J}
\end{bmatrix}
\cdot
\begin{bmatrix}
|z_1| c_1 \lambda_1^0 \\
|z_2| c_2 \lambda_2^0 \\
\cdot \\
\cdot \\
|z_I| c_I \lambda_I^0
\end{bmatrix}
\tag{14}
$$

or

$$\overline{\Lambda} = \overline{\overline{T}} \cdot \overline{\overline{K}} \cdot \overline{c} . \tag{15}$$

The meaning of the terms is:

$\overline{\Lambda}$ Vector containing measured conductivities at N different temperatures;

$\overline{\overline{T}}$ Matrix with the temperature information;

$\overline{\overline{K}}$ Matrix with the polynomial coefficients $k_{i,j}$;

$\overline{c}$ Vector with the ion concentrations multiplied by the parameters $|z| \cdot \lambda_i^0$;

The question rises whether it is possible to calculate the vector $\overline{c}$ (containing the concentration information) from a known $\overline{\overline{T}}$ matrix and a measured $\overline{\Lambda}$ vector. When the $\overline{\overline{K}}$ matrix is assumed to be known, which implies a chosen set of ions, eq 14 can be written as

$$
\begin{bmatrix} \Lambda_1 \\ \Lambda_2 \\ . \\ . \\ \Lambda_N \end{bmatrix}
=
\begin{bmatrix}
\sum_{j=0}^{J} k_{1,j}(T_1 - T_0)^j & \sum_{j=0}^{J} k_{2,j}(T_1 - T_0)^j & . & \sum_{j=0}^{J} k_{1,j}(T_1 - T_0)^j \\
\sum_{j=0}^{J} k_{1,j}(T_2 - T_0)^j & \sum_{j=0}^{J} k_{2,j}(T_2 - T_0)^j & . & . \\
. & & . & . \\
\sum_{j=0}^{J} k_{1,j}(T_N - T_0)^j & . & . & \sum_{j=0}^{J} k_{1,j}(T_N - T_0)^j
\end{bmatrix}
\cdot
\begin{bmatrix} |z_1| c_1 \lambda_1^0 \\ |z_2| c_2 \lambda_2^0 \\ . \\ |z_i| c_i \lambda_i^0 \end{bmatrix}
\tag{16}
$$

where the $\overline{\overline{T}} \cdot \overline{\overline{K}}$ matrix is written as a single one. The elements in this matrix are polynomials for the n-th temperature (rows) and the i-th ion (columns). If the $\overline{\overline{T}} \cdot \overline{\overline{K}}$ matrix has an inverse, the concentrations will follow from

$$\overline{c} = \left(\overline{\overline{T}} \cdot \overline{\overline{K}} \right)^{-1} \cdot \overline{\Lambda} . \tag{17}$$

The first condition for having an inverse is that the matrix is square, so the minimal number of necessary experiments is equal to the number of ions to fit ($N = I$). For $N > I$ an estimator must be used. The second condition is that the determinant is not equal to zero. This is true when the coefficients are different for every ion and the order of the polynomial is equal or larger than $I - 1$.

So, it is possible to find the concentrations of individual ions in a solution under the following conditions:

- The measured conductivity scan must be a linear combination of the temperature responses for the individual ions. This means that
  - → Every ion which is significantly present in the electrolyte conductivity must be represented in the calculations;
  - → No two ions may have the same temperature dependency (which will probably never be the case);
  - → When I types of ions have to be calculated, the conductivity of the electrolyte must be measured at least $N = I$ temperatures;

→ The order of the used polynomials is equal to or larger than the number of different ions to fit minus one ($J = I - 1$);

• The coefficients $k_{i,j}$ of the individual ions must be known. The coefficients for the third order fit are given by Harned and Owen *(15)* for nine types of ions;

Improvement by estimation

The calculation of ion concentrations using the basic algorithm of eq 17 requires conductivity measurements at as much as temperatures as the number of ions to be fitted. It is more accurate to do more measurements *(N > I)* and use an estimation method. In order to find the best fit from the measurements, some theory concerning parameter estimation is necessary. The method introduced here is a matrix based algorithm for minimizing the mean square error of the estimation *(16)*.

Consider the generalized system

$$\overline{\Lambda} = \overline{\overline{B}} \cdot \overline{c} + \overline{v} \tag{18}$$

with

$\overline{\Lambda}$ the vector containing the observations,

$\overline{\overline{B}}$ a matrix representing the system (in our case equal to $\overline{\overline{T}} \cdot \overline{\overline{K}}$),

$\overline{c}$ the input vector to be estimated and

$\overline{v}$ the noise or error in the measurement.

The aim is to find an estimate $\hat{\overline{c}}$ for the vector $\overline{c}$ satisfying the observed vector $\overline{\Lambda}$. The theoretical description of this estimation is beyond the scope of this chapter. Under certain particular conditions a relatively simple estimation can be derived.

The linear minimum variance unbiased estimate $\hat{\overline{c}}$ of $\overline{c}$ given data $\overline{\Lambda}$ under these conditions is according to Gauss-Markoff theorem equal to

$$\hat{\overline{c}} = \left(\overline{\overline{B}}^T \overline{\overline{B}}\right)^{-1} \overline{\overline{B}}^T \cdot \overline{\Lambda} \tag{19}$$

which reduces to eq 17 for a square matrix $\overline{\overline{B}}$. Because the variances σ_v^2 of the vector $\overline{v}$ are assumed to be equal, the moment matrix $\overline{\overline{C}}_{\overline{v}}$ can also be eliminated, and the error matrix can be expressed as

$$\overline{\overline{C}}_e = \left(\overline{\overline{B}}^T \overline{\overline{B}}\right)^{-1} \cdot \sigma_v^2 \tag{20}$$

which contains all covariances of the fitted vector $\hat{\overline{c}}$ in its entries. So, the numbers on the diagonal of this error matrix are the variances of the fitted parameters. Using this knowledge, the standard deviation of the whole fit can be defined as

$$\sigma_c = \sqrt{\text{trace}\left(\overline{\overline{B}}^T\overline{\overline{B}}\right)^{-1}} \cdot \sigma_v \qquad (21)$$

where the trace function is the summing of the elements on the diagonal of a matrix.

Increasing the accuracy of estimation

Now the variance in the measurement error can be determined, the propagation of this error through the estimation algorithm can be evaluated. This was done for the situation where the number of applied temperatures is equal to the number of ions to be fitted *(N = I)*. However, by using more than N measurements, a decrease in the final error can be expected because of suppression of measurement noise.

In Figure 15 the normalized calculated standard deviation in the estimated concentration is represented for three different temperature ranges ($T_1..T_N$) and estimations for $N = 2$ to 11. For this numerical example, the coefficients $k_{i,j}$ for a 100 mM sodium chloride solution are used.

It can be seen that the accuracy of the estimation can be increased, either by increasing the number of measurements or the temperature range. The improvement in accuracy by increasing the temperature range is larger than the improvement obtained by using more measurements.

Summary

Reconsidering the original eq 16, describing the set of conductivity measurements at N temperatures, the ultimate method of determination should be

- Perform the measurements, by heating a solution while measuring the conductivity. Take more measurements (conductivities at known temperatures) than the number of ions to be fitted *(N > I)* since this will increase the accuracy of the estimation. Notice that the application of a large temperature sweep will increase the accuracy much more. Some conductivity meters have an automatic temperature compensation which must thus be switched off. Also the use of an auto range function will disturb the measurement since the variance in the measurement appeared to be constant per operational range;

- Assume a set of ions and calculate the matrix $\overline{\overline{B}}$ with the elements

$$\overline{\overline{B}}_{n,i} = \sum_{j=0}^{J} k_{i,j}\left(T_n - T_0\right)^j$$

  with T_0 the reference temperature, T_n the temperature of measurement n, J the order of the polynomial fit and $k_{i,j}$ the polynomial fit coefficients for the temperature dependency of the mobilities (which can be found in literature *(15)*);

- Create the conductivity vector $\overline{\Lambda}$, which is a column of N conductivity measurements at N different temperatures;

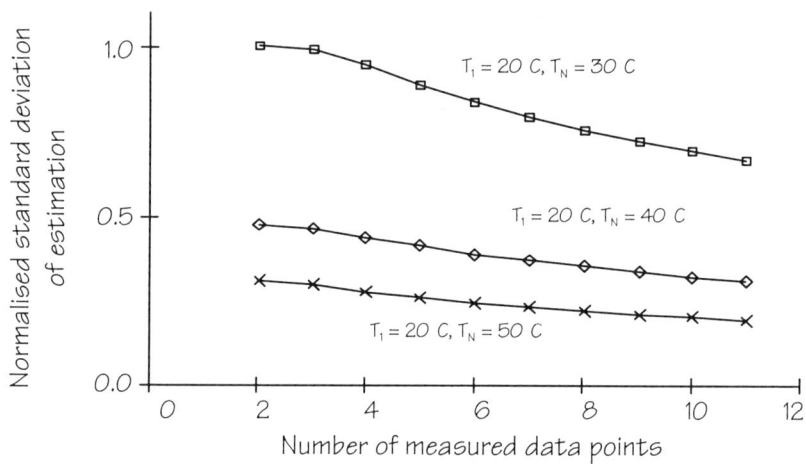

Figure 15. Effect of the temperature range, $T_1..T_N$, and the number of measurements on the estimation accuracy.

- Use eq 19 for finding the estimation vector $\hat{\bar{c}}$, which is a column of the $|z_i|c_i\lambda_i^0$ products for the I ions. Because the charge of the ions z_i and the limiting molar conductivity λ_i^0 are known, the concentrations of the individual ions can be calculated.

The size of the error in the estimated concentrations is represented by eq 21 in terms of the covariance of the measurement error.

Improvement by implementing the zero charge condition

Because in practical solutions the total charge is equal to zero, the condition

$$\sum_{\text{Cations}}|z_i|c_i = \sum_{\text{Anions}}|z_i|c_i \tag{22}$$

can be implemented in the algorithm in order to increase accuracy and to avoid useless answers like negative concentrations. This equation can be implemented in the model, eq 17, where the $\overline{\overline{T}} \cdot \overline{\overline{K}}$ matrix has now become the augmented $\overline{\overline{\overline{T}}} \cdot \overline{\overline{\overline{K}}}$ matrix. The elements in the augmented vector $\overline{\Lambda}$ and the augmented matrix $\overline{\overline{\overline{T}}} \cdot \overline{\overline{\overline{K}}}$ do not have an equal unity anymore. This does not make any difference for numerical evaluations, however.

This augmented model can be evaluated like the original model using the same estimation algorithm. However, the condition stated in eq 22 will have the same priority as every single measurement and so its importance will be suppressed with an increasing number of applied temperatures N. Two options are available for increasing its priority:

- The first option is to give the last row of the augmented $\overline{\overline{\overline{T}}} \cdot \overline{\overline{\overline{K}}}$ matrix a weight factor equal to the number of measurements.
- Another possibility is obtained by first reducing the measured information and then implementing the zero charge condition.

The further evaluation of these two options is treated elsewhere *(13)*. The error coefficient matrix of eq 20 has now become meaningless, and so the standard deviation of the fit (eq 21), since these are only valid when having equal variances $\overline{V}$ for all the measurements. The variances are not equal anymore, since the zero charge condition will have another variance than the actual conductivity measurements.

Implementation and results

The operational principle requires heating of the electrolyte. It is the benefit of a μ-TAS that local heating in a small measuring volume can be applied to rapidly heat the electrolyte locally. For this purpose, a dedicated EC-and-temperature sensor / thermal actuator was designed and developed *(17)*. The device is shown in Figure 16 and consists of a 1x1 mm² platinum structure on a glass substrate. A photograph of this device, mounted on a small piece of printed circuit board is shown in Figure 17. By applying an AC-current to one of the Pt-meanders, as shown at the left-hand part of Figure 16, the electrolyte can be heated locally. An increase of about 10^0C in temperature is obtained after a few milliseconds. Directly after switching off the heating current, both the EC and the local temperature are measured, as shown at the right-hand part of Figure 16: the structure now functions both as an interdigitated EC probe and resistive Pt-film temperature sensor.

From the measured results, a conductivity versus temperature plot can be constructed, as shown in Figure 18. After some calculations, using the theory described above, a unique fit can be obtained from which the specific ion concentrations can be calculated. A result, showing the successful application of the method is shown in Figure 19, in which the separate concentrations of 3 different ions in an electrolyte are determined. In a solution containing 25 mM NaCl, five different concentrations of KCl were added, as shown on the x-axis of Figure 19. The concentrations, as estimated by the algorithm are plotted on the y-axis and follow the added amount of KCl, whereas the estimated [Na$^+$] remains constant, as expected.

In conclusion, from a non-selective conductivity measurement, it is possible to find specific ion concentrations by recording the conductivity at different temperatures. The key to this is that every ion has its own specific limiting molar conductivity which depends uniquely on temperature. This method needs an assumed set of ions: the electrolyte conductivity is a linear combination of the specific ionic conductivities of these ions.

Additionally, every ion which is significantly present in the solution must be included in the calculation, since the method is based on the conductivity being a linear combination of all the separate ionic conductivities.

Summarizing, it can be said that the advantage of this method is that it introduces selectivity by smart data interpretation, and not by the sensor itself.

Concluding remarks

Research groups can continue to pursue the development of the ideal chemical sensor, or they can use the existing devices and micromachining techniques favorably in Total Analysis Systems. The flaws of existing detectors can thereby be circumvented. Amongst several other advantages, it is shown in this chapter, that μ-TAS's can offer a type of sample inlet, the microdialysis technique, which prevents severe fouling of the detector. Furthermore, it is shown that the function of calibration can be integrated in a μ-TAS by using electrochemically driven pumps.

Selection in EC sensing can be obtained by local heating, temperature- and EC-measuring. All these functions are shown to be performed with a device, comparably simple as an EC probe itself: properly shaped metal films. Only because of the small dimensions of a μ-TAS and the use of smart signal processing, the specific ion concentration determination by EC sensing is made possible.

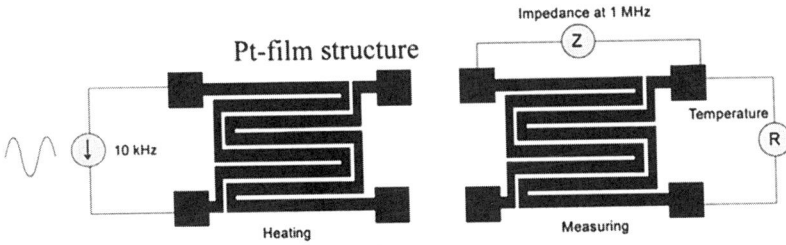

Figure 16. Integrated sensor-actuator device for local heating and temperature/conductivity measurement.

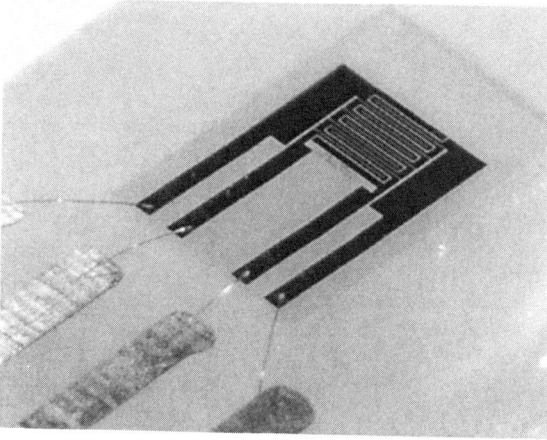

Figure 17. Photograph of the integrated sensor-actuator device on a glass substrate, mounted on a piece of PCB. The connection wires and strips are protected by transparent epoxy.

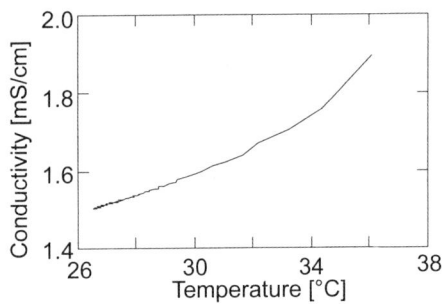

Figure 18. Measured conductivity versus temperature plot.

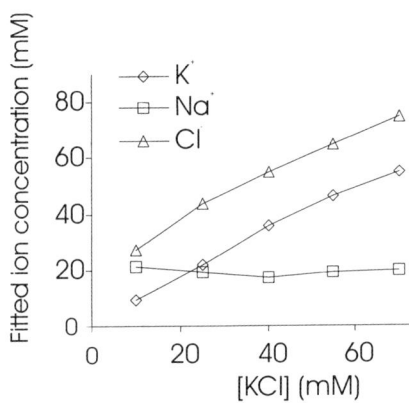

Figure 19. Fitted ion concentrations in five solutions with 25 mM NaCl and several KCl concentrations.

References

1. Manz, A.; Graber, N.; Widmer, H.M., *Sensors and Actuators* **1990**, *B1*, 244-248.
2. Manz, A.; Verpoorte, E.; Raymond, D.E.; Effenhauser, C.S.; Durggrat, N.; Widmer, H.M., In *Proc. µTAS '94;* van den Berg, A.; Bergveld, P., Eds.; Kluwer Academic Publ.: Dordrecht, 1995, 5-27.
3. de Boer, J.; Plijter-Groendijk, H.; Visser, K.R.; Mook, G.A.; Korf, J., *Eur. J. Appl. Physiol.*, **1994** *69*, 281-286.
4. Crank, J., *The mathematics of diffusion,* Oxford University Press: London, 1964.
5. Böhm, S.; Olthuis, W.; Bergveld, P., In *Proc. µTAS '98;* Harrison, D.J.; van den Berg, A., Eds.; Kluwer Academic Publ.: Dordrecht, 1998, 31-34.
6. Böhm, S.; Pijanowska, D.; Olthuis, W.; Bergveld, P., In *Proc. Dutch Sensor Conference*; van den Berg, A.; Bergveld, P., Eds; Kluwer Academic Publ.: Dordrecht, 1998, 91-95.
7. Bergveld, P.; Böhm, S.; Olthuis, W., International Patent Application, PCT/NL99/00057, 1999.
8. Böhm, S.; Olthuis, W.; Bergveld, P., *J. Biomed. Microdev.* **1999**, *1:2,* 121-130.
9. Leistiko, O.; Jensen, P.F., In *Proc. µTAS '98;* Harrison, D.J.; van den Berg, A., Eds.; Kluwer Academic Publ.: Dordrecht, 1998, 291-294.
10. Roulet, J.-C.; Fluri, K.; Verpoorte, E.; Völkel, R.; Herzig, H.-P.; de Rooij, N.F.; Dändliker, R., In *Proc. µTAS '98;* Harrison, D.J.; van den Berg, A., Eds.; Kluwer Academic Publ.: Dordrecht, 1998, 287-290.
11. Darling, R.B.; Yager, P.; Weigl, B.; Kriebel, J.; Mayes, K., In *Proc. µTAS '98;* Harrison, D.J.; van den Berg, A., Eds.; Kluwer Academic Publ.: Dordrecht, 1998, 105-108.
12. Fielden, P.R.; Baldock, S.J.; Goddard, N.J.; Pickering, L.W.; Prest, J.E.; Snook, R.D.; Treves Brown, B.J.; Vaireanu, D.I., In *Proc. µTAS '98;* Harrison, D.J.; van den Berg, A., Eds.; Kluwer Academic Publ.: Dordrecht, 1998, 323- 326.
13. Langereis, G.R., Ph.D thesis, University of Twente, Enschede, 1999.
14. Langereis, G.R.; Bergveld, P.; Olthuis, W., European Patent no. 98202690.8-2204, 1998.
15. Harned, S.; Owen, B., *The physical chemistry of electrolytic solutions*, Reinhold Publishing Corporation: New York, 1958.
16. Liebelt, P.B., *An introduction to optimal estimation*, Addison-Wesley Publishing Company: Reading, MA, 1967.
17. Langereis, G.R.; Bergveld, P.; Olthuis, W., *Sensors and Actuators* **1999**, *B 53/3*, 197-203.

ENZYME- AND PROTEIN-BASED SENSORS

Chapter 6

Fluorescent Biosensing Systems Based on Analyte-Induced Conformational Changes of Genetically Engineered Periplasmic Binding Proteins

Lyndon L. E. Salins, Suresh Shrestha, and Sylvia Daunert

Department of Chemistry, University of Kentucky, Lexington, KY 40506

Periplasmic binding proteins from bacteria such as *Escherichia coli* (*E. coli*) are important biomolecules which participate in the transport of substrates between the periplasm and the cytoplasm of the cells. In the presence of their respective substrates, the two globular domains of the proteins undergo a hinge motion completely engulfing the ligands. The short peptide links, which connect the two domains, act as the hinge during this structural rearrangement. These conformational changes that proteins such as the phosphate binding protein (PBP), sulfate binding protein (SBP), and galactose/glucose binding protein (GBP) undergo upon binding to their respective ligands can be used as the basis for the development of an optical sensing system for phosphate, sulfate, and glucose, respectively.

The wild-type forms of these proteins lack cysteine residues in their primary structure. Using the polymerase chain reaction (PCR), single cysteines were introduced via site-directed mutagenesis. These mutant proteins were expressed in the periplasm and released by osmotic shock. Novel purification methods were developed whereby the proteins of interest were isolated from the crude extract in a single step using perfusion anion exchange chromatography. The unique cysteine residues were labeled with various environment-sensitive fluorophores such as N-[2-(1-maleimidyl)ethyl]-7-(diethylamino)coumarin-3-carboxamide (MDCC), 6-acryloyl-2-dimethylaminonaphthalene (acrylodan), 5-((((2-iodoacetyl)amino)ethyl)amino)naphthalene-1-sulfonic acid (1,5-IAEDANS), and N-((2-iodoacetoxy)ethyl)-N-methyl)amino-7-nitrobenz-2-oxa-1,3-diazole (IANBD ester). Site-specific labeling ensures the reporting of conformational changes that the proteins undergo upon ligand binding. The changes in the fluorescence properties of the conjugates were monitored and related to the amount of ligand present. These steady-state fluorescence studies clearly indicated that the hinge motion and binding properties of these proteins could be utilized to develop a fluorescence-based biosensing system for phosphate, sulfate, and glucose.

Introduction

There is a growing need for screening and monitoring applications that are rapid, reliable, cost-effective, selective, and able to detect low levels of environmental pollutants and biomolecules. Chemical devices such as biosensors can fulfill this need since they are capable of determining a particular analyte. Modern advancements in the fields of molecular biology, protein chemistry, and genetic engineering have resulted in the emergence of sensing systems based on biological molecules that can be used as vital tools in the development of biosensors.

The development of an assay to determine the concentration of inorganic phosphate (P_i) and sulfate concentrations in a defined volume of water would be extremely useful from an ecological standpoint. Such a test, when performed in a natural body of water, might indicate localized pollutant dispersal into the surrounding environment (1, 2). P_i and sulfate pollution can be harmful to the natural inhabitants of the area (3, 4). In addition to a need for detection of phosphate on earth, there is also a growing need for such methods in future space stations where the recycling of water will be a vital task. Analytical methods, which can accurately determine these analytes are needed because of their importance not only in the environment (5) but in the function of living organisms as well (6). The purity and stability of natural bodies of water, including drinking water, is highly dependant on P_i and sulfate concentrations while the monitoring of blood glucose levels is vital in patients suffering from diabetes. The need for a high-throughput screening method for these analytes in the environmental and medical field calls for the development of smarter sensors that can recognize specific molecules.

The possibility of using some of the members of the periplasmic family of binding proteins such as the phosphate binding protein (PBP), sulfate binding protein (SBP), and galactose/glucose binding protein (GBP), as the biorecognition element in a biosensing scheme for the detection of P_i, sulfate, and glucose respectively, have been explored. The selectivity of these proteins originates from their natural role which, in Gram-negative bacteria such as *E. coli*, is to serve as an initial receptor for the highly specific translocation of ligands to the cytoplasm. The high specificity for their ligands can be taken advantage of in the analysis of complex mixtures. The tertiary structure of the proteins consist of two globular domains connected by three short peptide links that serve as a hinge. The ligand binding site is located deep within the cleft between the two domains. Each periplasmic protein selectively binds a single analyte molecule. The ligand induces a hinge motion in the protein, indicated by the movement of α-helices, β-sheets, and strands that are not restricted by tertiary packing (7). The resultant conformational change constitutes the basis of the sensor development.

PBP, the product of the *phoS* gene (8), is responsible for the highly specific transport of phosphate (9) to the cytoplasm. The mature protein consists of 321 amino acids with a molecular weight of 34,422 Da (10). From kinetic plots and the resin method for assaying binding activity, it has been shown that one protein molecule binds to one molecule of phosphate (11). PBP can bind to both monobasic

($H_2PO_4^-$) and dibasic (HPO_4^{2-}) phosphate (12). It binds P_i tightly (13), K_d of 0.1 μM, and rapidly (1.36 × 10^8 M^{-1}s^{-1}) (6). The protein consists of a single polypeptide chain that is folded into two globular domains connected via three peptide segments which serve as a flexible hinge (14). When P_i is absent, the two domains are far apart with the cleft accessible to solvent molecules (15, 16). The substrate binding site is located deep within this cleft. In the bound-form structure, the domains undergo a large relative angle change (16) and are closer to each other engulfing and burying the P_i substrate (17). The conformational change that the protein undergoes is represented in Figure 1. The exclusion of water molecules from the binding pocket enables efficient hydrogen-bonding interactions between P_i and several polarizable main chain amino groups and hydroxyl side chains in the active site. Along with van der Waals forces, these interactions are responsible for stabilizing the negative charge of the phosphate ligand (14).

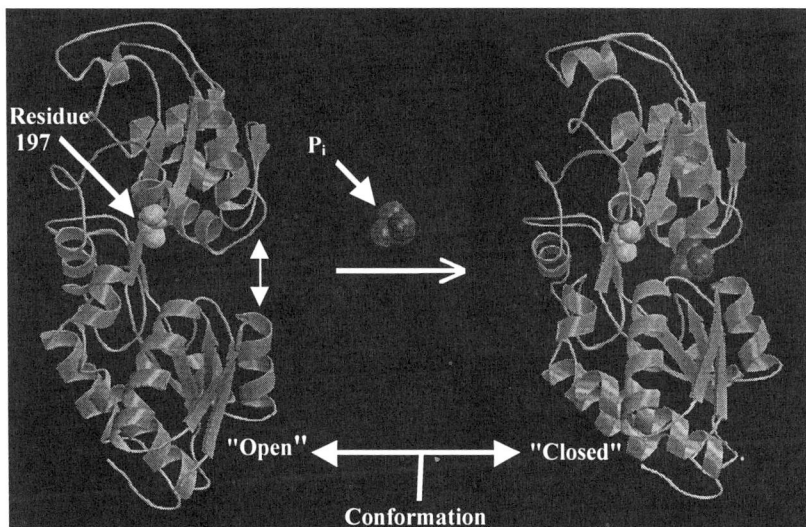

Figure 1. Crystal structures of PBP in the absence and presence of phosphate.

SBP exhibits similar tertiary features to the hinge-bending PBP (18). It consists of two distinct domains each arising from both the N- and C-terminals of the polypeptide chain. However, the active site in SBP is designed to bind the fully ionized form of sulfate, while the site in PBP recognizes the proton(s) of the weak acid phosphate (12). SBP has a molecular weight of 35 kDa and binds sulfate with a K_d of 0.12 μM (19). The anion is bound in the active site and its negative charge stabilized via hydrogen bonds with seven donor groups along with van der Waals forces (20, 21).

Galactose and glucose uptake in *E. coli* is mediated by GBP (*22*). The mature protein consists of a single polypeptide chain comprised of 309 amino acids with a molecular weight of 33,310 Da (*23*). Like SBP, GBP is ellipsoidal in shape with two different but similarly folded domains connected by three peptide segments that serve as a flexible hinge. The two globular domains form a central binding pocket for the substrate. The protein binds to D-galactose and D-glucose with dissociation constants, K_d, of 0.4 and 0.2 μM, respectively (*24*). The aspartic acid residue at position 14 forms hydrogen bonds with the the hydroxyl group on carbon 4 of the sugar when the latter is either in the axial (D-galactose) or equatorial position (D-glucose). This explains the fact that there is such negligible difference in the affinity of GBP for the two epimers. In the binding site, the sugar ligand is sandwiched between the two aromatic residues, phenylalanine at position 16 and tryptophan at position 183 (*23*).

In order to take advantage of the selectivity of these proteins for their analytes in the biosensing scheme, a mode of signal transduction that is capable of sensing the protein-analyte interactions must be developed. To achieve this, genetically altered mutants of PBP, SBP, and GBP, containing a unique cysteine residue, were constructed by performing PCR-site directed mutagenesis. Wild-type PBP, SBP, and GBP lack cysteine residues in their primary structure; therefore, the replacement of an amino acid residue with a cysteine ensures a site-selective labeling position. The residues chosen were not involved in ligand binding and are located at sites where they might experience a change in the surrounding micro-environment. This renders them sensitive to the changes induced by the protein-ligand interactions. Tagging a fluorescent probe to the protein via the cysteine residues, enables the performance of fluorescence measurements where the change in the intensity of the emitted signal can be monitored. The accompanying conformational change can be thus assessed by monitoring perturbations in the fluorescence properties of the fluorophore. Calibration plots relating the fluorescence intensity with the concentration of ligand present in the sample solution can then be constructed.

The mutant proteins were expressed in *E. coli*, released from the periplasmic space by osmotic shock, and purified in a single step by perfusion chromatography using an anion-exchange column. After the purity of the proteins were verified, they were then labeled with several environment-sensitive fluorophores (Figure 2), such as 6-acryloyl-2-dimethylaminonaphthalene (acrylodan), 5-((((2-iodoacetyl)amino)ethyl)amino)naphthalene-1-sulfonic acid (1,5-IAEDANS), N-[2-(1-maleimidyl)ethyl]-7-(diethylamino)coumarin-3-carboxamide (MDCC), and N-((2-iodoacetoxy)ethyl)-N-methyl)amino-7-nitrobenz-2-oxa-1,3-diazole (IANBD ester), through the sulfhydryl groups of the cysteine residues. All the probes used in this study are known to experience a change in their spectral properties based on their surrounding environments. The excess fluorophores were separated from the conjugates employing size-exclusion chromatography using a Sephadex column. The labeled proteins were finally dialysed in the appropriate buffers before fluorescent measurements were obtained.

In this chapter we outline the development of genetically engineered biosensing systems based on ligand-induced conformational changes of periplasmic binding proteins. Strategies employed in selecting appropriate labeling sites and fluorescent reporter probes in the rational design of three sensing systems are discussed.

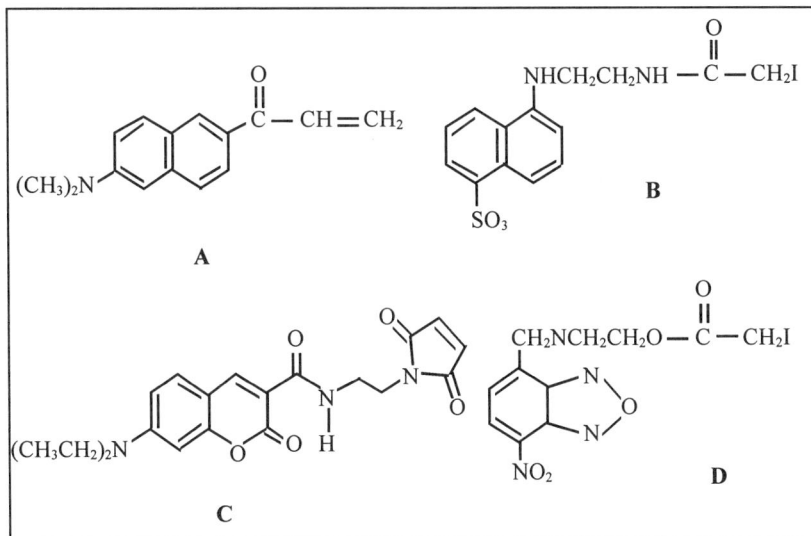

Figure 2. Sulfhydro-specific fluorescent probes used in the conjugation of binding proteins via unique cysteine moeities. A. Acrylodan (λ_{ex} 351 nm, λ_{em} 510 nm); B. 1,5-IAEDANS (λ_{ex} 336 nm, λ_{em} 490 nm); C. MDCC (λ_{ex} 425 nm, λ_{em} 470 nm); D. IANBD ester (λ_{ex} 472 nm, λ_{em} 536 nm).

Experimental Procedures

Bacterial Strains and Plasmids.

The *E. coli* strain ANCC75, which contains the plasmid pSN507 carrying the *phoS* gene (*25*), was kindly provided by Dr. H. Shinagawa (Osaka University). Plasmid pSD501 was constructed in our laboratory from pSN507 by PCR-site directed mutagenesis. Specifically, the MluI-PstI restriction fragment, which carries the *pstC*, *pstA*, *pstB*, *phoU*, and *phoS* genes in pSN507, was excised and a new fragment containing the *phoS* gene with the desired alanine to cysteine mutation at position 197 was inserted in its place. DNA sequence analysis verified the presence of the mutation.

Since the gene sequence which codes for SBP is known (*26*), primers were designed and the PCR methodology was employed to extract the *sbp* gene from JM109 chromosomal DNA. Plasmids containing cysteine mutations at positions 55, 90, 171, 181, and 186 were constructed using the pET-17b vector obtained from

92

Novagen (Madison, WI). They were then transformed into the BL21(DE3) strain of *E. coli*, with the expression now under the control of the T7 promoter.

Utilizing the known sequence of GBP (*22*), primers were constructed to isolate the *mglB* gene directly from JM107 genomic DNA using PCR. The gene was then incorporated into the pUC(E)8-19 vector yielding pSD503. Overlap extension PCR site-directed mutagenesis was performed on the *mglB* gene in pSD503 to introduce single amino acid mutations to cysteine at positions 148, 152, and 182.

Materials.

Luria-Bertani (LB) media and restriction endonucleases were purchased from GibcoBRL (Gaithersburg, MD). Primers used in PCR were bought from Operon Technologies (Alameda, CA). Tris buffer ([tris(hydroxymethyl)aminomethane]) was obtained from VWR Scientific (S. Plainfield, NJ). The antibiotics, tetracycline and ampicillin, as well as bovine serum albumin (BSA), were bought from Sigma (St. Louis, MO). Ethylenediaminetetraacetic acid (EDTA) as well as all organic and inorganic salts, including dithiothreitol (DTT), were obtained from either Fisher Scientific (Fair Lawn, NJ), VWR Scientific (S. Plainfield, NJ), or Sigma (St. Louis, MO). SBP and GBP expressions were induced with isopropyl-β-D-thiogalactopyranoside (IPTG), purchased from GibcoBRL (Gaithersburg, MD). The bicinchoninic acid (BCA) protein micro assay reagent kit from Pierce (Rockford, IL) was used to determine the concentration of purified proteins. The fluorophore, N-[2-(1-maleimidyl)ethyl]-7-(diethylamino)coumarin-3-carboxamide (MDCC), which was conjugated to the mutant proteins, was synthesized in our laboratory following the method of Corrie (*27, 28*). The fluorescent probes, acrylodan, 1,5-IAEDANS, and IANBD ester were purchased from Molecular Probes (Eugene, OR). Size-exclusion chromatography was utilized to separate conjugated protein from excess fluorophore with a Sephadex G-25 column from Sigma (St. Louis, MO).

Apparatus.

Gene amplification and mutagenesis were performed on a Gene Amp PCR System 2400 by Perkin Elmer (Norwalk, CT). Bacterial colonies were grown on agar plates at 37 °C in a Fisher Scientific Incubator (Fair Lawn, NJ). Cell cultures were grown in an Orbital Shaker from Forma Scientific (Marietta, OH) and pelleted using a Beckman J2-MI Centrifuge (Palo Alto, CA). Unpurified protein fractions were filtered with a 0.2 μm filter from Nalgene (Rochester, NY).

The BioCAD SPRINT Perfusion Chromatography System by PerSeptive Biosystems (Cambridge, MA) was used for protein purification. Liquid protein samples were lyophilized using the VirTis Bench Top 3 Freeze Dryer (Gardiner, NY). Proteins were dialysed against the proper buffer using a 12-14,000 MWCO SPECTRA/POR molecular porous membrane by Spectrum Medical Industries (Los Angeles, CA). Protein absorbances in the BCA assay were determined with the Hewlett Packard 8453 UV-Visible System (USA).

Expression and Purification of Mutant Proteins.

A detailed description of the expression of mutant PBP in *E.coli* is provided in reference 24. For the expression of the SBP and GBP mutants, the plasmids of interest were transformed into *E. coli* and grown at 37 °C in LB media containing the antibiotic ampicillin. When the cultures attained an OD_{600} of about 0.6, IPTG was added to induce protein expression. The cultures were allowed to incubate for an additional 3.5 h (for SBP) or 17 h (for GBP). The cells were harvested and the proteins released from the periplasmic space via osmotic shock (*8*). The crude extracts were lyophilized, dialyzed against 10 mM Tris-HCl, 1 mM $MgCl_2$, 1 mM DTT, pH 8.0, and filtered using a 0.2 μm syringe filter.

The proteins were purified in a single step using perfusion chromatography with an anion exchange column composed of quarternized polyethyleneimine (HQ) functional group. The purity of the fractions were verified by sodium dodecyl sulfate polyacrylamide gel electrophoresis (SDS-PAGE) using the PhastSystem from Pharmacia Biotech (Uppsala, Sweden).

Conjugation of Mutant Proteins with fluorophores.

Purified proteins were first dialyzed against 10 mM Tris-HCl, 1 mM DTT (pH 7-8) to eliminate disulfide linkages and free up the cysteine groups. The samples were dialyzed against 10 mM Tris-HCl (pH 7-8) to remove any excess DTT and then introduced into reaction vials. A five-fold molar excess (compared to the proteins) of fluorophores were added to the vials in small increments while the solutions were being stirred. The reactions were allowed to proceed in the dark at 4 °C for 4-5 h. The conjugates were then loaded onto a Sephadex G-25 size-exclusion column to separate the labeled proteins from the unbound fluorophores. The samples were eluted with 10 mM Tris-HCl (pH 7-8).

Steady State Fluorescence Measurements.

All fluorescence studies were performed on a Fluorolog-2 spectrofluorometer, Spex Industries (Edison, NJ), equipped with a 450-Watt Xenon arc lamp. The excitation and emission monochromator slit widths were both set at 2 mm. The excitation and emission wavelengths of the various fluorophores used are indicated in Figure 2. All data were acquired at room temperature using quartz cuvettes with sample volumes of 1.5 mL.

Dilutions of labeled PBP and SBP were made in a solution of 10 mM Tris-HCl, pH 8.0 and 7.7, respectively, each containing 0.01 % BSA to prevent adsorption of the protein to the walls of the cuvette. The GBP conjugates were diluted in the same buffer, but at a pH of 7.0 and in the absence of BSA. The labeled proteins were allowed to incubate with K_2HPO_4, Na_2SO_4, and D-glucose for 15, 20, and 15 min, respectively, on a shaker at 400 rpm at 4 °C. Calibration plots were constructed by adding aliquots of different analyte concentrations to the buffered solution containing the fluorescently-labeled proteins and monitoring changes in the signal intensity.

Selectivity studies were conducted to measure the response of PBP-MDCC towards cations such as carbonate, nitrate, sulfate, chloride, arsenate, tartrate, and perchlorate. To assess the stability and lifetime of the conjugate, the fluorescence signal of MDCC on the PBP was monitored under the given parameters at various points during extended periods of time.

Results and Discussion

Conformational changes are naturally occurring phenomena that take place when the periplasmic family of binding proteins associate with their respective substrates. In Gram negative bacteria such as *E. coli*, these proteins are found located in the periplasm, the region between the outer and the inner membranes. We have sought to use as a tool, that nature has provided, the structural rearrangements that these proteins undergo upon binding as the basis of the sensor development. The mechanism of interaction between an analyte and the labeled protein was investigated in our laboratory by studying the static and dynamical behavior of the latter through steady-state and time-resolved fluorescence spectroscopy (*29*). The research outlined here describes our initial efforts in the development of biosensors for the detection of phosphate, sulfate, and glucose.

The periplasmic binding proteins of bacteria such as *E. coli* serve as the receptors in an active transport system for a specific ligand (*30*). When the ligand has passed into the periplasm non-specifically through the outer membrane, it binds to the periplasmic binding protein with high affinity. The resulting bound complex interacts with the membrane-bound transport proteins of the system, at which point the ligand is released and transferred into the cytoplasm with the help of ATP hydrolysis (*31*). Since these proteins bind their ligands in a very specific manner (K_d in the micromolar range), we have chosen to use them as the biorecognition elements in a fluorescence-based sensing system. The binding mechanism functions in a similar way in all periplasmic binding proteins: the two globular domains enclose the ligand, closing the cleft which contains the binding site, with the three peptide chains that connect those domains acting as a hinge. This conformational change constitutes the origin of the sensing system. An environment-specific fluorophore conjugated to the protein can be expected to experience a change in its microenvironment as a result of the conformational change in the protein. The fluorescence signal is affected by this alteration in the environment of the fluorophore, and is either enhanced or quenched; it is difficult to characterize the existing microenvironment at each protein residue specifically enough to predict which effect will ensue.

The rational design of sensing systems involves selecting amino acids for modification that are in close proximity to the binding cleft by looking at either the NMR or the 3-dimensional crystal structure of the protein of interest (*32, 33*). Since these proteins do not contain any cysteine residues in their structure, introduction of this residue ensures a specific site for the attachment of a fluorescent probe. With the help of genetic engineering techniques such as PCR-site directed mutagenesis, we were able to modify single amino acid residues of PBP, SBP, and GBP, site-selectively, for the specific attachment of environmentally-sensitive fluorophores. The sites for modification were not involved in ligand binding and were chosen in close proximity to the binding cleft, since residues at these locations can be expected to experience a change in their local environment on substrate binding.

The first step in the development of a sensing system for P_i involved site-directed mutagenesis using PCR to encorporate a unique cysteine residue in PBP. Since neither of the three wild-type proteins of interest contain the amino acid cysteine in their primary structure, the introduction of this residue enables the attachment of a sulfhydro-specific probe. Using the criteria mentioned in the previous paragraph, the alanine at position 197 was chosen to be replaced by a cysteine. The plasmid pSD501 was constructed from pSN507 by deleting the *Mlu*I-*Pst*I fragment, containing the *pho* family of genes, and replacing it with a mutant *phoS* gene which carried the alanine to cysteine mutation. The mutation was verified by DNA sequence analysis. Protein expression was induced under conditions of P_i starvation, which prompts the bacteria to synthesize higher levels of mutant PBP to help in the scavenging of this essential nutrient. Periplasmic proteins were released by osmotic shock using sucrose and EDTA and followed by the addition of a low osmotic pressure liquid. The resultant swelling of the cells disrupts the outer membrane, thereby releasing the proteins. The sample of interest was isolated in a single step using anion-exchange perfusion chromatography. The purity of PBP was verified by running it on a SDS-PAGE and developing the gel with silver stain.

The lone cysteine residue was then conjugated with the sulfhydro-specific fluorophores, acrylodan and MDCC. The former binds cysteine via its thiol group forming stable thioether bonds. The probe's spectral intensity is sensitive to the protein's conformational changes or association with a substrate (*34*). In contrast to most commercially available maleimido probes, where the maleimido moeity is directly attached to the aromatic ring, in MDCC an aliphatic spacer is incorporated between the two groups (*27, 28*). This renders the environment-sensitive fluorophore with unrestricted motion: the two-carbon spacer arm between the maleimide and the coumarin moeity allows flexibility of the fluorophore, preventing any interference with the conformational change of the protein or its interaction with the ligand. The purified PBP was dialyzed against 10 mM Tris-HCl, pH 7-8 to get rid of any DTT which was used to reduce disulfide linkages. The molar concentration of the fluorophores in the reaction mixture were five times that of the protein. They were added in small increments to the PBP solution at 4 °C, while the latter was being stirred to ensure equal exposure of the cysteine residue on all protein molecules. The reaction was allowed to proceed in the dark for 5 h in the case of acrylodan and for 4 h with MDCC. Excess fluorophore was separated from the conjugated protein by running the mixture through a Sephadex G-25 size-exclusion column.

A considerable change in fluorescence was observed upon the PBP-acrylodan conjugate binding to P_i. Results indicated about a 50 % enhancement of the fluorescence signal upon binding of micromolar concentrations of the ligand (*29*). This suggests that a sensitive detection system for P_i can be developed using this genetically engineered protein. However, for sensing applications, the fluorescence labeling protocol should be site-specific. To verify that our protocol was selective for only the cysteine at position 197, the fluorophore was also conjugated to wild-type PBP under similar conditions. From the steady-state emission spectrum of this conjugate, no measurable fluorescence could be observed, clearly indicating that the protein had not been labeled non-specifically. With the help of this acrylodan-labeled protein we have also studied the dynamics of the system in order to better understand the signal transduction process. Differences in the mechanism(s) of dipolar relaxation are responsible for the observed steady-state profiles of the labeled protein when free and when bound to the substrate. However, local structural changes influence the local state surrounding the mutation site, thus affecting the mechanism(s) of dipolar relaxation. Time-resolved anisotropy measurements demonstrate that global structural changes take place upon ligand binding due to the closure of the binding pocket. A 5 ns decrease in the rotational correlation time of the global motion was observed, indicating that the volume/structure of the rotating body had decreased. This event induces a series of changes that are responsible for the observed changes in the local behavior of the fluorophore. Changes in the rate of local motion of acrylodan about its axis were observed: a slight decrease from 0.68 ± 0.17 ns to 0.25 ± 0.11 ns was measured upon association of the protein with P_i. An increase in the semiangle of acrylodan when the protein binds its ligand was also measured. These results indicate that, although the binding pocket is closed when the labeled protein binds P_i, the fluorophore rotates faster and gains a greater range of motion.

Fluorescence quenching techniques allow for the quantification of the degree of accessibility of the fluorophore to the quenching agent through the Stern-Volmer relationship:

$$\tau_0/\tau = 1 + \tau_0 <k_q>[Q]$$

In this equation, τ_0 and τ are the mean fluorescence lifetimes of the fluorophore in the absence and presence of the quencher, respectively, [Q] is the concentration of the quenching species, and $<k_q>$ is the bimolecular quenching constant which serves as a measure for determining the degree of accessibility of acrylodan to the quencher molecule. Dynamic quenching measurements reveal that acrylodan becomes more accessible to iodide when the binding pocket of the PBP is closed. On binding phosphate, the bimolecular quenching constant increases from $(6.0 \pm 0.23) \times 10^8$ $M^{-1}s^{-1}$ in the ligand-free state to $(8.1 \pm 0.16) \times 10^8$ $M^{-1}s^{-1}$. These results suggest that local structural changes near the acrylodan are responsible for changes in the accessibility of the fluorophore to the iodide quencher. Additional details of these studies can we found in reference 29.

MDCC was also conjugated to mutant PBP and fluorescence studies in solution were performed to characterize the binding properties of the protein to evaluate the feasibility of using it as a biorecognition element in a biosensor. An incubation time study of PBP-MDCC with P_i, performed in the presence of BSA, indicated maximum fluorescence enhancement after 15 min of exposure to the ligand.[35] The was chosen as the incubation time for all further studies. Calibration curves were constructed with P_i at pHs of 6, 7, and 8 as seen in Figure 3. The best results were obtained at a pH of 8 when the phosphate was present in the dibasic form. The system was demonstrated to have a detection limit of 6×10^{-8} M, indicating excellent detection limits of the system to its ligand. There was a maximum fluorescence signal increase of 93 % using a protein concentration of 4.7×10^{-8} M. This enhancement in the fluorescence intensity upon ligand binding can be attributed to the positioning of the fluorophore in a less steric environment when PBP binds P_i. Dynamic quenching studies using iodide as the quencher have also suggested that the probe is more accessible to the quenching species when the protein is in the bound form compared to the ligand free structure (35).

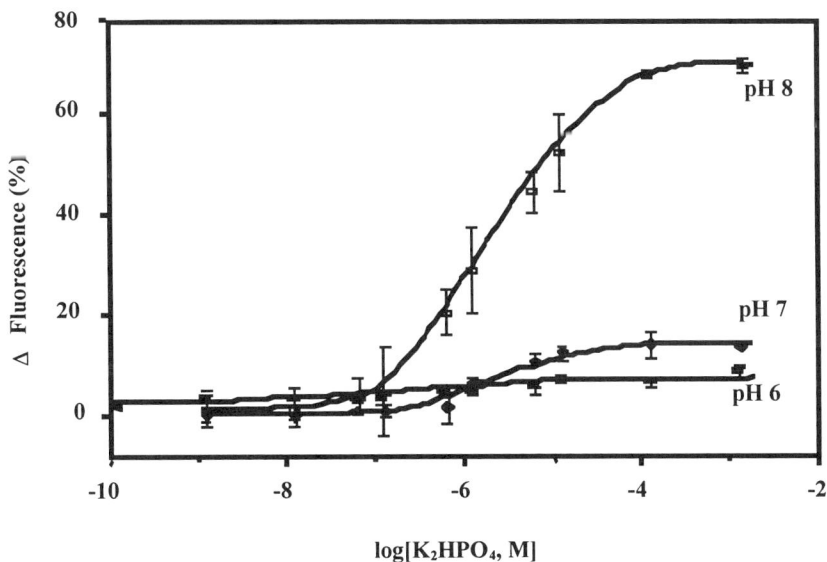

Figure 3. Calibration Plot of PBP-MDCC with P_i at different buffer pH..

Selectivity studies were then conducted with nonmetals such as arsenate, sulfate, nitrate, chloride, carbonate, perchlorate, and tartrate. Of the nonmetals tested, only incubation of PBP-MDCC with arsenate demonstrated fluorescence enhancement results similar in magnitude to those with P_i. This is to be expected, since arsenate is similar in structure in many ways. However, a comparison of the calibration curves indicate that the detection limit of PBP for arsenate is nearly two orders of magnitude higher than for P_i, indicating less specific response of the system to arsenate. The maximum increase in signal intensity with arsenate was about 55 % compared to the 93 % obtained with P_i (*35*). As seen in Table I, the conjugate system still responded to micromolar concentrations of P_i despite being saturated with the interfering anions. A stability study was conducted to observe the change in fluorescence signal of PBP-MDCC as related to the time since its original conjugation. The deterioration of the signal has not been sufficient to affect experimental results. The functionality of the protein itself has not been degraded by time, as evidenced by the fact that the fluorescence intensity change upon binding remained consistent for over a year (*35*).

These significant enhancement results of PBP-MDCC with P_i, coupled with the demonstrated preference in detection capabilities and magnitude of signal change which the system shows for the ligand of interest, indicate the possibility of the use of its use as a sensing system for phosphate. Initial studies like reducing the solution volume to microliter levels followed by assaying on a fluorescence plate reader and entrapping the conjugated protein in solution behind a membrane attached to a bifurcated fiber-optic tip, have demonstrated promise thus far (35), PBP may prove to be the basis of a very specific and sensitive detection system for P_i.

Table I. Change in fluorescence intensity of PBP-MDCC upon addition of 12.2 μM P_i in the presence of saturating concentrations of potential interferents.

Interfering Anion	P_i-Induced Signal Enhancement (%)
Carbonate	28.5
Nitrate	30.0
Sulfate	54.1
Chloride	64.4
Arsenate	No Change

Sulfate and glucose transport in E. coli are mediated by periplasmic SBP and GBP, respectively. Both these proteins entrap their ligands, in a highly specific manner, with the closure of the two domains in a hinge-like fashion. The conformational change that accompanies this binding event is the basis of the development of optical sensing systems for these analytes. The change in the fluorescence signal of the labeled protein can be related to the concentration of the analyte present. To achieve this purpose, suitable sites for site-specific labeling have been evaluated for attachment of the fluorophore. Owing to the lack of crystal structures of both SBP and GBP in the substrate-free form, amino acid sites were chosen by studying the crystal structure of the proteins in the substrate-bound form and also comparing them to the P_i-free and -bound forms of PBP. The rational behind doing this is because all binding proteins in this family have closely related tertiary structures and hence residues situated close to the binding cleft in one protein maybe close to the cleft in another protein as well.

Five sites were chosen for replacement of the native amino acids with a cysteine via site-directed mutagenesis. These residues are located on loops between α-helices located near the active site. Hence, they should be susceptible to changes in micro-environment upon association of the protein with sulfate. Mutant SBP expression was induced using IPTG, with similar isolation and purification protocol as used for PBP. The protein was labeled with MDCC and a SBP concentration of 2.6×10^{-8} M was used for the fluorescence studies. The conjugates were allowed to incubate with sulfate for 20 min. Detection limits for sulfate in the order of 1×10^{-7} to 1×10^{-6} M were obtained using these mutants. The changes in the signal intensity of the protein conjugates upon ligand binding are summarized in Table II. From the data, residue 90 clearly experienced a change in its surrounding environment upon the association of the protein with sulfate as indicated by the enhancement of the fluorescence signal. Residues 171 and 181 also experienced a change in their respective environments during the binding event; however, in these cases the signals were quenched. Whether the fluorescence intensity of the fluorophore was enhanced or quenched is dependant on the hydrophobic or hydrophilic nature of the surrounding solvent molecules or amino acid residues.

Table II. Amino acid residues selected for the rational design of the sensing system. The change in intensity upon sulfate binding is depicted for the MDCC-labeled SBP.

Mutation Sites	Residues Changed	Fluorescence Change (%)
55	Gly → Cys	No Change
90	Ser → Cys	+ 21
171	Ser → Cys	– 19
181	Val → Cys	– 14
186	Gly → Cys	No Change

Since GBP binds glucose tightly (K_d of 0.2 µM) (24), we have chosen to use it in the biosensing scheme to develop a sensor for the sugar. The gene of interest that codes for GBP was isolated from genomic DNA and amplified by PCR (36). The mutant protein was labeled with several fluorophores at the three sites where unique cysteine residues were introduced via overlap extension PCR. The locations chosen for labeling were all located around the cleft of the binding pocket. As seen in Table III, all the mutants showed different degrees of quenching of the fluorescence signal depending on the properties of the individual probes. The reason for such significant quenching may lie in the location of the conjugated fluorophore on the protein. As described earlier, a fluorophore on the lip of the protein may become buried in the hydrophobic regions of the protein when the globular domains engulf the ligand, and thus, would have far less exposure to solvent molecules, and a resultant less degree of change in signal.

Table III. Percentage of fluorescence quenching of mutant GBP labeled with various fluorophores upon the addition of glucose.

GBP Mutant	MDCC	Acrylodan	IANBD	IAEDANS
M182C	12	7		
H152C	30		5	
G148C	17			7

In summary, we have taken advantage of the structural changes that periplasmic binding proteins undergo in the presence of their respective substrates to develop biosensing systems that are selective and sensitive to submicromolar concentrations of their respective analyte. Through the use of modern protein engineering technology, we were able to modify a single amino acid residue for the site-selective covalent attachment of environmentally-sensitive fluorophores. In addition to the three proteins mentioned in this chapter, our laboratory is also involved in the development of biosensors for calcium (32, 33) and the phenothiazine

100

family of drugs (*37*) using genetically engineered calmodulin as the biorecognition element for biosensing applications.

Acknowledgments

We would like to thank the National Aeronautics and Space Administration (NCCW-60) and the Department of Energy (DE-FG05-95ER62010) for their funding for this research, as well as the Kentucky Research Challenge Trust Fund for the academic fellowship provided to L.S. S.D. is a Cottrell Scholar and a Lilly Faculty Awardee.

Literature Cited

1. Mayewski, P. A.; Spencer, M. J. *Atmos. Environ.* **1987**, *21(4)*, pp 863-869.
2. Klemow, K. M.; Tarutis, W.; Walski, T.; *J. Penn. Acad. Sci.* **1995**, *68*, p 182.
3. Whiting, D. *Fertilizer-Phosphorous and Water Pollution*; No. 282; Univ. of Minn. Dept. of Horticulture and Soil Sciences, 1997.
4. Abrams, R. *Acid Rain*; Committee on Environment and Public Works United States Senate, 1982; pp 170-171.
5. Mancy, K. H.; Weber, W. J.; *Analysis of Industrial Wastewaters*; Part III; Wiley-Interscience: New York, NY, 1971; Vol. 2.
6. Brune, M.; Hunter, J. L.; Corrie, J. E. T.; Webb, M. R. *Biochem.* **1994**, *33*, pp 8262-8271.
7. Wodak, S. J.; Janin, J. *Biochem.* **1981**, *20*, pp 6544-6552.
8. Willsky, G. R.; Malamy, M. H. *J. Bacteriol.* **1976**, *127*, pp 595-609.
9. Willsky, G. R.; Bennett, R. L.; Malamy, M. H. *J. Bacteriol.* **1973**, *113*, pp 529-539.
10. Magota, K.; Otsuji, N.; Miki, T.; Horiuchi, T.; Tsunasawa, S.; Kondo, J.; Sakiyama, F.; Amemura, M.; Morita, T.; Shinagawa, H.; Nakata, A. *J. Bacteriol.* **1984**, *157*, pp 909-917.
11. Pardee, A. B.; Prestidge, L. S.; Whipple, M. B.; Dreyfuss, J. *J. Biol. Chem.* **1966**, *241*, p 3962.
12. Wang, Z.; Choudhary, A.; Ledvina, P. S.; Quiocho, F. A. *J. Biol. Chem.* **1994**, *269*, pp 25091-25094.
13. Luecke, H.; Quiocho, F. A. *Nature.* **1990**, *347*, pp 402-406.
14. Ledvina, P. S.; Yao, N.; Choudhary, A.; Quiocho, F. A. *Proc. Natl. Acad. Sci.* **1996**, *93*, pp 6786-6791.
15. Sack, J. S.; Saper, M.A.; Quiocho, F. A. *J. Biol. Chem.* **1989**, *206*, pp 171-191.
16. Spurlino, J.; Lu, G.-Y.; Quiocho, F. A. *J. Biol. Chem.* **1991**, *266*, pp 5202-5219.
17. Quiocho, F. A. *Kidney Intl.* **1996**, *49*, pp 943-946.
18. Copley, R. R.; Barton, G. J. *J. Biol. Chem.* **1994**, *242*, pp 321-329.
19. Quiocho, F. A.; Jacobson, B. L. *J. Mol. Biol.* **1988**, *204*, pp 783-787.
20. Jacobson, B. L.; He, J. J.; Vermersch, P. S.; Lemon, D. D.; Quiocho, F. A. *J. Biol. Chem.* **1991**, *266*, pp 5220-5225.

21. Pflugrath, J. W.; Quiocho, F. A. *J. Mol. Biol.* **1988,** *200*, pp 163-180.
22. Scholle A.; Vreeman, J.; Blank, V.; Nold, A.; Boos, W.; Manson, M. *Mol. Gen. Genet.* **1987,** *208*, pp 247-253.
23. Mahoney, W. C.; Hogg, R. W.; Hermodson, M. A. *J. Biol. Chem.* **1981,** *256*, pp 4350-4356.
24. Miller, D. M.; Olson, J. S.; Pflugrath, J. W.; Quiocho, F. A. *J. Biol. Chem.* **1983,** *258*, pp 13665-13672.
25. Amemura, M.; Shinagawa, H.; Makino, K.; Otsuji, N.; Nakata, A. *J. Bacteriol.* **1982,** *152*, pp 692-701.
26. Hellinga, H. W.; Evans, P. R. *Eur. J. Biochem.* **1985,** *145*, pp 363-373.
27. Corrie, J. E. T. *J. Chem. Soc. Perkin Trans.* **1990,** *1*, pp 2151-2152.
28. Corrie, J. E. T. *J. Chem. Soc. Perkin Trans.* **1994,** *1*, pp 2975-2982.
29. Lundgren, J. S.; Salins, L. L. E.; Kaneva, I.; Daunert, S. *Anal. Chem.* **1999,** *71*, pp 589-595.
30. Ames, G. F. -L. *Ann. Rev. Biochem.* **1986,** *55*, pp 397-425.
31. Ames, G. F. -L; Mimura, C.; Holbrook, S.; Shyamala, V. *Adv. Enzymol.* **1992,** *65*, pp 1-47.
32. Schauer-Vukasinovic, V.; Cullen, L.; Daunert, S. *J. Am. Chem. Soc.* **1997,** *119*, pp 11102-11103.
33. Salins, L. L. E.; Schauer-Vukasinovic, V.; Daunert, S. *Proc. SPIE-Int. Soc. Opt. Eng.* **1998,** *3270*, pp 16-24.
34. Haugland, R. P. In *Handbook of Fluorescent Probes and Research Chemicals*; Spence, M. T. Z., Ed.; Molecular Probes, Inc.: Eugene, OR, 1996; p 55.
35. Salins, L. L. E.; Daunert, S., University of Kentucky, unpublished data.
36. Salins, L. L. E.; Ware, R. A.; Ensor, C. M.; Daunert, S., University of Kentucky, unpublished data.
37. Douglass, P. M.; Salins, L. L. E.; Daunert, S., University of Kentucky, unpublished data.

Chapter 7

Study of Bacterial Metal Resistance Protein-Based Sensitive Biosensors for Heavy Metal Monitoring

Ibolya Bontidean[1], Jon R. Lloyd[2], Jon L. Hobman[2], Nigel L. Brown[2], Bo Mattiasson[1], and Elisabeth Csöregi[1,3]

[1]Lund University, Center for Chemistry and Chemical Engineering, Department of Biotechnology, P.O. Box 124, S–221 00 Lund, Sweden
[2]The University of Birmingham, School of Biological Sciences, Edgbaston, Birmingham B15 2TT, United Kingdom

A capacitive signal transducer was used with metal-resistance (SmtA) and metal regulatory (MerR) proteins to construct sensitive biosensors for monitoring heavy metal ions. The proteins were overexpressed in *E. coli*, purified and immobilized on a gold electrode modified with self-assembled thiol layers. The protein-modified electrode was used as the working electrode in an electrochemical cell placed in a flow injection system. Both the metallothionein from the cyanobacterium *Synechococcus* PCC 7942 and the MerR regulatory protein from transposon Tn*501* enabled monitoring of Cu^{2+}, Hg^{2+}, Zn^{2+}, and Cd^{2+} starting from femtomolar concentrations. The metal ions were bound differentially and the shape of the binding curves may sense conformational changes related to the biological roles of the proteins.

Heavy metal ions are extremely toxic, and hence their determination at trace levels is an important task, not only in analytical chemistry, but also in other fields, such as environmental monitoring, clinical toxicology, wastewater treatment, animal husbandry and industrial process monitoring.

Classical methods characterized by high sensitivity and selectivity e.g., atomic absorption and emission spectroscopies (*1*), inductively coupled plasma mass spectroscopy (*1, 2*) etc., can be used for heavy metal measurements, but they often require sophisticated instrumentation and qualified personnel.

[3]Corresponding author

102

Electrochemical methods, such as ion selective electrodes, polarography etc. are also often used, since they require simpler instrumentation (*3*), but these are not useful at low concentrations.

Biosensors, known to monitor various analytes both selectively and sensitively, have also been reported for heavy metal detection. Several electrode configurations using whole cells, enzymes or apoenzymes have been designed (*4-6*). The main advantage of such biosensors is that samples often require little pretreatment and the *bioavailable* concentration of the toxic heavy metal is measured, rather than the *total* concentration. However, a limited selectivity and quite low sensitivity characterize these sensors described in the literature.

Metal-binding proteins are synthesized by many bacteria in response to the presence of specific heavy metals (e.g., silver, bismuth, cadmium, cobalt, copper, mercury, nickel or zinc). Some of these are normal components of the cellular machinery (such as metal uptake proteins for copper or zinc), others are part of specific mechanisms of resistance to antimicrobial metals (e.g. to cadmium or mercury). In each case the heavy metal induces expression of the genes for metal-binding proteins under the control of a specific regulator responsive only to that metal, or a few related metals. The specificity of prokaryotic metallothionein (SmtA, expressed and used as a fusion with glutathione-S-transferase) (*7*) and the mercury-responsive regulator MerR (*8*) was exploited as the biological recognition element of a capacitive biosensor.

Experimental Section

Chemicals.

The fusion protein GST-SmtA and the regulatory protein MerR, were isolated and purified as described elsewhere (*9, 10*). Gold rods, 99.99%, used as electrode material, and 1-dodecanethiol were from Aldrich Chemicals (Milwaukee, WI), thioctic acid, bovine serum albumin (BSA) and HEPES buffer were purchased from Sigma (St. Louis, MO) and 1-(3-dimethylaminopropyl)-3-ethyl-carbodiimide (EDC) was obtained from Fluka AG (Buchs, Switzerland). Dithiothreitol (DTT) was from Chemicon (Malmö, Sweden), and H_3BO_3, $Na_2B_4O_7$ and heavy metal salts $CuCl_2 \cdot 2H_2O$, $ZnCl_2$, $HgCl_2$ and $Cd(NO_3)_2 \cdot H_2O$ were all purchased from Merck (Darmstadt, Germany). All reagents were of analytical grade. Solutions were prepared with water obtained from a Milli-Q system preceded by a reverse osmosis step, both from Millipore (Bedford, MA), if not otherwise specified. To remove all traces of metal ions, all glassware was soaked for 3 days in 3 M HNO_3 and 1 day in water before use.

Biosensor Design.

A detailed description of biosensor construction was previously published (*9*); here only a brief description of the procedure is given. The proteins were dissolved in phosphate-buffered saline (70 mM NaCl, 1.3 mM KCl, 5 mM Na_2HPO_4, 0.9 mM KH_2PO_4; pH 7.3) containing 50% (v/v) glycerol to a final concentration of 1 mg/ml and immobilized via a thiol monolayer to the gold electrodes using EDC coupling. First, the storage buffer for the proteins was exchanged with 100 mM borate coupling buffer, pH 8.75, by ultrafiltration and readjustment of the concentration to approximately 0.04 mg/ml with borate buffer. Next, the gold electrodes were polished, sonicated and plasma cleaned, and treated with thioctic acid to form a self-assembled monolayer. The thiol monolayer was activated with a solution of 1% EDC in dried acetonitrile for 5 h, washed with 100 mM borate buffer, pH 8.75, and placed in the protein solution at 4°C for 24 h. Finally, the electrodes were washed with borate buffer and immersed in a solution of 1-dodecanethiol for 20 minutes just before taking measurements. Electrode preparation was performed at room temperature, unless otherwise specified.

Instrumentation

The protein-based biosensors were placed as the working electrode in a three (four)-electrode electrochemical cell (with a dead volume of 10 μL) connected to a fast potentiostat (Zäta Elektronik, Lund Sweden) (9, 11). A Pt foil served as the auxiliary electrode, and a Pt wire together with a second Ag/AgCl (0.1 M KCl) served as quasi and real reference electrodes. The second reference electrode (Ag/AgCl) was placed in the outlet stream (11). The cell was arranged in a flow injection (FI) system as presented in Figure 1. Buffer solutions (10 mM borate or HEPES, both containing 0.02% sodium azide) were pumped with a peristaltic pump (Alitea AB, Stockholm, Sweden) at a flow rate of 0.5 ml/min. Samples were injected into the flow via a 250 μl sample loop. The carrier buffers were filtered through a 0.22 μm Millipore filter and degassed before use.

Results and Discussion

Biorecognition Elements

Two different proteins were used in this work, the metallothionein SmtA and the metalloregulatory protein MerR. Metallothioneins are low molecular weight proteins, which are characterized by a high cysteine content and a selective capacity to bind heavy metal ions. They have been isolated from a wide range of organisms and their properties have been studied using a great variety of analytical methods (12-14). However, there has been little use of electrochemistry in the study of the characteristics of these kind of molecules and no application for biosensor design was reported until recently (9, 15, 16). The fusion protein GST-SmtA used in this work was selected for its ability to bind several ions (Cu, Hg, Cd, and Zn), and hence was not anticipated to allow determination of individual metal ions. The amino acid sequence of this protein is known, and its

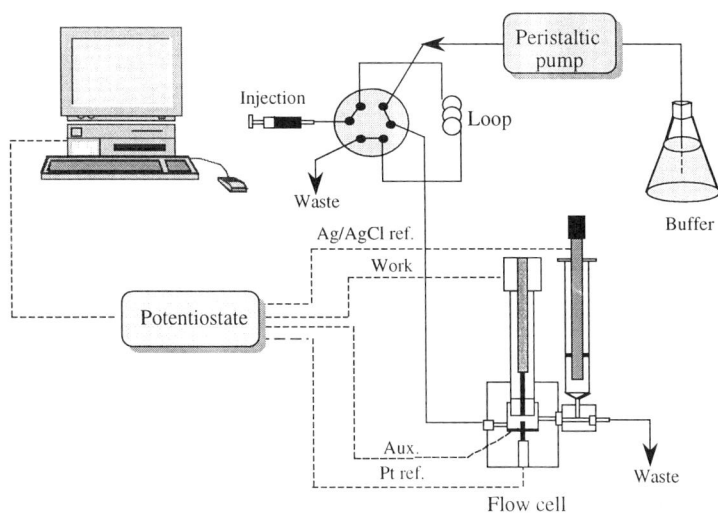

Labels visible in figure:
Peristaltic pump
Injection
Loop
Waste
Buffer
Ag/AgCl ref.
Work
Potentiostate
Aux.
Pt ref.
Waste
Flow cell

Figure 1. Schematic drawing of the experimental set-up.

3D structure is assumed to be similar to a domain of human metallothionein. The MerR regulatory protein is encoded by the *mer* operon of Tn*501* (*8*) and is highly specific for Hg^{2+} in its biological response. The amino acid sequence is known, but not the structure, and 3 cysteine residues are responsible for binding Hg^{2+} in the dimeric MerR protein (*17*).

Capacitive Transducer

A recently described capacitive sensor was shown to be able to detect antigens with high sensitivity (*11*) and was therefore considered to be have potential use as signal transducer for heavy metal detection. It was previously demonstrated that it is able to transduce the conformational change which occurs when metal ions bind to the protein (*9, 15, 16*). The principle of detection involves capacitance measurements made by applying a potential pulse of 50 mV and recording the current transients following the potential step according to eq:

$$i(t)=u/R_s exp(-t/R_s C_1)$$

where i(t) is the current at time t, u is the amplitude of the potential pulse applied, R_s is the resistance between the gold and the reference electrodes, C_1 is the total capacitance over the immobilized layer and t is the time elapsed after the potential pulse was applied. The working electrode had a rest potential of 0 mV *vs.* the Ag/AgCl reference electrode. The current values were collected with a frequency of 50 kHz and the first ten values were used for the evaluation of the capacitance.

Protein Biosensors

Both protein biosensors responded to heavy metal ions over a broad concentration range, while electrodes prepared identically using bovine serum albumin (BSA) did not respond (see Figure 2.). A detailed description of the characteristics (detection limit, stability, selectivity, etc.) of the biosensors based on the two above mentioned proteins were recently presented (*9*). The GST-SmtA electrode binds a variety of metal ions as part of its biological function (*18*), while the Tn501 MerR is highly specific in its biological function for Hg^{2+}, having an in vitro response to Hg^{2+} in the initiation of the transcription about 103-fold more sensitive than its response to Cd^{2+}. Since the MerR protein is oxygen sensitive, these electrodes were prepared under nitrogen atmosphere, but attempts were made to stabilize the protein by adding excess DTT to the protein solution, and preparing the electrodes in air. Results presented here were obtained with electrodes prepared in air with the protein solutions containing 2mM DTT. Comparing the relative responses of the MerR biosensor (Figure 3B) to those obtained for the GST-SmtA-based one (Figure 3A) it can be seen that MerR electrodes have higher selectivity for Hg^{2+} than the GST-SmtA ones which senses all four heavy metal ions. Capacitance changes are shown relative to the

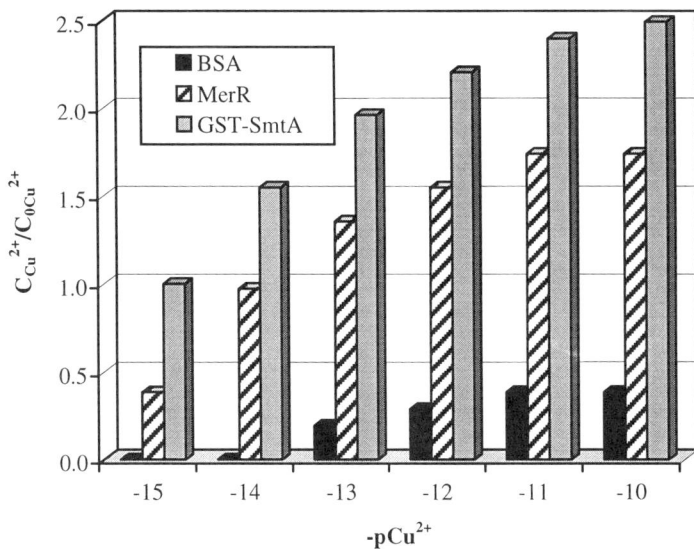

Figure 2. *Capacitance changes for electrodes based on different proteins at injection of Cu^{2+} relative to the capacitance change of GST-SmtA electrode for 1 fM Cu^{2+} (C_{0Cu}^{2+}) considered as unit.*

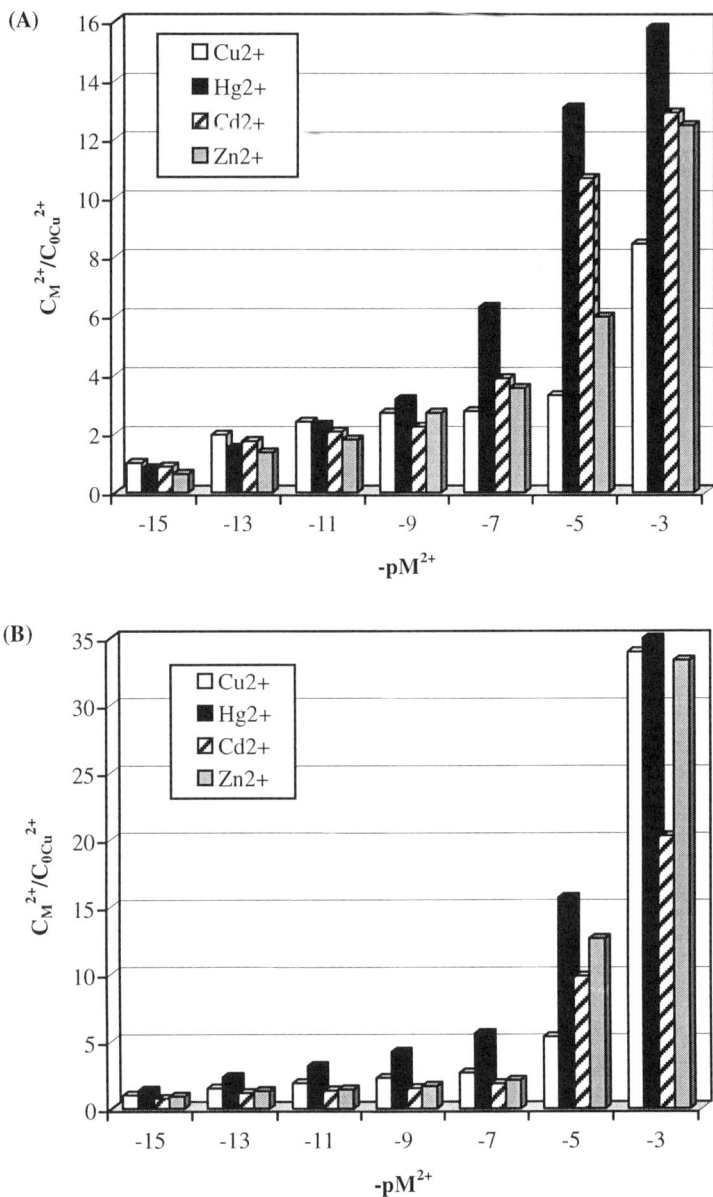

Figure 3. Capacitance changes obtained for GST-SmtA (A) and MerR (B) based electrodes respectively, for injection of Cu^{2+}, Hg^{2+}, Cd^{2+} and Zn^{2+} relative to the capacitance change for 1 fM Cu^{2+} (C_{0Cu}^{2+}) considered as unit.

capacitance change obtained with GST-SmtA electrodes for injection of 1 fM Cu^{2+} (this being considered as a detection unit).

In order to elucidate the binding mechanism, the responses of both biosensors were studied to the four metal ions (Cu, Hg, Cd, and Zn) across the concentration range 10^{-15}-10^{-3} M, and the effects of pH and the nature of the buffer were assessed. Although both protein-based biosensors showed a similar shape of the capacitance vs. concentration dependence for each metal ion, the corresponding curves are displaced.

The GST-SmtA electrode is most sensitive for Cu^{2+} at concentrations below 10^{-11} M (Cu^{2+}>Cd^{2+}>Hg^{2+}>Zn^{2+}) but becomes more sensitive to Hg^{2+} at concentrations above 10^{-10} M, and to Cd^{2+} and Zn^{2+} above 10^{-7} M (Figure 3A). The slope of the capacitance curve changes at 10^{-7} M for Hg^{2+}, at 10^{-5} M for Cd^{2+} and Zn^{2+}, and at 10^{-3} M for Cu^{2+}. It is assumed that below 10^{-10} M each metal ion titrates the metal binding sulphhydryl and histidine groups of SmtA at different rates, the sum of the titration events for each metal ion being reflected in the capacitance curves. The dramatic capacitance change observed at higher concentrations is assumed to represent the folding of the SmtA region forming the typical metallothionein cage structure.

The MerR electrodes (Figure 3B) were also responsive to all four metal ions at low concentrations (<10^{-9} M) but show an accentuated sensitivity for Hg^{2+} and a sensitivity pattern of Hg^{2+}>Cu^{2+}>Zn^{2+}>Cd^{2+}. The change in capacitance occurs at 10^{-7} M for Hg^{2+} and Cd^{2+}, at 10^{-6} M for Zn^{2+} and at 10^{-5} M for Cu^{2+}. This is compatible with a model in which there is a change from non-specific binding of metal to cysteine and histidine to specific binding of the metal ions to the specific binding site in MerR, which is associated with a conformational change related to its gene activation function ([10, 19]). The capacitance changes caused by metal ion concentrations might be only partially explained conformational changes, as this occurs i.e. at 10^{-7} M for Hg^{2+} and at 10^{-6} M for Zn^{2+}, whereas the *in vitro* response for activation of transcription shows at least 10^5-fold difference ([19]). The conformational changes caused by the metals may contribute partly, but not exclusively to the specificity of transcriptional activation by metal binding. Ongoing experiments using other metal-responsive regulators related to MerR, such as ZntR ([20]) are expected to clarify the observed phenomenon.

The capacitive signals of the GST-SmtA electrode for the four metal ions in different buffers (HEPES and borate) at different pH values (7, 8, and 8.75) on exposure to Cu^{2+} and Hg^{2+} (see Figure 4) were determined in order to determine the effect of pH on binding. The capacitance changes obtained in HEPES were approximately two-fold lower than those obtained in borate and a decrease of one unit of pH in both buffers caused an approximately 30-35% signal decrease. Thus, borate buffer seemed to give the optimum response. However, while electrodes could not be regenerated with EDTA in borate at pH 8.75 when exposed to 1 mM Cu^{2+}, this could be done in both buffers at pH 8.0, confirming our assumption that the detailed mechanisms of binding might differ at different pH values.

In order to assess the robustness of the GST-SmtA electrodes the regeneration of the electrode was studied using 1 mM EDTA following injection of 10^{-4} M Hg^{2+} in both borate and HEPES at pH 8.0. Results (not shown) indicated that the absolute capacitance in HEPES was relatively constant across

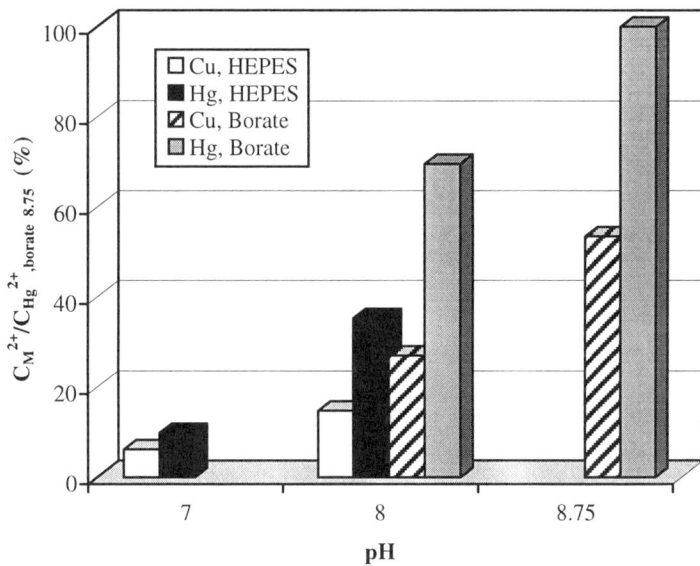

Figure 4. Dependence of relative capacitance changes on the nature and pH of the carrier buffer, where the capacitance change for Hg^{2+} in borate buffer with pH 8.75 was considered as 100%.

three exposure cycles to Hg and EDTA, whereas those obtained in borate drifted slightly upwards. However, both buffer conditions allowed the use of the electrodes for several measurements. The full extent of electrode reusability, effects on electrode regeneration, and optimal storage conditions have to be elucidated more rigorously, but we have already shown that if the electrode was stored in the presence of heavy metal ions and reactivated with EDTA just before use, it displayed a stable response for 10 days (9).

The data reported here indicate that not only can protein-based biosensors monitor heavy metal ions at low concentrations and across a wide concentration range, but they can provide useful information on the response of a protein to the metal, which may have a biological significance. However, the results should be interpreted with caution, as oxidation or other denaturation of the protein may alter the observed response, as was the case with MerR electrodes prepared in DTT solution compared with those prepared under nitrogen.

Conclusions

The electrodes described here are not specific for a single heavy metal, but they are yielding information in the specificity of metal binding, and may suggest ways in which one can tailor protein-based capacitive biosensors to specific metals. Recently new members of the MerR family of proteins have been identified, which will be considered for further biosensor development. ZntR from *E.coli* has a similar sequence to MerR but selectively responds to Zn^{2+} and PbR from *Alcaligenes eutrophus* CH34, regulates the expression of lead resistance genes. This is highly specific *in vivo* for Pb(II) and does not respond to Zn(II), Cd(II), Cu(II) or Hg(II). Biosensor development using these new proteins is planned. Future work also targets testing of optimal electrode designs in both mixture of metal ions and real samples.

Acknowledgements

The European Commission (project ENV4-CT95-0141), the Swedish Natural Research Council/NFR, the Swedish Institute, MISTRA (Coldren project), and the UK Biotechnology and Biological Sciences Research Council (No. G07943) supported this work. Prof. N. J. Robinson (Newcastle) is kindly acknowledged for the gift of plasmid pGEX3X-*smtA*, and K.J. Jakeman for technical assistance in protein purification.

Literature Cited

1. Jackson, K. W.; Chen, G. *Anal. Chem.* **1996,** *68*, 231R-256R.
2. Burlingame, A. L.; Boyd, R. K.; Gaskell, S. J. *Anal. Chem.* **1996,** *68*, 599R-651R.

112

3. Anderson, J. L.; Bowden, E. F.; Pickup, P. G. *Anal. Chem.* **1996**, *68*, 379R-444R.
4. Wittman, C.; Riedel, K.; Schmid, R. D. In *Handbook of Biosensors and Electronic Noses*; Kress-Rogers, E., Ed.; CRC: Boca Raton, FL, 1997, pp 299-332.
5. Ögren, L.; Johansson, G. *Anal. Chim. Acta* **1978**, *96*, 1-11.
6. Mattiasson, B.; Nilsson, H.; Olsson, B. *J. Appl. Biochem.* **1979**, *1*, 377-384.
7. Shi, J. G.; Lindsay, W. P.; Huckle, J. W.; Morby, A. P.; Robinson, N. J. *FEBS Lett.* **1992**, *303*, 159-163.
8. Lund, P. A.; Ford, S. J.; Brown, N. L. *J. Gen. Microbiol.* **1986**, *132*, 465-480.
9. Bontidean, I.; Berggren, C.; Johansson, G.; Csöregi, E.; Mattiasson, B.; Lloyd, J. R.; Jakeman, K. J.; Brown, N. L. *Anal. Chem.* **1998**, *70*, 4162-4169.
10. Parkhill, J.; Ansari, A. Z.; Wright, J. G.; Brown, N. L.; O'Halloran, T. V. *EMBO J.* **1993**, *12*, 413-421.
11. Berggren, C.; Johansson, G. *Anal. Chem.* **1997**, *69*, 3651-3657.
12. *Metallothionein;* Kägi, J. H. R.; Nordberg, M., Eds.; Birkhäuser Verlag: Basel, 1979.
13. *Metallothionein II*, Kägi, J. H. R.; Kojima, Y., Eds.; Birkhäuser Verlag: Basel, 1987.
14. *Metallothionein: Synthesis, Structure and Properties of Metallothionein Phytochelatin and Metal-thiolate Complexes*, Stillman, M. J.; Shaw III, C. F.; Suzuki, K. T., Eds.; VCH: Verlagsgesellschaft, Germany, 1992.
15. Brown, N. L.; Lloyd, J. R.; Jakeman, K.; Hobman, J. L.; Bontidean, I.; Mattiasson, B.; Csöregi, E. *Biochem. Soc. Trans.* **1998**, *26*, 662-665.
16. Bontidean, I.; Lloyd, J. R.; Hobman, J. L.; Wilson, J. R.; Csöregi, E.; Mattiasson, B.; Brown, N. L. *J. Inorg. Biochem.* **1999**, *in press*.
17. Wright, J. G.; Tsang, H.-T.; Penner-Hahn, P.; O'Halloran, T. V. *J. Am. Chem. Soc.* **1990**, *131*, 2434-2435.
18. Daniels, M. J.; Turner-Cavet, J. S.; Selkirk, R.; Sun, H. Z.; Parkinson, J. A.; Sadler, P. J.; Robinson, N. J. *J. Biol. Chem.* **1998**, *273*, 22957-22961.
19. Ralston, D.; O'Halloran, T. V. *Proc. Natl. Acad. Sci. USA* **1990**, *87*, 3846-3850.
20. Brocklehurst, K. R.; Hobman, J. L.; Lawley, B.; Blank, L.; Marshall, S. J.; Brown, N. L.; Morby, A. P. *Molec. Microbiol.* **1999**, *31*, 893-902.

Chapter 8

Cellobiose Dehydrogenase and Peroxidase Biosensors for Determination of Phenolic Compounds

Annika Lindgren, Tautgirdas Ruzgas, Leonard Stoica, Florentina Munteanu, and Lo Gorton

Department of Analytical Chemistry, Lund University, P.O. Box 124, SE–22100 Lund, Sweden

Phenolic structures are very common in environmentally hazardous compounds, justifying the development of biosensors targeted at measuring phenolic compounds. Common phenol biosensors are based on tyrosinase, laccase or tissues containing these enzymes. These enzymes have shown to have a rather narrow substrate selectivity. In this study we report on sensors based on two more general group-specific enzymes, not equally commonly used to construct phenol biosensors. In the first sensor the enzyme is peroxidase. In the presence of hydrogen peroxide the enzymatic oxidation of phenols forms phenoxy radicals that can be electrochemically re-reduced at the electrode (-50 mV *vs.* Ag/AgCl), resulting in a reduction current proportional to the phenol concentration. The other sensor is based on cellobiose dehydrogenase (CDH). In this recently developed sensor, diphenolic compounds are oxidized at the electrode surface at +300 mV *vs.* Ag/AgCl. The diphenols are then regenerated by the adsorbed CDH in the presence of cellobiose. The CDH biosensor results in very low detection limits (<5 nM). The peroxidase biosensor can be used to determine a wide range of phenolic compounds, whereas the CDH electrode only can be used for diphenols, and thus efficiently discriminates between diphenolic and monophenolic compounds.

Introduction

A large number of organic pollutants, widely distributed throughout the environment, have a phenolic structure. These phenols and substituted phenols are used in many industrial processes, *e.g.* in the manufacture of plastics, paper, dyes, drugs, pesticides and antioxidants *(1)*. Phenols are also breakdown products from natural organic compounds such as humic substances, lignins and tannins *(2)*. The phenols and related aromatic compounds are highly toxic, carcinogenic and allergenic *(1)*. Due to the toxic effects of the phenols their determination in the environment is of great importance. Biosensors have offered selective and sensitive detection of phenolic compounds. The most commonly used amperometric biosensors for phenol determination are based on phenol oxidases (tyrosinase or laccase) or tissues containing these enzymes (ref. in *(3)*). Another enzyme that can be used for phenol sensors is peroxidase (POD) *(4-8)*. Although the enzymatic reaction of phenol oxidases and peroxidases is different, they work in a similar manner on the electrode surface, see Figure 1a *(3)*. The enzyme is oxidized by an oxidizing agent, molecular oxygen (O_2) in the case of phenol oxidases and hydrogen peroxide (H_2O_2) for POD. The enzyme is then re-reduced to its native state by electrons donated from a phenolic compound forming a phenoxy radical or a quinone that in turn can be reduced at the electrode surface at potentials below 0 V *vs.* Ag/AgCl, resulting in a current proportional to the phenol concentration.

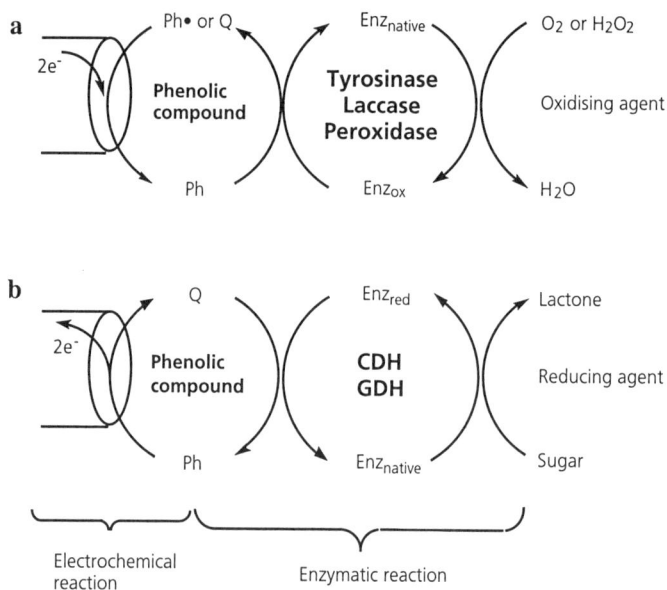

Figure 1. Reaction scheme of enzyme-modified electrodes for the determination of phenolic compounds. CDH is cellobiose dehydrogenase and GDH is PQQ-dependent glucose dehydrogenase. Ph, Ph• and Q are the phenolic molecule, its phenoxy radical, and its quinone, respectively.

The other less common type of phenol biosensor involves enzymes like PQQ (pyrroloquinoline quinone)-dependent glucose dehydrogenase (GDH) *(9-12)* or cellobiose dehydrogenase (CDH) *(13)*, for which the detection reaction sequence is based on the reverse mechanism, see Figure 1b. The native form of these enzymes is the oxidized form. In the presence of a sugar the enzyme is reduced, the enzyme can then be reoxidized to its native state by an electron acceptor such as a quinone. At an anodic potential *ortho-* and *para*-diphenolic compounds can be oxidized forming quinonic compounds *(14)* that can serve as electron acceptors for GDH and CDH. A glucose oxidase biosensor for chlorophenols is based on a similar reaction sequence as the CDH and the GDH biosensors. However, in this case the chlorophenols were initially oxidized in batch using chloroperoxidase or bis(trifluoroacetoxy)iodobenzene to form electroactive species, which in turn were recycled between glucose oxidase and the electrode surface, serving the basis for an amplified response reaction cycle *(15,16)*.

The main difference between the two types of sensors is that the first type (based on phenol oxidases or peroxidase) uses a low applied potential (-100 - 0 mV *vs.* Ag/AgCl) *(3)*, whereas the second type (based on CDH or GDH) needs a higher potential (300 - 400 mV *vs.* Ag/AgCl) *(9,11,13)* to be able to oxidize the phenols and therefore the risk of electrooxidizing interfering compounds in the sample is higher. Tyrosinase, laccase and peroxidase can all be used for both phenolic and diphenolic compounds, however, for laccase and tyrosinase the sensitivity is much higher for catecholic compounds *(3)*. CDH and GDH require a quinone, therefore a diphenol or aminophenol is needed.

Tyrosinase was shown to have a relatively narrow substrate selectivity *(3)*. In this study we focus on one sensor of each type, namely on POD and CDH biosensors, since these are more general group-specific enzymes compared with tyrosinase and laccase.

Both POD *(17,18)* and CDH *(13,19,20)* are among the few enzymes that are able to communicate with the electrode through direct electron transfer (ET) *(21)*. However, it is the mediated electron transfer that forms the basis of the use of these enzymes for phenol determination *(5-7,13)*. The common electron acceptors and donors of the enzymes are summarized in Table I.

Table I. The prosthetic group, common electron acceptors and donors for the used enzymes *(22-25)*.

Enzyme	Catalyze the transport of electrons		Prosthetic
	from	*to*	*group*
POD	*phenols, aromatic amines,* ferrocyanide	H_2O_2, small organic hydroperoxides	heme
CDH	*cellobiose,* lactose, maltose	*quinone,* cyt *c,* ferricyanide, oxygen	heme - FAD

NOTE: The acceptors/donors in italics are the ones utilized in this study

Peroxidases (POD) catalyze the oxidation of phenolics, whereas CDH catalyzes the reduction of quinones (which is generated electrochemically from diphenols). The prosthetic group of peroxidase is heme *(22)*, whereas CDH contains two prosthetic groups one heme and one flavin (flavin adenine dinucleotide, FAD) *(23-25)*, see Figure 2. The oxidation of sugars and the subsequent reduction of two-electron

116

acceptors, *e.g.* quinones, take place at the FAD domain of CDH *(26,27)*. The function of the heme part is not well understood, however, it is needed when one-electron acceptors are used and also in direct electrochemical communication with an electrode *(19,20)*.

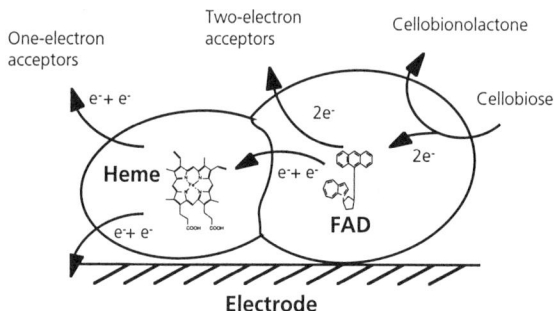

Figure 2. Schematic picture of the two domain structure of CDH also illustrating proposed catalytic and electron transfer paths through/within the enzyme.

Experimental

Cellobiose dehydrogenase (CDH) from *Phanerochaete chrysosporium* (0.84 g l^{-1} in 50 mM ammonium acetate buffer, pH 5) *(23,26)* was kindly provided by Dr. G. Henriksson and Dr. G. Pettersson (Department of Biochemistry, Uppsala University, Sweden) and horseradish peroxidase (HRP) was purchased from Boehringer-Mannheim, Mannheim, Germany (cat. no. 814407, RZ 3.2-3.3, approx. 1000 U /mg (ABTS)).

The biosensors were prepared by placing a droplet (5-10 µl) of enzyme solution (0.5-5 g l^{-1}) onto the polished circular end of a rod of spectroscopic graphite (SGL Carbon, Werk Ringsdorff, Bonn, Germany, type RW001, Ø 3.05 mm). The enzyme was let to adsorb for 20 h (4°C). For HRP it was found that an immobilization time of 30 min was fully sufficient, however, for CDH an adsorption time of 20 h was needed to get stable electrodes with high activity *(13)*. After the adsorption step, the electrodes were thoroughly rinsed with water. If the electrodes were not used immediately they were stored in buffer at 4°C *(5,13)*. For HRP covalent coupling using carbodiimide or crosslinking using glutaraldehyde was tested but without giving any increased response or stability *(5)*.

The enzyme modified electrode was fitted into a Teflon holder and inserted into a flow-through wall-jet amperometric cell *(28)*. The enzyme electrode was used as the working electrode, an Ag/AgCl (0.1 M KCl) electrode as reference electrode, and a platinum wire served as the auxiliary electrode. The electrodes were connected to a three-electrode potentiostat (Zäta Elektronik, Lund, Sweden), controlling the applied potential of the working electrode, and the current was recorded on a recorder (Kipp and Zonen, Delft, The Netherlands). The electrochemical cell was connected to one of the flow injection (FI) systems illustrated in Figure 3. As can be seen from Figure 1, a continuous supply of the enzyme substrate (POD: H_2O_2; CDH: cellobiose) is needed to use these biosensors for determination of (di)phenols. For the CDH biosensor, cellobiose was added directly to the flow carrier, see Figure 3b. In the case of POD modified electrodes the enzyme substrate, H_2O_2, was produced on-line by pumping glucose through an immobilized glucose oxidase/mutarotase reactor (7), because of problems with the decomposition of H_2O_2 (5). The phenolic compounds were injected in a second carrier stream to avoid unwanted reactions and peak broadening in the reactor, see Figure 3a.

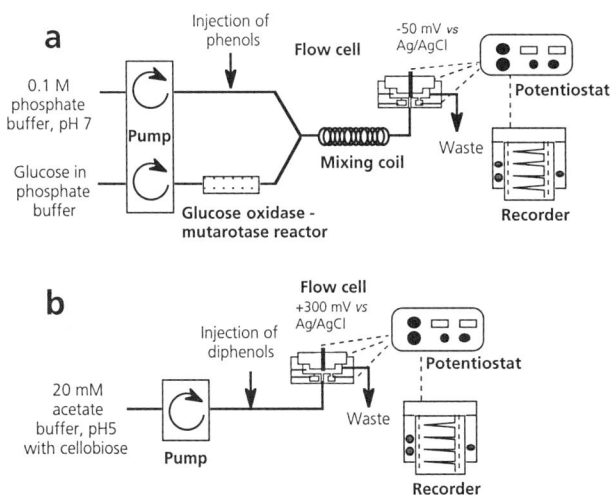

Figure 3. FI system for phenolic determination using (a) POD or (b) CDH modified electrodes. The flow rate through the electrochemical cell was 0.5 ml min^{-1} for both the POD and the CDH system.

Results and Discussion

Principle of the Measurements

An example of a measurement is shown in Figure 4. Each experiment started by pumping only carrier buffer through the FI-system. When a stable baseline was achieved the enzyme substrate (POD: H_2O_2; CDH: cellobiose) was added to the carrier buffer and a steady-state current due to the direct ET was registered, see Figure

118

4. If phenols are injected in the POD system, enzymatic oxidation of the phenols results in radicals that can be electrochemically re-reduced at the electrode (-50 mV *vs*. Ag/AgCl), resulting in a reduction current proportional to the phenol concentration. In the CDH system the injected diphenolic compounds are oxidized at the electrode surface at +300 mV *vs*. Ag/AgCl, forming quinones that serve as electron acceptors for the reduced FAD of CDH, thus restoring the diphenols. The diphenolic compounds can thus be recycled between the enzyme and the electrode (see Figure 1b). The amplification, caused by the recycling, can be described by the amplification factor. It is calculated as the amplified response divided by the response caused by direct oxidation of the diphenol (injections of diphenol in the absence of cellobiose result in small oxidation peaks due to the direct oxidation).

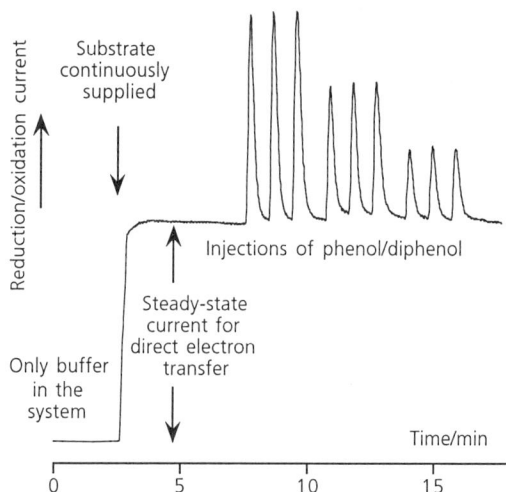

Figure 4. Example of a measurement. The y-axis is reduction current for POD and oxidation current for CDH

Optimizations

Parameters such as applied potential, immobilization time, storage conditions, flow rate, and suitable substrate concentration were optimized in previous publications *(5-7,13)*. The values of the optimized parameters are given in Figure 3, and in Experimental. For the optimized system sensitivity, linearity, operational, and storage stability were studied.

In this type of system the substrate concentration is very crucial, since the steady-state current due to the direct electron transfer is always present and the noise is highly dependent on this continuous background current. Too high a substrate concentration will also result in decreased activity with time. However, too low a concentration will limit the reaction with (di)phenols, resulting in a more restricted linear range. As a compromise 100 µM H₂O₂ and 200 µM cellobiose, respectively, were chosen as suitable substrate concentrations. Using these concentrations, calibration plots for a variety of (di)phenolic compounds were registered. HRP-modified electrodes gave calibration curves reflecting a Michaelis-Menten like dependence ($I=I_{max} c/(c+K_m^{app})$), see Figure 5a, with a linear range of approximately one order of magnitude *(7)*, whereas the CDH-modified electrodes showed a linear range extending over three orders of magnitude, Figure 5b.

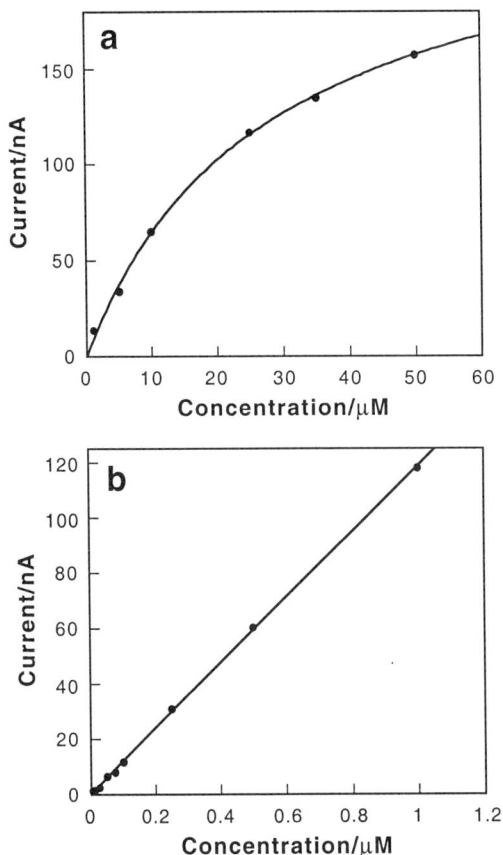

Figure 5. Calibration graph for catechol using (a) HRP and (b) CDH biosensor. The calibration data using HRP are fitted to the Michaelis-Menten equation. For experimental conditions see Figure 3, [H₂O₂]: 100 µM, [cellobiose]: 200 µM.

Stability, Repeatability and Reproducibility

The operational stability of the HRP and CDH biosensors was studied by repeated injections of (di)phenols (phenol and dopamine, respectively) with a steady-state supply of substrate (H_2O_2 or cellubiose), see Figure 6. During one hour the response decreased about 2-3% for both biosensors.

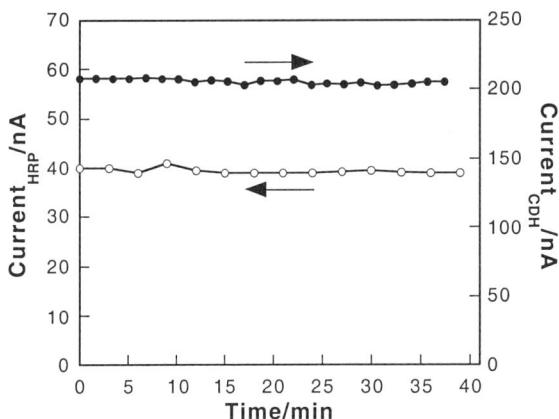

Figure 6. Operational stability of the (O) HRP and (●) CDH modified electrodes studied by repeated injections of 10 µM phenol for HRP and 1 µM dopamine for CDH. For experimental conditions see Figure 3

An unused CDH electrode can be stored refrigerated at 4°C for some weeks with retained activity, however, if the electrode has been used, the response decreased to half the initial value after 1 week storage in refrigerator. The HRP electrode has better storage stability. During two weeks almost no decrease in response was observed, however, after two weeks the decrease was faster and after 3 weeks the response was 60% of the initial response. It is not clear whether the decrease in current seen during the long term stability tests is caused by desorption or by inactivation of the enzyme. However, the operational stability proves that the very simple electrode preparation method chosen, gave very stable electrodes with diminutive desorption and/or inactivation.

The relative standard deviation (RSD) between injections was very good for both biosensors; 0.85% for HRP (n=16, phenol) and 0.60% for CDH (n=28, hydroquinone (13)). The reproducibility between electrodes, prepared and used different days, was about 30% for both HRP and CDH biosensors.

Selectivity for Various (Di)phenols

The selectivity profiles for a variety of phenols and diphenols obtained using HRP and CDH modified electrodes are presented in Table II. HRP showed similar responses to both mono- and diphenols. The responses to aromatic amines are generally higher compared with the corresponding phenols. The aromatic amines, however, showed a calibration graph with a lower Michaelis-Menten constant, K_m^{app}, than for the other compounds. CDH showed high response for the tested diphenols, and virtually no response for the monophenols.

Table II. Selectivity presented as relative response (%) of HRP and CDH modified electrodes.

Analytes	Relative response/%	
	HRP	CDH
Diphenols		
catechol	100	100
hydroquinone	n.d.	181
dopamine	n.d.	218
3,4-dihydroxybenzylamine	n.d.	69
3,4-dihydroxybenzaldehyde	41	84
3,4-dihydroxybenzoic acid	39	37
3,4-dihydroxyhydrocinnamic acid	n.d.	54
3,4-dihydroxyphenylacetic acid	n.d.	33
Monophenols		
phenol	103	<0.1
p-cresol	112	<0.1
vanillin	85	<0.01
guaiacol	17	<0.1
4-chloro-3-methylphenol	106	n.d.
2-chlorophenol	n.d.	<0.1
4-chlorophenol	73	<0.1
2,4-dichlorophenol	24	n.d.
Amino(mono)phenols and aromatic amines		
o-aminophenol	349	n.d.
p-aminophenol	316	28
aniline	46	n.d.
o-phenylenediamine	437	n.d.
m-phenylenediamine	155	n.d.
p-phenylenediamine	487	n.d.

NOTE: For HRP electrodes, the concentration of phenolic compounds in buffer solution was 20 μM, whereas 1 μM of phenolics was used for the CDH electrode (for the monophenols the concentration was 10-50 μM). The current response for catechol was 72 nA for HRP and 116 nA for CDH (20 and 1 μM catechol, respectively).

n.d.: not determined

As can be seen in Table II the responses within a group of related compounds differ. This may be due to differences in enzyme selectivity but also in how efficiently the (di)phenols are oxidized/reduced at the electrode surface, especially in the case of the CDH biosensor where the applied potential is rather close to the formal potential of several of the diphenols. The differences are probably also caused by differences in the stability of the products formed in the electrochemical reduction/oxidation and/or enzymatic reduction. Another possible explanation is the formation of by-products that cannot be recycled between the enzyme and the electrode.

The detection limits of the CDH electrodes were generally very good, below 5 nM for six of the eight tested diphenols. The sensitivity and detection limit for catechol (1.6 A M^{-1} cm^{-2} and 1.7 nM, respectively) (13) can be compared with the best reported values for catechol with a tyrosinase-modified solid graphite electrode (1.8 A M^{-1} cm^{-2} and 2.3 nM, respectively) (3,29). The CDH biosensor gave also a similar (or slightly lower) detection limit for dopamine (2.5 nM (13)) as the related GDH biosensor (3 nM (10)). For the GDH sensor a lot of work was carried out to optimize the immobilization protocol and the electrode configuration (9-12); in this study, a very simple adsorption procedure was used, which nevertheless gave stable electrodes. There is most likely capacity to optimize the electrode configuration for the CDH electrodes to further improve the detection limits and sensitivities. Another opportunity to improve the sensor is to replace the *Phanerochaete chrysosporium* CDH with CDH from another fungus, *e.g. Humicola insolens*, having a completely different pH-optimum (30). An advantage over the GDH biosensor is that the CDH electrode does not need to be incubated with the cofactor before the measurements to reconstitute the holoenzyme, a step necessary for a GDH biosensor (10-12).

The HRP biosensor showed response to all the tested phenols. This makes the biosensor suitable to be used as a detection unit in liquid chromatography for determination of phenols. However, the detection limits using HRP was slightly too high for direct environmental measurements, since they were in the 0.1-4 μM range. The sensitivities and detection limits for some aromatic amines and phenols have been studied using other plant peroxidases, *e.g.* tobacco and peanut peroxidase electrodes, and tobacco peroxidase showed as low detection limits as 10 nM for several of the aromatic amines (7). The results of the tobacco and peanut peroxidases give hope for future improvements in detection limits by screening some more plant peroxidases. Preliminary experiments in the laboratory have shown that sweet potato peroxidase is a good candidate.

Both biosensors have in earlier publications been applied to humic containing river waters (6,13). Neither of the two biosensors showed any significant difference in response if the calibration solutions were prepared in river water instead of in carrier buffer. However, the buffer composition needed to be slightly adjusted to accomplish detection of phenols in river water. For the CDH sensor it was possible to detect hydroquinone down to a concentration of 15 nM (a detection limit one order of magnitude higher compared with that in buffer solution). No detection limit was calculated for the HRP sensor, but the response to 10 μM 2-amino-4-chlorophenol was more than 10 times higher than the background response. The enzymes seem to withstand the exposure to the humic containing waters. The CDH sensor showed no decrease in response after 30 injections of river water. For the HRP sensor a slight decrease in response was noted, most probably because of adsorption of humics to the electrode surface.

123

Conclusions

The CDH sensor showed very good detection limits and operational stability. HRP showed also stable responses, however, the detection limits need to be improved, for example by replacement by some other peroxidases. The selectivity profiles of the two sensors differ; HRP covers both mono- and diphenols whereas CDH only works for diphenols. This discrimination can be used by combining the two biosensors in a sensor array.

Acknowledgments

The authors thank the Swedish Natural Science Research Council (NFR), the Crafoord Foundation, and the TEMPUS program (S JEP 09227-95) for financial support. The authors also thank Dr. G. Henriksson and Dr. G. Pettersson for donating CDH.

Literature Cited

1. Deichmann, W. B.; Keplinger, M. L. In *Patty's Industrial Hygiene and Toxicology*; Clayton, G. D., Clayton, F. E., Eds.; John Wiley & Sons, Inc.: New York, USA, 1981; Vol. 2A; pp 2567-2627.
2. Marko-Varga, G. A. In *Environmental Analysis: Techniques, Applications and Quality Assurance*; Barceló, D., Ed.; Elsevier: Amsterdam, The Netherlands, 1993; Vol. 13; pp 225-271.
3. Marko-Varga, G.; Emnéus, J.; Gorton, L.; Ruzgas, T. *Trends Anal. Chem.* **1995**, *14*, 319-328.
4. Kulys, J.; Bilitewski, U.; Schmid, R. D. *Bioelectrochem. Bioenerg.* **1991**, *26*, 277-286.
5. Lindgren, A.; Emnéus, J.; Ruzgas, T.; Gorton, L.; Marko-Varga, G. *Anal. Chim. Acta* **1997**, *347*, 51-62.
6. Ruzgas, T.; Emnéus, J.; Gorton, L.; Marko-Varga, G. *Anal. Chim. Acta* **1995**, *311*, 245-253.
7. Munteanu, F.-D.; Lindgren, A.; Emnéus, J.; Gorton, L.; Ruzgas, T.; Csöregi, E.; Ciucu, A.; van Huystee, R. B.; Gazaryan, I. G.; Lagrimini, L. M. *Anal. Chem.* **1998**, *70*, 2596-2600.
8. Kane, S. A.; Iwuoha, E. I.; Smyth, M. R. *Analyst* **1998**, *123*, 2001-2006.
9. Eremenko, A.; Makower, A.; Jin, W.; Rüger, P.; Scheller, F. *Biosens. Bioelectron.* **1995**, *10*, 717-722.
10. Lisdat, F.; Wollenberger, U.; Paeschke, M.; Scheller, F. W. *Anal. Chim. Acta* **1998**, *368*, 233-241.
11. Lisdat, F.; Wollenberger, U.; Makower, A.; Hörtnagl, H.; Pfeiffer, D.; Scheller, F. W. *Biosens. Bioelectron.* **1997**, *12*, 1199-1211.
12. Lisdat, F.; Ho, W. O.; Wollenberger, U.; Scheller, F. W.; Richter, T.; Bilitewski, U. *Electroanalysis* **1998**, *10*, 803-807.

13. Lindgren, A.; Stoica, L.; Ruzgas, T.; Ciucu, A.; Gorton, L. *Analyst* **1999**, *124*, 527-532.
14. Sternson, A. W.; McCreery, R.; Feinberg, B.; Adams, R. N. *J. Electroanal. Chem.* **1973**, *46*, 313-321.
15. Saby, C.; Luong, J. H. T. *Electroanalysis* **1998**, *10*, 7-11.
16. Male, K. B.; Saby, C.; Luong, J. H. T. *Anal. Chem.* **1998**, *70*, 4134-4139.
17. Yaropolov, A. I.; Tarasevich, M. R.; Varfolomeev, S. D. *Bioelectrochem. Bioenerg.* **1978**, *5*, 18-24.
18. Ruzgas, T.; Csöregi, E.; Emnéus, J.; Gorton, L.; Marko-Varga, G. *Anal. Chim. Acta* **1996**, *330*, 123-138.
19. Larsson, T.; Elmgren, M.; Lindquist, S.-E.; Tessema, M.; Gorton, L.; Henriksson, G. *Anal. Chim. Acta* **1996**, *331*, 207-215.
20. Lindgren, A.; Larsson, T.; Ruzgas, T.; Gorton, L. **1999**, *unpublished*.
21. Ghindilis, A. L.; Atanasov, P.; Wilkins, E. *Electroanalysis* **1997**, *9*, 661-674.
22. *Peroxidases in Chemistry and Biology*; Everse, J.; Everse, K. E.; Grisham, M. B., Eds.; CRC Press: Boca Raton, FL, 1991; Vol. 2.
23. Henriksson, G. PhD Thesis, Uppsala University, Uppsala, Sweden, 1995.
24. Morpeth, F. F. *Biochem. J.* **1985**, *228*, 557-564.
25. Ander, P. *FEMS Microbiol. Rev.* **1994**, *13*, 297-312.
26. Henriksson, G.; Pettersson, G.; Johansson, G.; Ruiz, A.; Uzcategui, E. *Eur. J. Biochem.* **1991**, *196*, 101-106.
27. Henriksson, G.; Johansson, G.; Pettersson, G. *Biochim. Biophys. Acta* **1993**, *1144*, 184-190.
28. Appelqvist, R.; Marko-Varga, G.; Gorton, L.; Torstensson, A.; Johansson, G. *Anal. Chim. Acta* **1985**, *169*, 237-247.
29. Ortega, F.; Domínguez, E.; Burestedt, E.; Emnéus, J.; Gorton, L.; Marko-Varga, G. *J. Chromatogr. A* **1994**, *675*, 65-78.
30. Igarashi, K.; Verhagen, M. F. J. M.; Samejima, M.; Schülein, M.; Eriksson, K.-E. L.; Nishino, T. *J. Biol. Chem.* **1999**, *274*, 3338-3344.

Chapter 9

Organophosphate Biosensors Based on Mediatorless Bioelectrocatalysis

Plamen Atanasov, Melissa Espinosa, and Ebtisam Wilkins

**Department of Chemical and Nuclear Engineering,
University of New Mexico, Albuquerque, NM 87131**

Organophosphorus compounds are significant major environmental pollutants due to their intensive use as pesticides. The modern techniques based on inhibition of cholinesterase enzyme activity are discussed. Potentiometric electrodes based on detection of cholinesterase inhibition by analytes have been developed. The detection of cholinesterase activity is based on the novel principle of molecular transduction. Immobilized peroxidase acting as the molecular transducer, catalyzes the electroreduction of hydrogen peroxide by direct (mediatorless) electron transfer. The sensing element consists of a carbon based electrode containing an assembly of co-immobilized enzymes: cholinesterase, choline oxidase and peroxidase.

Organophosphorus compounds (OPCs) are significant major environmental and food chain pollutants *(1,2)* due to their intensive use as pesticides in agriculture. Other important sources of such pollutants are manufacturing sites, spills during their transportation, and inappropriate use and storage. Chemicals of this group are also the basis for several different chemical weapons (Sarin, VX, etc), a potential source of serious environmental problems due to deliberate use, accidents or improper disposal. The destruction of OPC-based chemical weapons mandated by international agreements or as part of routine operations also leads to problems in environmental control and protection, and also stringent monitoring requirements in the destruction of weapons manufacturing plants.

Pollutants of this type are found to be present in many sampled soils, streams, ground and waste waters. One of the most important preventive measures in this case is to rapidly determine the source of the pollutant and magnitude of the threat using on-site measurements. Analysis of low levels of OPC pollutants in foods is another potential application of the sensor.

Chromatographic techniques are generally the most reliable methods for determination of OPCs. These techniques allow selective and quantitative determination. However, they have a number of disadvantages: the currently available equipment is complex and expensive, it require highly trained technicians, and it is unsuitable for rapid analyses under field conditions; the pretreatment and assay procedures are lengthy, hence fast analyses are impossible.

Environmental issues require far more sensitive, selective and quantitative methods, capable of low-level of pollutant detection in stream, ground, waste waters, in the soils and plants and in food as well *(1)*.

Organophosphorus Pesticides Assay

The most common approach for determination of OPCs is based on their inhibition of the activity of cholinesterase enzymes *(3, 4)*. cholinesterases are hydrolases catalyzing the hydrolysis reaction of a particular choline ester (butyryl-choline, acetylcholine, etc) to the corresponding carboxylic acid with the release of choline:

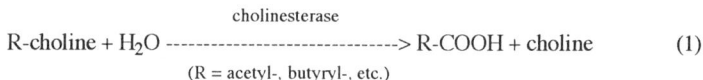

$$\text{R-choline} + H_2O \xrightarrow{\quad\text{cholinesterase}\quad} \text{R-COOH} + \text{choline} \qquad (1)$$

(R = acetyl-, butyryl-, etc.)

The presence of low concentrations of inhibitors strongly and specifically affects enzyme activity. Therefore, by measuring the enzyme activity the concentration of the organophosphorus compounds can be assayed.

The activity of the cholinesterases can be determined directly using traditional spectrophotometric methods and also electrochemical techniques. Electrochemical methods for cholinesterase activity assay that are based on pH-shift potentiometry have been described *(5-11)*. Conventional pH electrodes *(7-11)* and pH sensitive field effect transistors *(5, 6)* were employed as transducers coupled with cholinesterase enzymes. The main disadvantage of the pH-shift based method is a strong requirement for low buffer capacity of the sample. In addition, the sensitivity of pH based analytical techniques, in general, is less than that based on amperometric assay. The theoretical threshold of pH based assay methods is as low as 58 mV per decade of analyte concentration. Ion-selective membranes *(12)* and mediator-assisted potentiometry *(13)* have also been proposed for assays of cholinesterase inhibitors.

A number of papers describe techniques for determination of cholinesterase activity based on amperometric measurement of products formed as a result of enzymatic hydrolysis (equation 1). In this case, artificial (butyryl or acetyl thiocholine) cholinesterase substrates are used. Thiocholine, formed as a result of cholinesterase-catalyzed hydrolysis can be measured amperometrically on a platinum electrode *(14, 15)* or mercury electrode *(16)*. Analyses based on thiocholine determination employing an electrode modified by cobalt phthalocyanine *(17-22)* or cobalt tetraphenylporphyrin *(23)* have been described. Enzymatic hydrolysis of aminophenyl acetate leads to formation of aminophenol. The technique of determination of cholinesterase activity based on sensitive amperometric detection of aminophenol has been described elsewhere *(24, 25)*.

A popular method of determination of cholinesterase activity is based on coupling a cholinesterase enzyme with a choline electrode *(26-39)*. This coupling results in two consecutive enzyme reactions, first catalyzed by cholinesterase (equation 1), and second, catalyzed by choline oxidase:

$$\text{choline} + O_2 \xrightarrow{\text{Choline Oxidase}} \text{betaine} + H_2O_2 \tag{2}$$

The choline electrode usually consists of an amperometric transducer and immobilized choline oxidase. The most frequently used electrochemical transducers are hydrogen peroxide electrodes *(26-28, 33-36)*. The amperometric signal in this case is due to electrooxidation of hydrogen peroxide, which is the co-product of the enzymatic choline oxidation (equation 2). Oxygen amperometric sensors (Clark-type electrodes) have been also used as basic transducers for choline electrode construction *(29, 32, 37)*. The signal in this case is based on the reduction of molecular oxygen which is the co-reactant in reaction (equation 2). Redox mediators: hexacyanoferrate *(26)*, ferrocene derivatives *(38)* and tetracyanoquinodimethane *(39)* have also been used in the construction of choline electrodes.

In order to facilitate hydrogen peroxide detection, a third enzyme, horseradish peroxidase has been employed as a catalyst for hydrogen peroxide reduction with the enzymatic oxidation of an electrode donor substance. Horseradish peroxidase has been used as a catalyst for hydrogen peroxide reduction with the enzymatic oxidation of a electrode donor (AH):

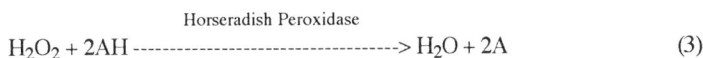

$$H_2O_2 + 2AH \xrightarrow{\text{Horseradish Peroxidase}} H_2O + 2A \tag{3}$$

Redox mediators can be used in this case for cathodic detection. An amperometric sensor for choline based on electron transfer between horseradish peroxidase and a redox polymer has been described *(19, 20)*. The use of redox mediators facilitate electron exchange but leads to system complication.

Choline electrodes coupled with acetylcholine esterase have been described as sensor for several acetylcholine esterase inhibitors: NaF, butoxycarboxime, trichlorfon, dimethoate *(25)*, malathion *(23)* paraoxon, aldicarb *(21, 22)*. All these studies used model toxic compounds and consisted of preliminary laboratory investigations aimed at environmental pesticide control.

During the last few years an alternative approach for organophosphates detection has been proposed *(40, 41)*. It relays on the monitoring of the reaction of enzymatic hydrolysis of OPCs by a specific microbial enzyme, organophosphate hydrolase (OPH). OPH-based biosensors have been shown to be useful for direct detection of various OPCs *(40-43)*, as well as for discrimination between different classes of neurotoxins *(42)*. OPH has been coupled with several different transducers: potentiometric *(41, 44-46)*, amperometric *(40, 47)* or fiber optic *(48, 49)*. Sensors utilizing the purified OPH enzyme *(41, 43, 46, 47)* as well as microbial sensors *(40, 44, 45, 48)* have been under development. The technology has been brought to the level of field demonstrations *(50)*.

Trends for future development of OPC sensors can be summarized as: improvement of sensor performances (decrease of detection limit and assay time); simplification of assay procedure; development of disposable sensing elements and development of inexpensive assays. This chapter summarizes our experience in developing a biosensor assay for OPCs based on the unique potentiometric principle and formatted for a disposable sensor technology.

Detection Based on Direct Electron Transfer

A potentiometric electrode based on direct mediatorless bio-electrocatalysis for determination of choline and butyryl-choline is developed. The electrode consists of a carbon based material and enzymes: peroxidase and choline oxidase co-immobilized on the electrode surface. Choline oxidase catalyzes the reaction of choline oxidation accompanied with hydrogen peroxide formation. Hydrogen peroxide acts as a substrate of the enzyme peroxidase.

In this system peroxidase catalyzes the reaction of hydrogen peroxide electro-reduction. The mechanism of this reaction is mediatorless, based on direct electron transfer from electrode to substrate molecule *via* the enzyme active center *(51)*

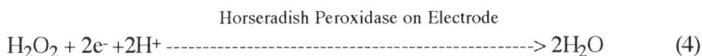

$$H_2O_2 + 2e^- + 2H^+ \xrightarrow{\text{Horseradish Peroxidase on Electrode}} 2H_2O \qquad (4)$$

Therefore, the enzyme peroxidase acts as an electrocatalyst (bio-electrocatalyst) eliminating the over-voltage for hydrogen peroxide reduction at the electrode. As result a significant increase of the electrode potential (ΔE) - anodic shift towards the equilibrium potential of the redox couple H_2O_2/H_2O occurs. The rate of this increase ($\Delta E/\Delta t$) is proportional to the hydrogen peroxide production rate.

The coupling of horseradish peroxidase acting as a hydrogen peroxide detector, with enzymes forming hydrogen peroxide as a co-product of analyte oxidation, has been used in designs of potentiometric sensors for determination of glucose *(52)* and lactate *(53)*.

A potentiometric electrode based on direct mediatorless bio-electrocatalysis for determination of choline and butyryl-choline has been described in our earlier work *(54)*. Using the two enzymes peroxidase and choline oxidase in a coupled system allows determination of choline concentration as a result of the consecutive enzyme reactions. In the system choline concentration assay is based on the measurement of the rate of electrode potential increase ($\Delta E/\Delta t$).

The ability of the electrode to detect choline over a wide concentration range allows its use for measurement of butyryl-choline concentration in assay systems employing coupled butyrylcholinesterase. The butyrylcholinesterase causes choline production due to enzymatic hydrolysis of butyryl-choline.

Butyryl-choline detection in such a system can be realized by two different coupling approaches: (i) co-immobilization of butyrylcholinesterase on a choline bi-enzyme electrode, which results in a tri-enzyme electrochemical sensing system; (ii) use of solubilized butyrylcholinesterase in a coupled system with an electrode for choline determination, by addition of standard amounts of the dissolved enzyme to the measuring cell. In both cases butyrylcholinesterase activity affects the choline electrode response providing the presence of choline in the solution. Both coupling approaches are suitable for analysis of butyrylcholinesterase inhibitors such as organophosphorus compounds.

The sensing element based on direct electron transfer does not involve an enzyme membrane coupled with a transducer. The layers of immobilized enzymes form a biosensitive system and is the transducer. This permits a high degree of miniaturization and results in inexpensive disposable sensing elements. The sensing element based on highly dispersed carbon material permits production of sensing elements based on screen printing. Electrodes based on this principle will be manufactured using commercially available carbon materials (screen printed carbon layers on an inert substrate), allowing development of a disposable electrode assembly. The electrode does not require any low-molecular weight mediator and can be arranged as an 'all-solid-state' sensor. Such electrodes, being based on a potentiometric principle, are suitable for miniaturization and arrangement in multi-sensor array for simultaneous determination of several analytes in real samples. The electrode response is not determined by a change in pH which makes the signal relatively unaffected by the electrolyte composition of the sample. Both approaches to use butyrylcholinesterase co- immobilized on the choline electrode in a tri-enzyme system or to use this enzyme solubilized in the sample during the assay possess some specific advantages and disadvantages.

The use of solubilized butyrylcholinesterase has a disadvantage compared to the use of butyrylcholinesterase co-immobilized on the electrode. This disadvantage is associated with excess of enzyme expenditure. However, in the case of co-immobilization of butyrylcholinesterase the analysis of irreversible inhibitors leads to a limited number of usage per electrode. Such a system is most suitable for disposable electrodes in 'one-shot' analysis. On the other hand, choline oxidase and peroxidase (the basic enzyme system of the choline electrodes) are much less influenced by butyrylcholinesterase inhibitors. In the case of the using solubilized butyryl-cholinesterase, incubation of the enzyme solution with inhibitors could be performed as a preliminary stage, thus avoiding contact between the inhibitors and the choline electrode. As a result the contact time of the choline electrode with the inhibitors will not exceed the usual time for analysis of enzyme activity (2-3 minutes). An important advantage of the system using solubilized butyrylcholinesterase is the possibility to control the choline electrode activity by a preliminary procedure. This allows calibration of the device before conducting a series of inhibitor (OPC) analyses.

The advantages of this technology over other disposable electrochemical systems *(22, 23)* are associated first with the potentiometric method of signal generation which provides independence on the electrode surface area and resulting in significant improvement in reproducibilty of the individual sensor's response parameters. In addition to it, the bioelectrocatalytic signal transduction provides extreme sensitivity *(54, 55)* towards changes in cholinesterase activity resulting in much lower limits of detection. The derivative potentiometric protocol adopted for the study (measurements of $\Delta E/\Delta t$) contributes to the sensitivity of analysis and provides a mean to obtain results before reaching the steady-state, thus expediting the assay. This chapter describes the development of the sensing element for a biosensor prototype for environmental analysis of organophosphorus pesticides. It brings the concept idea to a technological solution addressing the issues of sensor design, optimization, technological and analytical characterization of the sensing elements, their storage lifetime and application in the assay of several common pesticides.

Results and Discussion

The biosensor employs the tri-enzyme system arranged in the form of a disposable miniature electrode manufactured by screen-printing technology. Electrodes were tested in a miniature electrochemical cell using a standard Ag/AgCl reference electrode. A conventional high impedance voltmeter was used for signal registration. Trichlorfon was used as a model analyte for the optimization of the inhibitor assay procedure. The procedure consists of three steps:

(i) incubation stage: pre-calibration of the sensing element by obtaining its response to a standard concentration of butyryl choline as $\partial E/\partial t\,|_{in}$;

(ii) inhibition stage: incubation of the sensing element in a solution containing the inhibitor (analyte);

(iii) detection stage: obtaining of the resulting sensor response after inhibition as $\partial E/\partial t\,|_{res}$ and normalizing it to the initial response to obtain the sensor response.

The sensor response (SR), as a function of the inhibitor concentration (C_{inh}) can be defined by the following equation as a relative inhibition response:

$$SR\,(C_{inh}) \;=\; \{(\partial E/\partial t\,|_{in}) - (\partial E/\partial t\,|_{res})\} \,/\, (\partial E/\partial t\,|_{in}) \tag{5}$$

Fig. 1 presents typical recordings of the consecutive measurements of $\partial E/\partial t\,|_{in}$ and $\partial E/\partial t\,|_{res}$ as described above. This Figure illustrates the character of the primary sensor responses which are then treated mathematically according to equation 5. The assay allows determination of OPCs in sub-micromolar concentration ranges. The overall assay time including all operation stages is 20 minutes. Technology for electrode fabrication was evaluated in terms of its reproducibility.

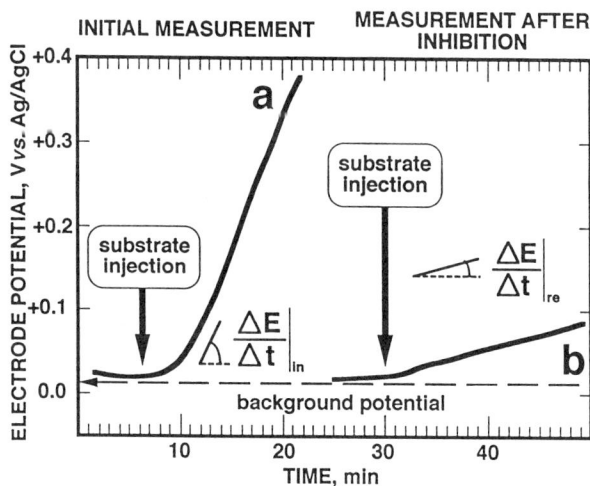

*Figure 1. Recordings of the electrode potentia shift (a) before inhibition [$\partial E/\partial t \mid_{in}$]
and (b) after inhibition [$\partial E/\partial t \mid_{res}$].*

Fig. 2 demonstrates the responses of 8 individual sensors, from the same
manufacturing batch, to a given standard concentration of trichlorfon (model OPC).
Both results for each individual OPC assay are presented (the consecutive
measurements of $\partial E/\partial t \mid_{in}$ and $\partial E/\partial t \mid_{res}$) to illustrate the reproducibility of the
individual steps of the assay. The error bars associated with the individual
measurements represent the measurement error. It can be seen that the distribution of
the individual responses is well grouped and the relative uncertainty of the assay does
not exceed 15 % (the 15 % of the average signal value interval is shown as shaded
zone in the graph which corresponds to the SD for the 8 individual measurements).

Series of sensors (manufacturing batches) were fabricated while varying the
number of the electrodes in the series (from 3 to 15) and the number of electrodes
which undergo simultaneously the electrochemical pre-treatment (depolarization steps
in the fabrication process). The limitations of the manufacturing process were assessed
demonstrating the advantages of larger batch production. The overall reproducibility of
the sensors analytical characteristics was in the acceptable level of 15 % deviation
from the expected sensitivity within a batch. Batch-to-batch reproducibility was found
to be within 25 % of the expected value.

Electrodes were incubated in solutions during the incubation stage and during the
inhibition stage at different pH and the sensor response was tested. Maximum
inhibition is observed for pH 10. Investigation of the dependence of pH of the
inhibitor containing solution on the inhibition effect of trichlorfon allowed to select
pH 10 for the OPC assay optimization at different temperatures.

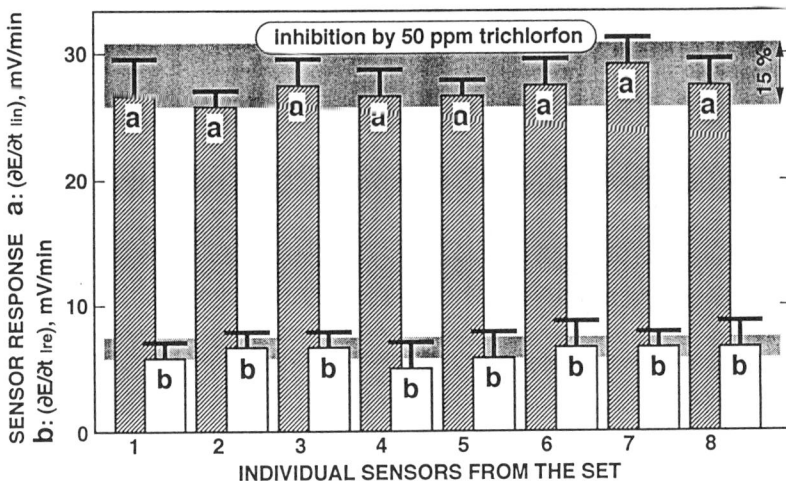

Figure 2. Intra-batch reproducibility: responses of 8 individual sensors, from the same manufacturing batch, to a given standard concentration of trichlorfon: (a) slopes before inhibition [$\partial E/\partial t \mid_{in}$] and (b) after inhibition [$\partial E/\partial t \mid_{res}$].

An increase in temperature of the inhibitor containing the solution enhances the inhibition effect. However, a high temperature above (45 °C) leads to thermal inactivation of the enzymes. Electrodes were incubated at different temperatures (in a thermostated cell) during the incubation stage and during the inhibition stage and the sensor response was tested at pH 10. It was found that the change in temperature during the incubation stage from ambient (23 °C) to elevated (43 °C) does not effect significantly the sensor response. The sensor response decreases with the increase in the temperature from 23 °C to 37 °C and demonstrates a local maximum at 40 °C. It should be noted from the practical point of view that the temperature dependence is expressed as a plateau of high sensor response from 32 °C to 40 °C which provides an advantage or relative insensitivity of the assay performance to temperature variations. When incubating the sensor at 40 °C for 10 min., a detection limit of 5 nM (1.3 ppb) concentration of trichlorfon was achieved.

The sensor shelf lifetime has been investigated in a series of experiments. Batches with uniformly fabricated sensors were kept in defined conditions: in refrigerator at 4 °C, at room temperature of 20 °C and in an incubating box at 40 °C. At particular time intervals sensors from those batches were tested with a model OPC (trichlorfon) to determine the sensitivity of the sensor response to the inhibitor. Having in mind that the sensors are disposable, the shelf lifetime of the electrodes made by this technology was determined as the period of time within which the sensor sensitivity obtained from such a test does not fall outside the deviation range of the intra-batch reproducibility (15 %). Dropping the sensitivity bellow 15 % from the average was considered as a manifestation of the decline in the sensor performance. The experimental results of this study are presented in Fig. 3. This Figure shows the relative value of the initial sensor response (compared to the first measurement made with a sensor from a given batch) as a function of the storage time. It can be seen from Fig. 3 that shelf lifetime of the sensors, when stored at elevated temperature (40 °C) were shorter then two days. During the first measurement, the sensor results showed a lower signal than the level of expectation. While stored in ambient temperature (20 °C) the sensors demonstrated a shelf lifetime of one week. Sensitivity of the sensors remained practically constant for more that 56 days when kept in refrigerated conditions (4 °C). Experiments with sensor batches stored at 4 °C are still in progress, showing more than two months of shelf lifetime. These results can be considered acceptable from the practical point of view as pertinent to a system where no special packaging was used for sensor preservation. Introducing packaging technologies as a part of the development of this assay system could significantly improve shelf lifetime of the sensors.

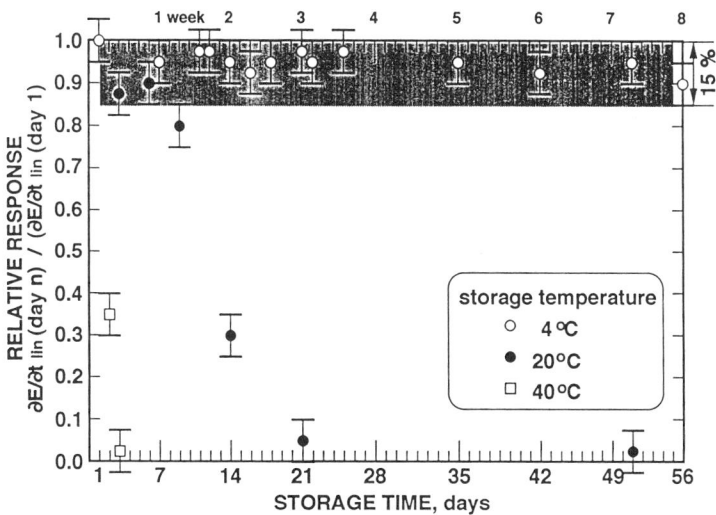

Figure 3. Shelf life-time of the sensors: dependence of the relative value of the initial sensor response (compared to the first measurement made with a sensor from a given batch) vs. the storage time at different temperatures.

There are two main characteristics of performance of OPC assay based on inhibition of cholinesterase activity: low detection limit and assay time. The assay time involves the stage of incubation of cholinesterase (solubilized or immobilized) with OPC containing solution. This stage is usually the longest stage of the assay. Obviously, longer incubation time leads to higher degree of enzyme inhibition and as a result in lower assay detection limit. However, practical analytical application requires rapid analysis. Therefore, there is a tradeoff between rapidity of the assay and the detection limit. The detection limits for different OPCs are different because of the difference in the ability to inhibit cholinesterase ranging from 10^{-6} to 10^{-9} M. The detection limits for different OPCs can differ by 3 orders (4). In general, very toxic OPCs inhibit cholinesterase stronger than less toxic ones and detection limit for more toxic OPCs is lower than that for less toxic OPCs.

Organophosphate pesticides studied in this work were: the model low-toxic OPC trichlorfon, and some common organophosphate pesticides: malathion, parathion, dichlorvos, and diazinon (Table I). Calibration curves for these pesticides (dependences of the sensor inhibition response on the analyte concentration) were obtained for all of these OPCs. These calibration curves were obtained under conditions (time of inhibition, pH and temperature) optimized with the model analyte trichlorfon. All of the pesticide calibration curves are similar and Fig. 4 illustrate the method by the example of malathion. The lowest concentration of pesticide samples assayed with 10 min. of incubation of the electrode in inhibitor containing solution was 5 ppb. This resulted in approximately 10 % of the relative inhibition signal. Fig. 4 predicts much better performance of our system compared with the literature data. For example, trichlorfon detection by means of ISFET had a reported limit of detection of ca. 250 ppb (5), while conductometric sensor assay registered trichlorfon at ca. 25 ppb (6), still an order of magnitude higher than the described sensor. An amperometric sensor was used to detect dichlorvos with a limit of detection of 350 ppb (21) and a potentiometric (pH-sensitive) sensor was shown to detect parathion at 39 ppm and diazinon at 35 ppb (9).

Table I. Characteristics the potentiometric biosensors assay.

Organophosphorus Compound	Low Limit of Detection (resulting in 10 % inhibition)	Upper Limit of the Range (resulting in 90 % inhibition)
trichlorfon	5	100,000
malathion	5	50,000
parathion	5	50,000
dichlorvos	5	1,000
diazinon	5	5,000

Incubation time: 10 minutes. Concentration units are ppb.

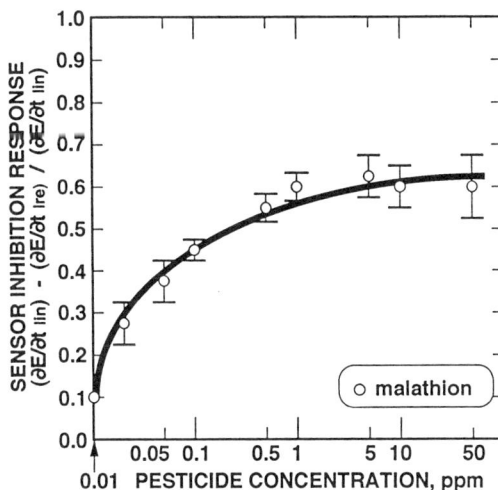

Figure 4. A typical calibration curve of the sensor assay for malathion. Each sensor response is an average of four individual measurements each obtained with different sensors from the same manufacturing batch (SEM are shown).

The overall assay time includes the incubation of the electrode in the analyte solution and the duration of measurements of electrode activity before and after the incubation step. The duration of the measurements of electrode activity does not exceed 5 min. for each assay, therefore, the overall assay time is about 20 min. Further development of electrode manufacturing in order to improve the reproducibility of electrode performance within one manufacturing set of electrodes will avoid the need for measurement of initial activity of each electrode. In this case, one electrode from the set can be used for determination of the initial activity of the whole set. This will result in only one measurement of electrode activity for inhibitor assay and will lead to decrease of the assay time.

Summary

This work demonstrates the potential for application of potentiometric enzyme electrodes based on mediatorless enzyme electrocatalysis for fast and sensitive assay of organophosphorus pesticides. The sensing element based on screen-printed carbon material permits mass fabrication of the electrodes at a low cost which is essential for the disposable sensor concept. The biosensor does not require any low-molecular weight mediator and can be arranged as an 'all-solid-state' device. Such electrodes, being based on a potentiometric principle, are suitable for miniaturization and arrangement in multi-sensor array for simultaneous determination of several analytes in real samples. This work presents the optimization of the assay parameters: it establishes the optimal pH and temperature range for maximal electrode response. The assay allows determination of OPCs in sub-micromolar concentration ranges with an overall assay time of 20 minutes. The enzyme sensor fabrication technology demonstrates very good intra-batch reproducibility of the main sensor analytical parameters. This is especially essential for a technology aimed to disposable sensors allowing calibration to be done by measurement of the performance of one sensor from the batch. The shelf lifetime of the disposable sensing elements is close to two months, while stored unpacked in refrigerated conditions. Applicability of the assay is illustrated by detection of several common of organophosphorus pesticides. This sensor technology provides a mean for estimation of the total OPC pesticide contamination of a given environmental sample, which is a typical feature of all techniques based on cholinesterase inhibition. It gives, however a promise for extremely low detection limits which makes this technology advantageous especially in the context of increasing environmental regulations and requirements for rapid detection of trace amounts of pesticides. Similarly to other disposable electrodes, reported elsewhere, this technology may find application in rapid screening and field site environmental monitoring, thus reducing the volume and the cost of the environmental analyses.

Acknowledgements

This research was supported in part by a grant from the DoE/Waste-Management Education and Research Consortium of New Mexico. Authors express their gratitude to Prof. Joseph Wang (Department of Chemistry and Biochemistry, New Mexico State University, Las Cruces, NM) for providing the basic (untreated) screen-printed electrodes for the sensor construction.

References

1. K.R. Rogers, *Biosens. Bioelectron.*, **1995**, *10*, 533.
2. C. Wittmann, K. Reidel, R.D. Schmid, in: E. Kress-Rogers (ed.), *Handbook of Biosensors and electronic Noses: Medicine , Food and the Environment*, CRC Press, Boca Raton, **1997**, p. 299.
3. P. Sklàdal, *Food Technol. Biotechnol.* **1996**, *34*, 43.
4. M. Trojanowicz, M.L. Hiuchman, *Trends Anal. Chem.*, **1996**, *15*, 38.
5. A.M. Hendji, N. Jaffrezic-Renaullt, C. Marlet, P. Clecht, A.A. Shul'ga, V.I. Strikha, L.I. Netchiporuk, A.P. Soldatkin, W.B. Wlodarski, *Anal. Chim. Acta*, **1993**, *281*, 3.
6. S.V. Dzydevich, A.A. Shul'ga, A.P. Soldatkin, A.M.N. Hendji, N. Jaffrezic-Renault, C. Martelet, *Electroanalysis*, **1994**, *6*, 752.

7. Y. Vlasov, A. Bratov, S. Levichev, Y. Tarantov, *Sens. Actuators B*, **1991**, 4, 283.
8. T. Danzer, G. Schwedt, *Anal. Chim. Acta*, **1996**, *318*, 275.
9. S. Kumaran, M. Morita,*Talanta*, **1995**, *42*, 649.
10. E.B. Nikol'skaya, G.A. Evtugyn, R.R. Iskanderov, *J. Anal. Chem.*, **1996**, *51*, 516.
11. G.A. Evtugyn, E.P.Rizaeva, E.E. Stoikova, V.Z. Latipova, H.C. Budnikov, *Electroanalysis*, **1997**, *9*, 1124.
12. T. Imato, N. Ishibashi, *Biosens. Bioelectron.*, **1995**, *10*, 435.
13. D.M. Ivnitski, J. Rishpon, *Biosens. Bioelectron.*, **1994**, *9*, 569.
14. R. Gruss, F. Scheller, M.J. Shao, C.C. Liu, *Anal. Lett.*, **1989**, *22*, 1159.
15. N. Mionetto, J.L. Marty, I. Karube, *Biosens. Bioelectron.*, **1994**, *9*, 463.
16. H.C. Budnikov, T.P. Medyantseva, S.S. Babkina, *J. Electroanal.Chem.*, **1991**, *310*, 49.
17. P. Sklàdal, *Anal. Chim. Acta*, **1991**, *252*, 11.
18. P. Sklàdal, *Anal. Chim. Acta*, **1992**, *269*, 281.
19. P. Sklàdal, M. Fiala, J. Krejci, *Intern. J. Environ. Anal. Chem.* **1996**, *65*, 139.
20. P. Sklàdal, G,S. Nunes, H. Yamanaka, M.L. Ribeiro, *Electroanalysis*, **1997**, *9*, 1083.
21. I.C. Hartley, J.P. Hart, *Analytical Proceedings*, **1994**, *31*, 333.
22. A.L. Hart, W.A. Collier, D. Janssen, *Biosens. Bioelectron.*, **1997**, *12*, 645.
23. Q. Deng, S. Dong, *Analyst*, **1996**, *121*, 1123.
24. C. Le Rosa, F. Pariente, E. Hernandez Lorenzo, *Anal. Chim. Acta*, **1994**, *295*, 273.
25. C. Le Rosa, F. Pariente, E. Hernandez Lorenzo, *Anal. Chim. Acta*, **1995**, *308*, 129.
26. B. Lopez Ruiz, E. Dempsey, C. Hua, M.R. Smyth, J. Wang, *Anal. Chim. Acta*, **1993**, *273*, 425.
27. E.N. Navera, M. Suzuki, K. Yokoyama, E. Tamiya, T. Takeuchi, I. Karube, J. Yamashita, *Anal. Chim. Acta*, **1993**, *281*, 673.
28. T.N. Nwosu, G. Palleschi, M. Mascini, *Anal. Lett.*, **1992**, *25*, 821
29. L. Doretti, P. Gattolin, S. Lora, *Anal. Lett.*, **1994**, *27*, 2455.
30. M.G. Garguilo, H. Nhan, A. Proctor, A.C. Michael, *Anal. Chem.*, **1993**, *65*, 523.
31. M.G. Garguilo, A.C. Michael, *Anal. Chim. Acta*, **1995**, *307*, 291.
32. S. Fennouh, V. Casimiri, C. Burstein, *Biosens. Bioelectron.*, **1997**, *12*, 97.
33. A. Cagnini, I. Palchetti, M. Mascini, A.P.F. Turner, *Microchim. Acta*, **1995**, *121*, 155.
34. M. Bernabei, C. Cremisini, M. Mascini, G. Palleschi, *Anal. Lett.*, **1991**, 24, 1317.
35. G. Palleschi, M. Bernabei, C. Cremisini, M. Mascini,*Sens. Actuators B*, **1992**, 7, 513.
36. I. Palchetti, A. Cagnini, M. del Carlo, C. Coppi, M. Mascini, A.P.F. Turner, *Anal. Chim. Acta*, **1997**, *331*, 315.

138

37. L. Campanella, M. Achilli, M.P. Sammartino, M. Tomassetti, *Bioelectrochem. Bioenerg.*, **1991**, *26*, 237.
38. S. Yabuki, F. Mizutani, T. Katsura, *Sens. Actuators B*, **1994**, *20*, 159.
39. D. Martorell, F. Cespedes, E. Martinez-Fabregas, S. Alegret, *Anal. Chim. Acta*, **1997**, *337*, 305.
40. E.I. Rainina, E.N. Efremenco, S.D. Varfolomeyev, A.L. Simonian, J.R. Wild, *Biosens. Bioelectron.*, **1996**, *11*, 991.
41. A.A. Karyakin, O.A. Bobrova, L.V. Luckachova, E.E. Karyakina, *Sens. Actuators B*, **1996**, 33, 34.
42. A.L. Simonian, E.I. Rainina, J.R. Wild, *Anal. Lett.*, **1997**, *30*, 2453
43. A.L. Simonian, B.D. diSioudi, J.R. Wild, *Anal. Chim. Acta*, **1999**, *389*, 189.
44. A. Mulchandani, P. Mulchandani, S. Chauhan, I. Kaneva, W. Chen, *Electroanalysis*, **1998**, *10*, 733.
45. A. Mulchandani, P. Mulchandani, I. Kaneva, W. Chen, *Anal. Chem.*, **1998**, *70*, 4140.
46. P. Mulchandani, A. Mulchandani, I. Kaneva, W. Chen, *Biosens. Bioelectron.*, **1999**, *14*, 77.
47. A. Mulchandani, P. Mulchandani, W. Chen, J. Wang, L. Chen, *Anal. Chem.*, **1999**, *71*, 2246.
48. A. Mulchandani, I. Kaneva, W. Chen, *Anal. Chem.*, **1998**, *70*, 5042.
49. A. Mulchandani, S.T. Pan, W. Chen, *Biotechnol. Progress*, **1999**, *15*, 130.
50. A. Mulchandani, P. Mulchandani, W. Chen, *Field Anal. Chem. Technol.*, **1998**, *2*, 363
51. A.L. Ghindilis, P. Atanasov, E. Wilkins, *Electroanalysis*, **1997**, *9*, 661.
52. A.L. Ghindilis, I.N. Kurochkin, *Biosens. Bioelectron.*, **1994**, *9*, 353
53. A.L. Ghindilis, A. Makower, F.W. Scheller, *Sens. Actuators B*, **1995**, *28*, 109.
54. J. Diehl-Faxon, A.L. Ghindilis, P. Atanasov, E. Wilkins, *Sens. Actuators B*, **1996**, *35-36*, 448.
55. A.L. Ghindilis, T.G. Morzunova, A.V. Barmin, I.N. Kurochkin, *Biosens. Bioelectron.*, **1996**, 11, 873.

Chapter 10

New Organically Modified Sol–Gel Glasses and Their Applications In Sensors Construction

Prem C. Pandey

Department of Chemistry, Banaras Hindu University, Varanasi 221 005, India

The preparation of organically modified sol-gel glasses (Ormosil) of three different types is reported. These ormosils are prepared for the development of three types of sensors namely: 1) Non-mediated glucose biosensor; 2) neutral carrier based ion sensor and 3) mediated biosensors. The non-mediated glucose sensor is developed using the ormosil precursors: 1) 3-aminopropyltriethoxysilane, 2-(3, 4-epoxycyclohexyl)-ethyltrimethoxy silane followed by addition of ditilled water, graphite powder, polyethylene glycol and hydrochloric acid. The development of ion sensor involve a base layer of ferrocene linked ormosil made from ferrocene carboxaldehyde, 3-aminopropyltriethoxysilane, 2-(3, 4-epoxycyclohexyl)-ethyltrimethoxy, distilled water and HCl followed by assembling of a neutral carrier incorporated plasticized PVC matrix membrane. The development of mediated biosensors involve a ferrocene encapsulated palladium-linked ormosil having the composition; Palladium chloride, glycidoxypropyltrimethoxy silane, ferrocene monocarboxalic acid, distilled water and HCl. The performance, characterization and applications of these ormosil are reported in the present chapter.

The sol-gel glasses modified electrodes prepared by introducing graphite particles (particle size 1-50 μm) have received great attention[1-3]. The dispersed carbon provides the electrical conductivity essential for electrochemical measurements. These carbon-ceramic electrodes (CCEs) can be bulk modifier by organic, inorganic, or biochemical species[4] which can be subsequently used in the preparation of Pd-modifier, enzyme–doped carbon-Ormosil (organically modified Sol-Gel glasses) and this made possible to cast silica-carbon matrices in virtually any desired geometrical configuration. Another way for the construction of modified CCEs which can be used for the large surface area amplification is the use of hydrophilic and hydrophobic silica-forming monomers, such as cyanoethyltrialkoxysilane as hydrophilic monomer and methyltrimethoxysilane as hydrophobic monomers. When hydrophobic silica forming monomers are used, the resulting electrodes reject water, leaving only segregated islands of carbon at the outermost surface in contact with electrolyte[5]. Thus the ratio of hydrophilic and hydrophobic monomers in organically modified sol-gel CCEs is crucial in sensor design. Additionally monomers containing a Si-C bond and an easily derivatized radicals through cross-linking such as an amino, vinyl-, epoxy-, or mercapto- groups can be used to prepare readily derivatized xerogels[5] and provides a convenient method for the production of organically modified surfaces[6] which can be used as covalent anchors for specific chelating agents. The lower degree of cross-linking provides an inherent strain relaxation pathway allowing thick (> 1μm) silica films to be prepared in one coating step[7]. Another way of increasing the exposed surface area (wetted section of sol-gel glass) is by incorporating readily leachable, water-soluble components in the matrix and dissolving them out by immersing the electrodes in an electrolyte solution. This leads wide open channels for the penetrating electrolyte, thereby increasing the wetted section inside the sol-gel matrix. A typical example is the use of poly (ethylene glycol)[4]. The present chapter describes the development of non-mediated glucose biosensor made from ormosil using 3-aminopropyltrimethoxysinale (relatively hydrophilic modifier) and 2-(3, 4-epoxycyclohexyl)-ethyltrimethoxy silane (relatively hydrophobic modifier) in the presence poly (ethylene glycol) and graphite powder.

The developments of solid-state potentiometric Ion-selective electrodes(ISEs) for both cations and anions have received increasing attention because of its simplicity in the construction and greater possibility of commercialization. The coated wire electrodes have gained attentions due to the simplicity involved in the construction of the ISEs avoiding the problems associated conventional ISEs. However, there still exists a need for further researches to overcome the problems associated to the drift and overshoot of the potential difference across liquid-ion exchanger/metal interface leading to the variable base line (steady-state potential) recovery during the potentiometric operation. A significant share of attention on these problems has been made during last few years[8-15]. The application of conducting polymers[8-15] has been demonstrated on these lines.The problem associated to the overshoot of the base line potential was reduced due to the electroactivity and cation exchange behavior of the Polypyrrole/polysulfonate (PPy/PSS) film resulting the formation of the concentrations cells also referred as symmetric cell (ions-in, ions-out) at the test solution/PVC and PVC/ (PPy/PSS) interfaces. We have recently developed solid-state ion-sensor based on a double layer of polyindole and dibenzo-18-crown-6 or valinomycin incorporated plasticized PVC matrix membrane[14,15]. The reproducible potential difference across interface was due to the presence of dissociated doped perchlorate anions and positively charged polymer domain within the polymer network. However the leaching out of the perchlorate anions while operating ion-sensor in aqueous solution results subsequent variation of potential difference across polyindole/metal interface in a regular pattern. Accordingly, there still exits a need to develop metal/new material/PVC interface which results reproducible potential difference across interface. The application of stable redox material between metal and PVC interface may overcome this problem. However, most of the redox material with well defined redox electrochemistry are relatively soluble in

queous solution that might cause variation in the potential difference across metal/redox
material interface. The present chapter describes a new interface resulting from a
errocene-linked sol-gel glass laid down over a Pt disk electrode followed by self
assembling of neutral carrier incorporated plasticized polyvinyl chloride matrix
membrane. The ferrocene carboxaldehyde is cross linked to one of the ormosil precursor
through the formation of schiff base linkage. The new interface provide reproducible
potential difference across metal/sol-gel interface while continuously using ion-sensor for
two month. The new ferrocene linked sol-gel glass is charaterized by scanning electron
micrographs and cyclic voltammetry. The ion-sensor for potassium ion is reported in this
chapter based on the application of dibenzo-18-crown-6 incorporated into plasticized PVC
matrix membrane over the ferrocene-linked sol-gel glass.

The encapsulation of redox material within sol-gel glass has gained significant share of
attention for sensors designing. Pankrata and Lev[16] reported tetrathiafulvalene mediated
carbon ceramic electrode (CCEs) with limited storage and in-use stability. Several other
reports on ferrocene encapsulated sol-gel glasses including those of Lev el al [2,17,18] are
available. Audebert et al[19] reported modified electrodes from organic-inorganic hybrid gels
containing ferrocene unit covalently bonded inside a silica network and modified
electrodes from organic-inorganic hybrid gels formed by hydrolysis-polycondensation of
some trimethoxysilylferrocenes[20]. Collinson et al[21] reported electroactivity of redox probes
encapsulated within sol-gel derived silicate film based on anionic i.e. $[Fe(CN))_6^{3-/4-}]$;
$[IrCl_6^{2-/3-}]$ and cationic i.e. ferrocenemethanol $[FcCH_2OH^{0/+}]$gel-doped probes. Other
reports on ferrocene based sol-gel sensors are also available[22]. There are great potential to
study ferrocene encapsulated/linked sol-gel glasses with reversible electrochemistry of
encapsulated ferrocene for sensors applications. The present chapter describes the
development of such ferrocene encapsulated Palladium linked ormosil.

EXPERIMENTAL

Materials and Methods. 3-Aminopropyltrimethoxy silane, trimethoxysilane, dibutyl
phthalate, dibenzo-18-crown-6, ferrocene carboxaldehyde, ferrocene monocarboxalic acid,
palladium chloride, poly (vinyl chloride) and graphite powder (particle size 1-2 μm) were
obtained from Aldrich; 2-(3, 4-epoxycyclohexyl)-ethyltrimethoxy silane and 3-
glycidoxypropyltrimethoxysilane were obtained from United Chemical Technologies, Inc.,
Petrarch[TM] Silanes and Silicones, Bristol, PA, USA; glucose oxidase, tetrabutyl
ammonium perchlorate (TBAP), and polyethylene glycol (Mol.wt. ~ 6000) were obtained
from Sigma. Tetraphenyl borate was obtained from the E-Merck, India. Ltd. The aqueous
solutions were prepared in double distilled deionized water. All other chemical employed
were of analytical grade.

Construction of Ormosil Based Glucose Biosensor. The electrode body used for the
preparation of composite sol-gel glass modified electrodes was similar to that described
in earlier[14]. The new material is developed with common sol-gel precursors; 3-
aminopropyltrimethoxy silane (70 μl), 2-(3, 4-epoxycyclohexyl)-ethyltrimethoxy silane

(20 µl) , 0.1 M HCl (5 µl) , glucose oxidase (4 mg), 700 µl saturated aqueous solution of poly (ethylene glycol) and 2 mg graphite powder. The resulting mixture in each cases is stirred thoroughly and desired amount (~ 70 µl) of the homogeneous solution is added to the well of the specially designed electrode body. The gelation is allowed to occur at 25°C for 30 h. A smooth very thin GOD immobilised film of organically modified sol-gel glass appeared on the Pt surface. The electrode obtained in this manner is washed with 0.1 M phosphate buffer, pH 7.0 several times and stored in 0.1 M phosphate buffer pH 7.0 at 4°C when not in use. Some of the glucose sensors made following such procedures were stored at room temperature (26°C).

Construction of Ion sensor Based on a Bilayer of Ferrocene-Linked Ormosil and Poly (vinyl chloride) matrix membrane. The electrode body used for the construction ferrocene-linked ormosil and subsequently ion-sensor based on ionophore impregnated PVC membrane was similar as described earlier[14]. The new material is prepared using optimum concentrations sol gel precursors 3-Aminopropyltrimethoxy (70 µl), 2-(3, 4-epoxycyclohexyl)- silane (20 µl), HCl (5 µl), Ferrocene carboxaldehyde (5 µl), double distilled water (50 µl).The resulting mixture is stirred thoroughly and desired amount of the homogeneous solution is added to the well of specially designed electrode body. The gelation is allowed to occur at 25°C for 12-30 h. A smooth very thin ferrocene linked sol-gel glass appeared on the Pt surface as evidenced by SEM and electrochemical measurement. The electrode obtained in this manner is washed with double distilled water and incubated overnight in 1 M KCl. The modified sol-gel glass electrode is then washed with double distilled water.

The solution of PVC casting membrane was made in dried THF of the composition; PVC fine powder (56 mg); dibenzo-18-crown-6 (1 mg); dibutyl phthalate (0.12 ml); tetraphenyl borate (0.75 mg); THF (1.3 ml). After complete dissolution of the membrane material; 70 µl of the solution was added to the recessed depth of the electrode body covered with ferrocene-linked sol-gel glass film . The solvent (THF) was allowed to evaporate slowly over a 20-h period at room temperature (25°C). On complete evaporation of the solvent a transparent smooth layer of the sensing membrane remained at the surface of modified sol-gel glass electrode. The resulting electrode was conditioned for 12 h in 1 M KCl solution.

Construction of Ferrocene Encapsulated Pd-Linked Ormosil. The electrode body for the construction of new ormosil was similar as described earlier[14]. The new material was prepared as follows : Palladium chloride (1 mg) and 4 mg ferrocene monocarboxylic acid is dissolved in 500 µl distilled water. This solution was added in 3-Glycidoxypropyltrimethoxy silane (70 µl) which results a black solution. The resulting reaction product is mixed with trimethoxy silane (30 µl), and 0.1 N HCl (5 µl). The resulting reaction mixture is stirred thoroughly up to 5 min at 25°C and desired amount of the homogeneous solution ranging between 70 µl is added to the well of specially designed electrode body. The gelation is allowed to occur at 25°C for 30 h. A smooth very thin ferrocene encapsulated Palladium-linked Ormosil appeared on the Pt surface as evidenced by electrochemical measurements.

Electrochemical Measurements. The electrochemical measurements were performed with a Solartron Electrochemical Interface (Solartron 1287 Electrochemical Interface). A one compartment cell with a working volume of 4 ml and a sol-gel glass modified working electrode, Ag/AgCl reference electrode and a platinum foil auxiliary electrode were used for the measurements. The cyclic voltammetry using GOD modified sol-gel glass electrode were studied between 0-1 V vs. Ag/AgCl. The amperometric measurements using GOD immobilized sol-gel modified electrode was operated at 0.70 V vs. Ag/AgCl. The experiments were performed in phosphate buffer (0.1 M, pH 7) employing both types of organically modified sol-gel glass electrodes.

The specific activity of glucose oxidase determined to be 80 U/mg before its immobilization on sol-gel matrix. The Scanning Electron Micrograph (SEM) measurements were made using JEOL- JSM 840A Scanning Electron Microscope.

Potentiometric Measurements. The potentiometric responses of the ion-sensor constructed above were carried out in 0.1 M Tris-HCl buffer, pH 7.0, and using a double junction calomel electrode with the cell assembly :

Calomel electrode/ KCl (sat.) / 0.1 M NH_4NO_3/test solution/modified sol-gel ion sensor

The ion-sensor together with double junction reference electrode was dipped in the stirred electrochemical cell with a working volume of 15 ml. The electrode potential was monitored with a Orion pH meter model SA 520 and recorded. At the steady-state potentiometric response, varying concentrations of the ionic solution (KCl, NH_4Cl, NaCl) were injected into the cell and the new steady-state potential was recorded. The measurements were made with the sol-gel modified electrode with and without the ion-sensing membrane.

RESULTS AND DISCUSSION

Ormosil Based Glucose Biosensor. Fig. 1a   which show the scanning electron microscopy of ormosil in absence of glucose oxidase whereas Fig.1b shows the SEM of ormosil with encapsulated glucose oxidase.The microscopy shows the smooth surface of ormosil in both cases. The present ormosils were prepared following a very simple one step gelation process and sol-gel precursors does not require sonication for the homogenisation of monomers and additives suspension as compared to relatively complex protocols of gelation reported for the preparation of sol-gel glasses by earlier workers[4,23,24]. The ormosil matrix inhibits intermolecular interactions of the encapsulated macromolecule[25] and the matrix functions as a solid solution of

Figure 1. (a) Scanning electron microscopy of ormosil made using 3-aminopropyltriethoxy silane, 2-(3, 4-epoxycyclohexyl)-ethyltrimethoxy silane and other ingredients; in absence of glucose oxidase, graphite powder and polyethylene glycol (a); in the presence of glucose oxidase and in absence of graphite powder and polyethylene glycol (b). (c) Scanning electron microscopy of ferrocene-linked ormosil. (d) Scanning electron microscopy of ormosil made using 3-glycidoxypropyltrimethoxy silane, Trimethoxy silane and other ingredient and in absence and the presence of ferrocene.

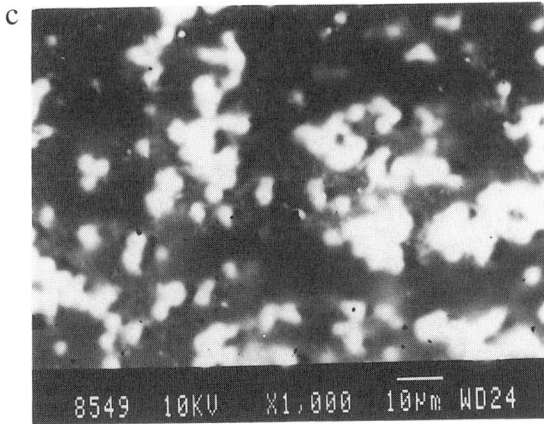

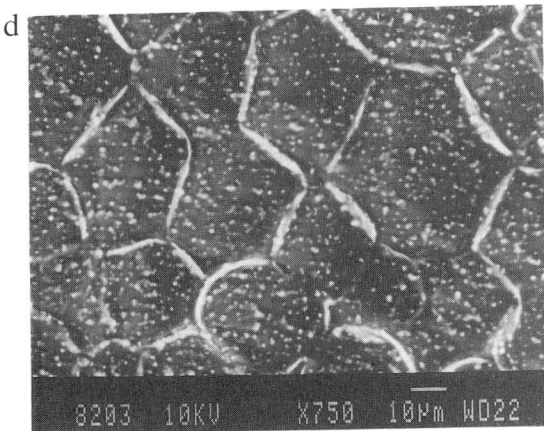

Figure 1. Continued.

Continued on next page.

e

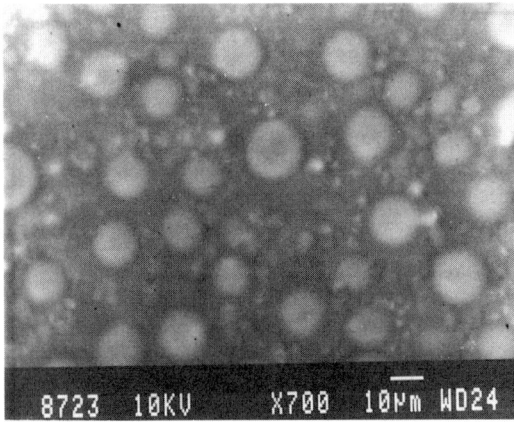

Figure 1. Continued.

encapsulated biomolecules. The present modified sol-gel film has been found to be strongly attached to the Pt surface and is not easily removable from the Pt surface.

The cyclic voltammetry results in the presence and absence of 50 mM in between 0-1 V vs. Ag/AgCl at the scan rate of 5 mV/s in 0.1 M phosphate buffer pH 7.0 at 25^0C shows large increase in anodic current on the addition of glucose corresponding to the oxidation of hydrogen peroxide. The data on amperometric response are shown in Fig.2. The curve shows the typical amperometric responses of the biosensor on subsequent addition of glucose. The inset (I) to Fig. 2 shows the calibration curve for glucose analysis. The response is also dependent on the composition of sol-gel glass precursors particularly in the presence and absence of polyethylene glycol and graphite powder. This is mainly due to relatively less concentration of oxygen at the site of enzymatic reaction required for the formation of hydrogen peroxide which can be explained from the consideration of exposed surface area of sol-gel glass to solution and water contact angle. It has been reported that neither highly hydrophobic nor totally hydrophilic sol-gel matrices are desirable for sensing application[4]. When chemical modifier such as metal dispersion, water soluble polymers and proteins are added to the materials, the resulting electrodes became more hydrophilic and subsequently alter the water contact angle which manifests the wettability. It has been reported[4] that a blank sol-gel electrode without any hydrophilic modifier shows a highest water contact angle (80 degree) and in turn lowest wettability (not amenable for nitrogen adsorption analysis) whereas the sol-gel electrode with all hydrophilic modifiers (carbon, poly(ethylene glycol), and Pd-GOD) shows lowest water contact angle (42 degree) and highest wettability (42 m^2 /g). An increase in wetted area increases the wetted conductive surface accessible to the solution and also the corresponding electrochemically active area and capacitive current. On the other hand unwetted area does not contribute to the capacitive or faradic currents. The results reported in Fig.2 follow the similar arguments. Recently Ingersoll and Bright have also reported the effect of dopant addition time and used oxygen as analyte to study sensor performance[26.] The variation of oxygen concentrations has also been determined experimentally based on direct electrochemical reduction of oxygen at CCEs and at blank Pt surface. These results are shown in insets (II) and (III) to Fig.2 respectively. The curves 1 and 2 of inset (II) to Fig.2 show the oxygen reduction at the surface of ormosil without and with polyethylene glycol and graphite powder. The results shows large cathodic peak current showing relatively increased rate of oxygen diffusion within the wetted section of ormosil made using polyethylene glycol and graphite powder. Finally we also studied the oxygen reduction at bare Pt electrode in same aqueous medium and the result recorded in inset (III) of Fig. 2 which two electron transfer as reported earlier[27,28].

The enzyme is caged within the ormosil hence occurrence of limited mass-transport kinetics at the solution/sol-gel glass (within the wetted surface area) is desirable which is also indicated by the results recorded vide infra. The formation of hydrogen peroxide, keeping all other components of the enzymatic reaction constant, depends on ;i) the concentration of dissolved oxygen diffusing solution across the wetted area of sol-gel glass and ii) glucose concentration within the sol-gel glass matrix. The results discussed

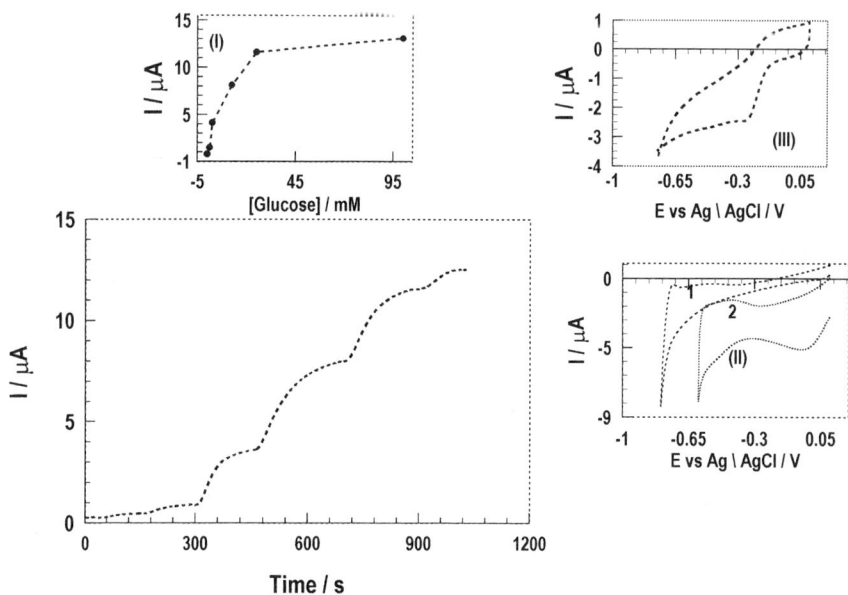

Figure 2. Typical amperometric response of glucose oxidase encapsulated ormosil. The recording shows typical response curve on the addition of subsequent increasing concentrations of glucose. The calibration curves for glucose analysis based on non-mediated electrochemical reaction from data are recorded in inset (I) to Fig. 2; the electrodes were held at 0.70o V vs. Ag/AgCl. Inset (II) shows the cyclic voltammograms of sol-gel modified electrodes showing the oxygen reduction in 0.1 M phosphate buffer pH 7.0 at the scan rate of 5 mV/S; curve 1 was recorded from graphite powder dispersed sol-gel modified electrode without glucose oxidase whereas curve 2 was recorded using the glucose encapsulated ormosil made in the presence of graphite powder and polyethylene glycol. The inset (III) shows the cyclic voltammogram showing oxygen reduction recorded using bare graphite disk electrode (dia. 3 mm).

above based on the contact angle and wetted area and also proved experimentally on oxygen reduction suggest that the diffusion of oxygen is a rate-limiting step. Although diffusion of glucose across the wetted area may also be a rate-limiting step however diffusion of glucose will continue until a steady state is reached. This might cause an increase in response time but not in magnitude of the amperometric response which is also supported by the experimental data discussed above. Additionally, the non-mediated detection of glucose requires the oxygen consumption to form hydrogen peroxide that can be electrochemically detected, the availability of oxygen at the site of enzymatic reaction is a rate-determining step. The results on amperometric responses recorded in Fig.2 support this conclusion. However, if there is a mediated electron exchange, by replacing oxygen using some mediator of well-defined electrochemistry, the dependence of oxygen concentration as a rate-limiting step may be eliminated. Accordingly, the construction of a mediated enzyme electrode which has been commercially implemented using ferrocene modified glucose oxidase[29] is of great interest. The biosensors based on ormosil requires the immobilization of glucose oxidase and mediator together within the sol-gel glass matrix. We tried to incorporate dimethyl ferrocene together with glucose oxidase within the sol-gel glass but we could not observe the mediated electrochemical signal. These findings indicate that the ferrocene molecule immobilized within the sol-gel glass does not allow the reaction between immobilized glucose oxidase and immobilized mediator which is related to the restricted mobility of the mediator as well as enzyme within the sol-gel matrix. In the present case we used soluble ferrocene to investigate the mediated response from the glucose oxidase which diffuses within the reactive surface area of sol-gel matrix and indeed we observed remarkable results. Actually the configuration of the biosensors based on the present approach does not fulfill the requirement of the mediated glucose biosensor, however we have got some very interesting observations on the mediated bioelectrochemistry using soluble ferrocene and glucose oxidase encapsulated within the ormosil. We studied the electrochemistry of ferrocene monocarboxyic acid at the surface of ormosil based glucose sensor. Fig. 3 shows the electrochemistry of soluble ferrocene monocarboxylic at different scan rates (3, 6, 10, 20, 50 & 100 mV/s) between –0.2-0.6 V vs. Ag/AgCl. The voltammograms of these systems were recorded in the same ferrocene solution. The inset in figures shows the plot of anodic peak current vs. square root of scan rates. The results shown in inset follow the same trends as described by Wang et al[30] that show a linear relation between peak current and square root of scan rates although the linear line does not pass through origin which shows that the system is not very well diffusion controlled. The peak separation increases with scan rates. Wang et al[30] has also reported the peak separation of 57 to 65 mV in two sol-gel systems and linear lines obtained from the plots of peak current vs square root of scan rate also do not pass through origin. The voltammograms shows reversible electrochemistry of ferrocene in each system with diffusion limited conditions in each case as evidenced from the results shown in insets. Subsequently we examined the mediated electrochemical response of glucose oxidase encapsulated ormosil in the presence and absence of 150 mM glucose. The cyclic voltammograms of CCEs in aqueous solution of ferrocene monocarboxylic acid between –0.2 to 0.6 V vs. Ag/AgCl at the scan rate of 5 mV/s are shown in Fig. 4 . Curve 1 of the voltammogram shows the

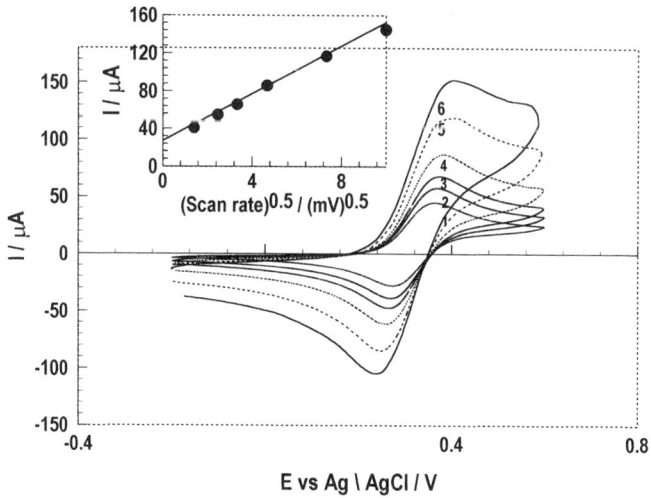

Figure 3. Cyclic voltammograms of 5 mM ferrocene monocarboxylic acid in 0.1 M phosphate buffer pH 7.1 on glucose oxidase encapsulated ormosil at 25°C and at the scan rate of 3, 6, 10, 20, 50 & 100 mV/s. The inset shows the plot of anodic peak current vs. Square root of scan rate.

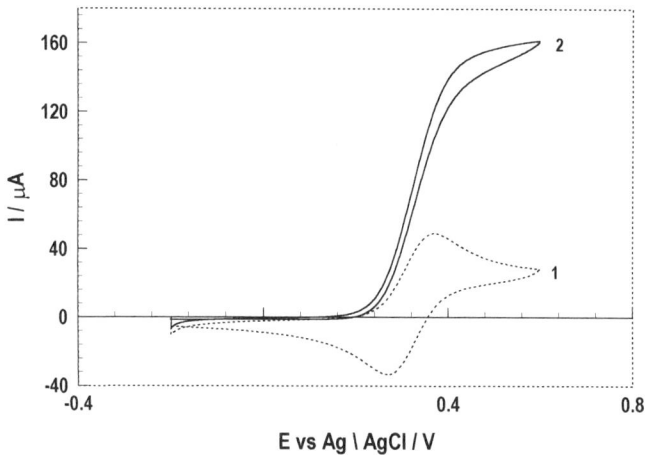

Figure 4. Cyclic voltammogram of glucose oxidase encapsulated ormosil in absence (1) and the presence (2) of 150 mM glucose acid in 0.1 M phosphate buffer pH 7.1 containing 5 mM ferrocene monocarboxylic acid at the scan rate of 5 mV/s.

recording in absence of glucose whereas curve 2 shows the recording in the presence of glucose. There is large increase in anodic current on the addition of glucose in each system showing the mediated electron exchange from the immobilized glucose oxidase. In order to have deeper insight of mediated electrochemical reaction between soluble ferrocene and immobilized glucose oxidase; we studied the amperometric responses of glucose oxidase encapsulated ormosil in the presence of constant concentration of soluble ferrocene monocarboxylic acid. The typical amperometric response curve glucose oxidase encapsulated ormosil at the constant potential of 0.35 V vs. Ag/AgCl shows the sensitivity of 2.5 µA/mM for glucose analysis with wide linearity based on mediated bioelectrochemistry. Thus incorporation of polyethylene glycol and graphite powder significantly affect the morphology of ormosil based biosensor that is related to the introduction of hydrophilic modifiers in order to control the thickness of the wetted section of the electrode (reaction layer thickness). Additionally the incorporation of graphite particles not only increases the wetted surface area of the electrode, it also facilitate the electron transfer within the sol-gel matrix as a result of increased electronic conductivity of the electrode.

The stability of ormosil based biosensors is determined under two conditions. In first case the CCE was stored in 0.1 M phosphate buffer pH 7.0 at 4°C whereas in second case the enzyme electrode was stored at room temperature in the same buffer. The stability of the ormosil based glucose sensor stored under first condition is relatively better, as compared to that of earlier sol-gel glass based glucose sensors under similar conditions, without loss of the amperometric response after 2 months. Under the second storage condition the response is consistent without loss for 20 days. The reproducibility of the sensor design was investigated using the ten sets of symmetrical teflon electrode body and using constant amount of sol-gel precursors and dopants. In each case 16 µl of homogenous precursors solution was added in to the well of electrode body with a recessed depth of 2 mm. This leads the production of ~0.3 mm thickness of sol-gel layer under 24 h drying at (25°C). Although the nature of individual events is random and the geometry and pore-size distribution of the product gel are difficult to determine, the nature of polymeric gel can be regulated to a certain extent by controlling the rates of the individual step[25].

Solid-State Ion Sensor Based on a Bilayer of Ferrocene-Linked Ormosil and Poly (viny chloride) Matrix Membrane. The physical status of new ferrocene-linked sol-gel glasses in the presence and absence of ferrocene were studied. A transparent smooth surface was obtained in the absence of ferrocene derivative whereas the color of ferrocene modified sol-gel glass was dark brown with smooth surface. The surface structures of the ferrocene-linked ormosil was studied from the scanning electron microscopy. Fig.1c shows the SEM of sol-gel glass linked to ferrocene. The gelation network is symmetrical with relatively better distribution of ferrocene through out the gel (Fig.1c). The ormosil made using these silanes are biocompatible and are good material for the immobilization of enzyme as proved by the amperometric response of glucose oxidase encapsulated ormosil. The cyclic voltammogramms of the ferrocene-linked ormosil shows diminished electrochemistry of ferrocene associated to restricted degree of translational motion within ormosil. The cyclic voltammograms recorded in between -0.2 - 0.6 V vs. Ag/AgCl do not show reversible electrochemistry as obtained using soluble ferrocene in aqueous solution. At lower frequency it shows capacitive CV whereas at higher frequencies it show redox behavior suggesting the presence of well behaved phase polymer.The open circuit potential of the ferrocene modified electrode was monitored using the following electrochemical cell :

Ferrocene modified sol-gel electrode / 0.01 M Tris-HCl buffer, pH 7.0/NH$_4$O$_3$ /SCE

The open circuit potential difference (ΔE) of the above cell was found to be -30 mV. This potential difference is highly reproducible for continuous measurement of the cell for two months which shows the non-polarizable behavior of the modified sol-gel electrode to be exploited in the construction of solid-state potentiometric ion-sensor.

The typical potentiometric response of the ion-sensor on the addition of varying concentrations of K$^+$ as KCl is studied. The response is very fast as compared to the response obtained with same neutral carrier assembled over polyindole modified electrode[14] with better detection limit (5.0 x 10^{-6} M). The order of selectivity is found to be K$^+$> Na$^+$> NH$_4$$^+$. This trend is similar as reported earlier for the same neutral carrier[14]. Fig. 5 shows the calibration curves for the analysis of K$^+$; Na$^+$ and NH$_4$$^+$· The performance of the sensor is better in all the respects i.e. response time, detection limit, slope and linearity.

The mechanism for yielding Nernstian response with reproducible potential difference across modified sol-gel glass/ metal surface is mainly attributed to the presence of ferrocene which generate into a stable asymmetric cell configuration (ion-in and electron out). The stability of ferrocene linked sol-gel glass provide reproducible response of the new ion-sensor for > 3 months.

A Novel Ferrocene Encapsulated Pd-Linked Ormosil. The electrochemistry of ferrocene-linked ormosil as described above shows capacitive CVs mainly due to restricted translational degree of freedom within the ormosil network. Subsequently we developed another ferrocene encapsulated ormosil using different ormosil precursors i.e. 3-glycidoxipropyltrimethoxy silane (hydrophilic silane) and trimethoxy silane(hydrophobic silane) together with other common ingrdient of gelation and ferrocene monocarboxalic acid. The electrochemistry of ferrocene encapsulated in the resulting ormosil is relatively better however still the CVs are not reversible and has no capability of mediated electrochemistry. The physical structure of this ormosil in the presence and absence of ferrocene monocarboxalic acid is shown in Fig. 1d and 1e respectively and indeed much ordered polymeric domains are recorded in both cases. In order to facilitate the reversibility of encapsulated ferrocene we modified one of the ormosil precursor (3-glycidoxypropyltrimethoxy silane) with Palladium chloride. On the

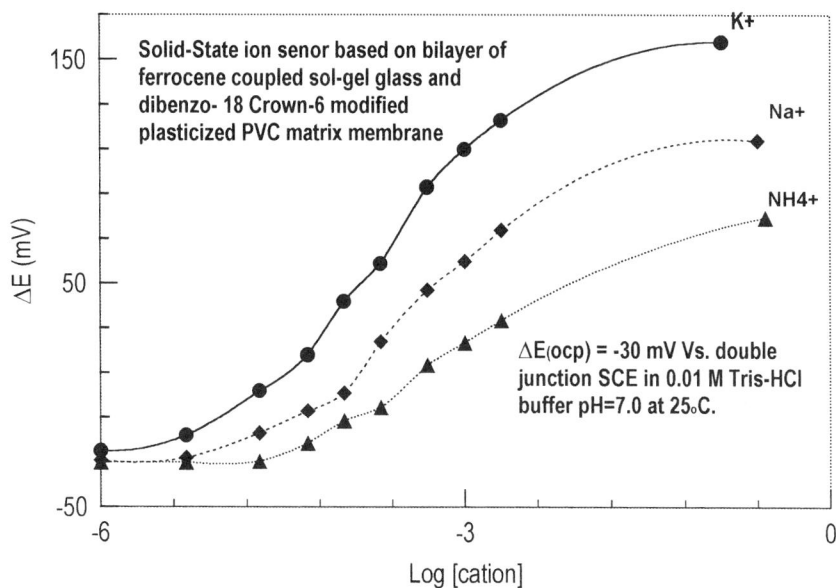

Figure 5. Calibration curves for the analysis of K^+, Na^+ and NH_4^+ ions using ion sensor made from bilayer of ferrocene-linked ormosil and dibenzo-18-crown-6 incorporated plasticized poly (vinyl chloride) matrix membrane.

addition of aqueous solution of Palladium chloride, a black solution is obtained. The Palladium is linked to two molecule of glycidoxypropyltrimethoxy silane. When this palladium-linked ormosil precursor is mixed with trimethoxy silane, ferrocene monocarboxalic acid and HCl, a novel ferrocene encapsulated Pd-linked ormosil is obtained. The electrochemistry of this novel material is studied and the cyclic voltammetry of the same is shown in Fig.6 at different sweep rates. The cyclic voltammetric data is nice with a peak separation of 57-59 mV and the electrode material seems promising. The inset to Fig.6 shows the relation of peak current vs. square root of scan rate. A linear relation passing throgh origin evidenced the novelty of the ferrocene encapsulated Pd-Ormosil material.

Subsequently we studied the mediated bioelectrochemistry of NADH at the surface of ferrocene encapsulated Pd-ormosil. Fig. 7 shows the typical mediated NADH oxidation at 0.35 V vs. Ag/AgCl on addition of increasing concentration of NADH. The inset to Fig.7 shows the calibration curve for NADH analysis. Further research work on this material is ongoing and will be published subsequently.

Conclusion. The new ormosil derived from optimized two silane out of four different typed of alkoxysilane precursors are reported particularly for the constuction of non-mediated glucose biosensor, a solid-state ion sensor and mediated biosensor and the new material are promising for the construction of other sensors/biosensor.

Acknowledgment . The author is thankful to UGC, New Delhi for financial assistance.

References
1. Tsionsky, M., Gun, J., Glezer, V. Lev, O. *Anal. Chem.,* **1994**, 66, 1747.
2. Gun, J., Tsionsky, M., and Lev, O., *Anal. Chim. Acta.* **1994**, 294, 351.
3. Gun, J., Rabinovich, L.,Tsionsky, M., Golan, Y., Rubinstein, I., Lev, O., *J. Electroanal. Chem.* **1995**, 395, 57.
4. Sampath, S., and Lev, O., *Anal. Chem.* **1996**, *68*, 2015.
5. Lev, O., Tsionsky, M., Ravinovich, L., Glezer, V., Sampath, S., Pankratov, L., and Gun. J., *Anal. Chem.* **1995**, 67,22A.
6. Schmidt, H., *Sol-Gel Sci. Technol.* **1994**,*1,217*.
7. Philipp, G., and Schmidt, H., *J. Non-Cryst. Solids.* **1984**,63 ,283.
8. Momma, T., Komaba, S., Yamamoto. M., Osaka, T., and Yamauch. S., *Sensors and Actuators, 1995,* B 24-25, 724-728.
9. Nikolskii, B. P., and Materova, E. A., *Ion Select. Electrode Rev,.***1995,** 7, 3-39.
10. Cadogan, A., Gao, Z., Lewenstam, A., Ivaska, A., and Diamond, D., *Anal. Chem.,* **1992**, *64,* 2496.
11. Campanella, L., Salvi, A. M., Sammartino, M. P., and Tomassetti, M., *Chim. Ind. (Milan),* **1986,** *68,* 71.

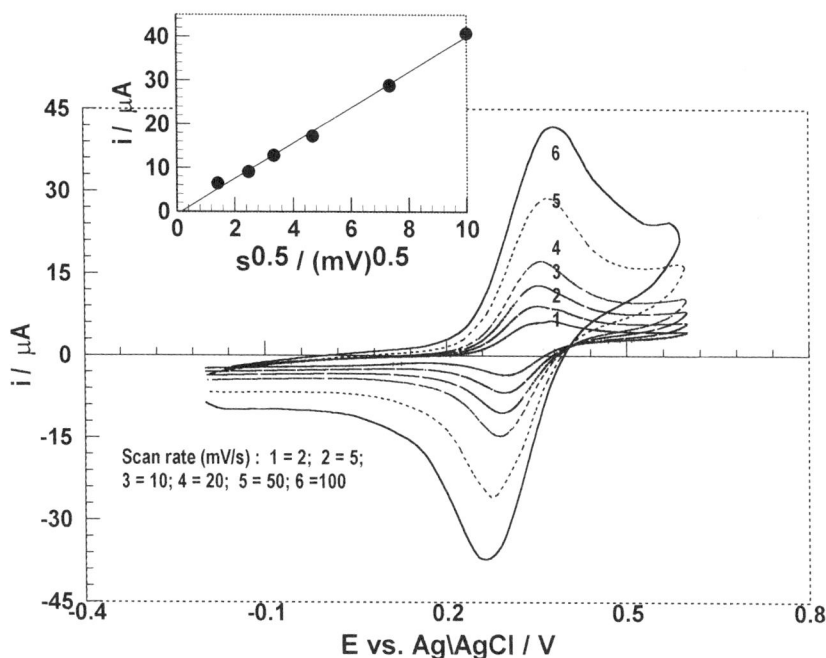

Figure 6. Cyclic voltammetry of ferrocene encapsulated Pd-ormosil in 0.1 M phosphate buffer pH 7.0 at the scan rate of 3, 6, 10, 20 and 100 mV/s. The inset shows the plot of peak current vs. square root of scan rate. Fig. 7. A typical amperometric response of ferrocene encapsulated Pd-ormosil electrode on the addition of varying concentration of NADH in 0.1 M phosphate buffer pH 7.0. The inset shows the calibration curve for NADH analysis based on mediated bioelectrochemistry.

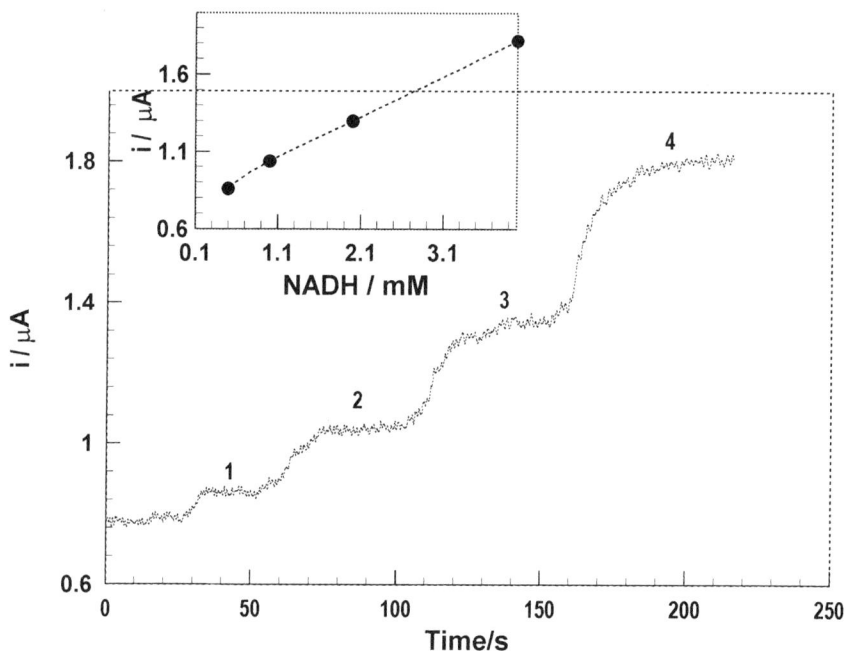

Figure 7. A typical amperometric response of ferrocene encapsulated Pd-ormosil electrode on the addition of varying concentration of NADH in 0.1 M phosphate buffer pH 7.0. The inset shows the calibration curve for NADH analysis based on mediated bioelectrochemistry.

12. Campanella, L., Mazzei, F., Morgia, C., Sammartinon, M. P., Tomassetti, M., Baroncelli, V., Battilotti, M., Colapicchioni, C., Giannini, I., and Porcelli, F., *Analusisim,* **1988**.*16* , 120.

13. Osaka, T., Fukuda, T., Kanagawa, H., Momma, T., Yamauchi, S., Komaba, M., and Yamamoto, M., *Sensors and Actuators,***1993***, B 24-25*, 724-728.

14. Pandey, P. C., and Prakash, R., *Sensors and Actuators.* **1998**, *46,* 61-65.

15. Pandey, P. C. and Prakash, R., *J. Electrochem. Soc..* **1998**, 145, 4103-4107.

16. Pankratov, B., and Lev, O., *J. Electroanal. Chem.,*.**1993**, 393, 35

17. Lev, O., Wv, Z., Bharathi, S., Glezer, V., Modestov, A., Gun, J., Ravinovich, L., and Sampath, S.,. *Chem. Mater.* **1997**,9(11),2354-2375.

18. Gun , J., and Lev. O., *Anal .Letters.* **1996**, *29,* 1933-1938. *Electroanal. Chem..,* **1994**, *372 ,* 275-278.

19. Audebert, P., Cerveau, G., Corriv, R. J. P., and Costa, N., *J. Electroanal. Chem.,* **1994**, *372,* 275

20. Audebert, P., Calas, P., Cerveau, G., Corriv, R. J. P., and Costa, N., *J. Electroanal. Chem.,* **1996**, *413, 89.*

21. Collinson, M. M., Rausch, C. G., and Voigt, A., *Langmuir,* **1997**, *13 ,* 7245-7251.

22. Chut, S. L., Li, J., and Tan, S., *Analyst..,* **1997**, *122 ,* 1431-1434.

23. Narang, U., Prasad, P. N., Bright, F. V., Ramanathan, K., Kumar, N. D., Malhotra, B. D., and Chandra, S., *Anal. Chem.* **1994**, *66*, 3139.

24. Sampath, S., and Lev, O., *Electroanalysis.* **1996**, *8*, 1112.

25. Dave, B. C., Dunn, B., Valentine, J. S. and Zink, J. I., *Anal.Chem.* **1994**, 66,1120.

26. Ingersoll, C. M. and Bright, F.V., *J. Sol-Gel Sci. Technol.*,**1998**,11,169.

27. Lowry, J. P., Boutelle, M. G., O'Neill, R. D., and Fillenz, M., *Analyst* **1996**,121, 761.

28. Zimmerman, B., and Wightman, R. M., *Anal. Chem.* **1991**,63,24.

29. Degani, Y., and Heller, A., *J.Am.Chem.Soc.***1988**,110,2615.

30. Wang, J., and Pamidi, P. V. A., *Anal. Chem.***1997**,69, 4490.

Chapter 11

Electrochemical Regeneration of Immobilized NADP+ on Alginic Acid with Polymerized Mediator

Shin-ichiro Suye, Hideo Okada, Makoto Nakamura, and Mikio Sakakibara

Department of Applied Chemistry and Biotechnology, Fukui University, 3–9–1, Bunkyo, Fukui 910–8507, Japan

Polymerized NADP+ (Alg-NADP+) was prepared and its application for electrochemical bioreactor was investigated. NADP+ has been covalently immobilized to carboxyl group of alginic acid using water soluble carbodiimide (EDC). Absorbance at 260 nm of Alg-NADP+s showed that 90% of carboxyl groups of alginic acid were coupled with NADP+. The coenzyme activity of immobilized NADP+ has reached 90 to 95% on each Alg-NADP+. A cathodic peak in the cyclic voltammogram of Alg-NADP+ appeared at -1.2 V (*vs.* SCE.) corresponding to the reduction wave of free NADP+. Polymerized vilogens (Alg-V, poly-(L-Lys)-V) were also prepared in a similar manner. The anodic wave of NADP dimer was not observed in the presence of methyl viologen or polymerized viologen derivatives and ferredoxin NADP+ reductase (FRD). Glutathione reductase was used as the catalyst for production of reduced form of glutathione from oxidized form of it, polymerized NADP+, polymerized viologen and FRD were used for the electrochemical regeneration of NADPH. The conjugated redox reaction was successful with Alg-NADP+ and poly-(L-Lys)-V. Under given conditions, the conversion ratio of GSH from GSSG reached 100% after 1 h of incubation at 37°C and the concentration of GSH accumulated was 4.0 mM of reaction mixture.

Introduction

Various useful compounds have been produced by the coupling of NAD(P)-linked dehydrogenase and the electrochemical reaction for regeneration of reduced form of pyridine nucleotide coenzyme, NAD(P)H (1-10). Conventionally, NAD(P)H has been produced by enzymatic, electrochemical, and photochemical methods. Electrochemical reduction of NAD(P)$^+$ has an advantage of not requiring specific substrate. In the initial studies, NAD(P)$^+$ was reduced electrochemically without mediator (11), however, inactive dimmers of NAD(P) are formed as by-products at high overpotential (-1.1 V $vs.$ SCE) (12, 13). A mediator for electrolytic regeneration of reduced form of NAD(P) is used to degrease overpotential and avoid useless consumption of substrate for the NAD(P)H regeneration, although an oxidereductase must be used to prevent dimerization of NAD(P) radical (14, 15). It is preferable to immobilize mediator, as well as coenzyme, on a polymer for separation of main products from these substances in a reaction mixture (16, 17). Biosensors based on NAD(P)-linked dehydrogenase have been investigated in recent years (18-25). It is also necessary to entrap coenzyme and mediator on an electrode surface for practical use. For this purpose many polymerized NAD(P)s (26-31) and redox polymers (20, 22, 32-36) as mediator were prepared and applied to bioelctrochemical reactor and biosensing system.

In our previous paper, the attempt to prepare polymerized NAD$^+$ was successful by means of coupling of amino groups of NADP$^+$ and carboxyl groups of alginic acid with water soluble carbodiimide (31). NAD$^+$ immobilized on alginic acid could be reduced to the normal NADH electrochemically.

In this study, polymerized NADP$^+$ and polymerized viologen derivatives were also prepared in a similar manner. We investigated their applications to conjugated redox reaction coupled with NAD(P)-linked enzyme reaction and with electrochemical coenzyme regeneration reaction.

Materials and Methods

Chemicals and apparatus

NADP$^+$ was obtained from Kojin Co. Ltd. (Tokyo, Japan). Sodium alginate (MW 20,000) and poly-(L-lysine) hydrochloride (MW 5,000-15,000) were purchased from Nacalai Chemical Co. (Kyoto, Japan). Glutathione reductase (GR, EC 1.6.4.2, from yeast), oxidized form of glutathione (GSSG), and reduced form of glutathione (GSH) were obtained from Oriental Co. Ltd., (Tokyo, Japan). Ferredoxine-NADP$^+$ reductase (FRD, EC 1.18.1.2, from spinach leaves) was purchased from Sigma Co. (MO, USA). All other reagents and compounds were analytical grade. Hokuto Denko potensiostatt (HA-301, Tokyo, Japan) and function generator (HB-104) with an X-Y recorder (Nippon Denshi Kagaku Co., model U-

335, Tokyo, Japan) were used to record voltammograms using a three electrodes system.

Electrochemical Measurement

A basal pyrolytic graphite (BPG)-electrode (3 mm ϕ) was obtained from Megachem Co.(Tokyo, Japan). A cell combined with a conventional three-electrodes system of a working electrode (BPG), a reference electrode (saturated calomel electrode, SCE), and counter electrode (Pt wire electrode) were used. Electrode potential values will be recorded against SCE. To remove oxygen from the solution in the cell, the solution was bubbled with high-purity nitrogen gas.

Preparation of Polymerized NADP+ and Polymerized Mediators

Polymerized NADP+ (Alg-NADP+) was prepared according to our previous paper (31). A 13.8 mg sodium alginate and 70 mmol 1-ethyl-3-(3-dimethylaminopropyl)carbodiimide hydrochloride (EDC) were dissolved in 15 ml water. The final pH was adjusted to 4.7 and the solution was stirred for 40 min at room temperature. Then 70 mmol NADP+ was added to the solution. After readjustment of pH to 4.7, resulting solution was stirred for 12 h at room temperature. The reaction mixture was dialyzed with 10 mM Tris-HCl buffer (pH 7.0) for 12 h and then dialyzed with water for 12 h at 5°C. Alg-NADP+ was recovered lyophilization. Lyophilized products were stored at -5°C in the dark.

Polymerized viologen compounds were synthesized as follows; 1-Methyl-1'-bromobutyl-4-,4'-bipyridinum iodide bromide (BrC$_4$V) was synthesized from 4,4-bipyridine (37). BrC$_4$V was attached on amino groups of poly(oxyethylene)diamine modified alginic acid or poly-(L-lysine) covalently.

Coenzyme Activity

The coenzyme activity of Alg-NADP+ was determined enzymatically by the glutathione reductase system (38). The assay mixture consisted of 250 μmol Tris-HCl buffer (pH 8.0), 2.7 μmol EDTA, 13.2 μmol GSSG, an appropriate amount of Alg-NADP+, and 20 units of GR in a total volume of 2.87 ml. The reaction was started by the addition of the enzyme and the reaction mixture was incubated at 30°C. Absorbance at 340 nm was measured with a double beam spectrophotometer. The reference contained all components except for Alg-NADP+. In the calculation of the amount of Alg-NADPH produced, a molar absorption coefficient for NADPH of 6.22 x 10^3 l·mol^{-1}·cm^{-1} was used.

Driving of the Conjugate on of compartment I, and desired amounts of MV was added to the buffer solution of compartment II. The cells were immersed in a water bath at 37 °C and continuously purged with nitrogen gas. The objective oxidoreductase reaction was conjugated with the electrochemical regeneration of NADPH at a constant potential, -0.9 V *vs*. SCE. The concentration of GSH produced in the reaction mixture was determined by the Ellman method (39).

Results and Discussions

Polymerized NADP⁺

NADP⁺ has been covalently immobilized to the carboxyl groups on alginic acid using EDC at room temperature for 12 h. Alg-NADP⁺ produced was recovered in a powdered form by the procedure described in Materials and Methods. The UV spectrum of Alg-NADP⁺ was similar to that of free NADP⁺. Absorbance at 260 nm of Alg-NADP⁺s showed that 90% of carboxyl groups on D-mannuronic acid residue of alginic acid were attached to NADP⁺. Increasing amount of NADP⁺ gave higher NADP⁺ density. This result suggests that the degree of NADP⁺ density of Alg-NADP⁺ depends on the content of NADP⁺ in reaction mixture. The coenzyme activity of immobilized NADP⁺ has reached 90 - 95 % on each Alg-NADP⁺ from the coenzyme test using glutathione reductase. Alg-NADP⁺ could be stored at -5 °C without any loss of coenzyme activity.

Electrochemical Behaviors of Alg-NADP⁺ and Polymerized Mediators

Electrochemical behavior of Alg-NADP⁺ was investigated using a BPG electrode. The potential scan range was between +1.0 V and +1.7 V *vs*. SCE with a rate of 50 mV·s⁻¹. A cathodic peak at -1.2 V in the cyclic voltammogram of Alg-NADP⁺ appeared corresponding to the reduction wave of free NADP⁺. On this polymer, however, the anodic wave of NADP dimer (at +0.7 V) which is inactive for enzymatic reactions was also observed. The cyclic voltammogram of the mixture of Alg-NADP⁺ (eq. 2 mM free NADP⁺) and 2 mM MV shows two pairs of cathodic-anodic wave assignable to MV. Height of both cathodic peaks at -0.72 and -1.15 V increased with the addition of FRD and an anodic peak ca. +0.8 V appeared and the anodic wave of NADP dimer was not observed. These indicate that NADP⁺ moiety on the polymer could be regenerated to the reduced form electrocatalytically. Polymerized viologens (Alg-V, Poly-(L-Lys)-V) have also similar CVgrams of MV.

Enzymatic Reaction Conjugated with Electrochemical Regeneration of Coenzyme

A conjugated reaction of electrochemical reduction of Alg-NADP⁺ and GSH

Table I. Conversions of redox reactions with various coenzymes and mediators

Mediator	Coenzyme/Enzyme	Conversion (%)
Alg-V	NAD^+/DI	7.5
MV	$NADP^+/FRD$	51.4
Alg-V	$NADP^+/FRD$	100
MV	$Alg-NADP^+/FRD$	100
Alg-V	$Alg-NADP^+/FRD$	3.5
Poly-(L-Lys)-V	$Alg-NAD^+/DI$	2.5
Poly-(L-Lys)-V	$Alg-NADP^+/FRD$	100

production was carried out by using FRD and MV for more accurate driving of the electrochemical NADPH regenerating reaction and GR (Scheme 1). The result was shown in Fig. 1. GSSG was completely converted to GSH after 1.0 h. Table I shows the effects of combination of Alg-NADP$^+$ and polymerized viologen on GSH production. The efficiently proceedings of conjugated reaction was observed in the case of the combination of NADP$^+$ and Alg-V, Alg-NADP$^+$ and Poly-(L-Lys)-V, and MV and Alg-NADP$^+$.

On the other hand, when the combination of Alg-NADP$^+$ and Alg-V was used for the conjugated reaction, the conjugated reactions were not proceeded, GSH was hardly produced for long time. Taking into account these data, the interaction between the polymerized mediator and polymerized NADP and the effects on conjugated reaction could be explained as follows. The ionic charge effects of support polymer for immobilization of mediator and coenzyme might be important for conjugate reaction. The viologen on the positive charged polymer was attached toward the NADP$^+$ on the negative charged polymer, so that the conjugated reaction could proceed effectively. Smooth progress of the electron transfer should be necessary for coenzyme regeneration.

In conclusion, we showed that alginic acid on which NADP$^+$ was immobilized and polymerized viologen can be used as an electrochemical regeneration. Recently, many types of biosensing systems with NAD(P)-linked dehydrogenase have been designed and developed for environmental analysis. Parellada et. al., used glucose dehydrogenase together with phosphorylase A, phosphoglucomutase for the determination of phosphate (40). Lactate dehydrogenase immobilized electrode used for detection of heavy metals (41). This biosensor is based on the effects of heavy metal salts on the catalytic activity of lactate dehydrogenase. In these methods coenzyme and mediator could not be immobilized on a electrode. In addition, some special NAD(P)-linked dehydrogenases, such as allylic alcohol dehydrogenase (42), 2, 3-dioxygene reductase (43) have been found. These enzymes also might be useful in a biosensing of environmental materials, as well as an enzymatic assay.

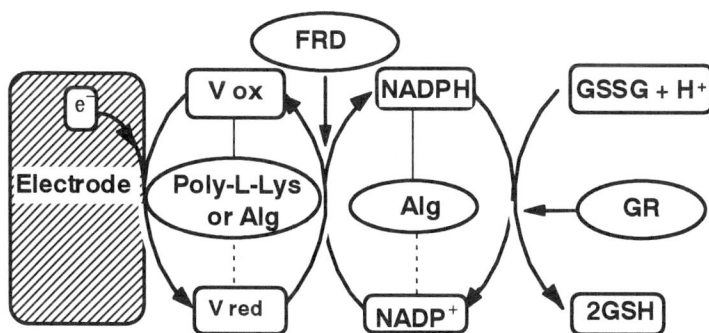

Scheme 1. Enzymatic production of GSH from GSSG with electrochemical regeneration of NADPH.

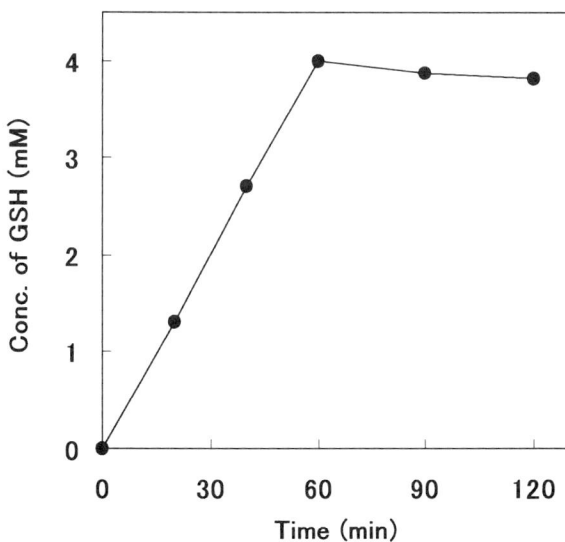

Fig. 1. Conjugate redox reaction driven by the electrochemical coenzyme regeneration on the carbon fabric electrode at -0.9 V.

Consequently, Alg-NAD(P)$^+$ and polymerized viologen can be applicable to the construction of various type of biosensors using NADP-linked dehydrogenase.

References

1. Davies, P.; Mosbach, K. *Biochim. Biophys. Acta* **1974**, *370*, 329-338.
2. Kelly, R. M.; Kirwin, D. J. *Biotechnol. Bioeng.* **1977**, *19*, 1215-1218.
3. Jaegfeldt, H.; Torstensson, A.; Johnsson, G. *Anal. Chim. Acta* **1978**, *97*, 221-228.
4. Mansson, M.-O.; Lasson, P.-O.; Mosbach, K. *FEBS Lett.* **1979**, *98*, 309-313.
5. Vandecasteele, J.-P. *Appl. Environ. Microbiol.* **1980**, *39*, 327-334.
6. Dicosimo, R.; Wong, C.-H.; Daniels, L.; Whitesides, G. M. *J. Org. Chem.* **1981**, *46*, 4622-4623.
7. Wandrey, C.; Fiolitakis, E.; Wichmann, U. *Ann. N. Y. Acad. Sci.* **1984**, *434*, 91-94.
8. Maeda, H.; Kajiwara, S. *Biotech. Bioeng.* **1985**, *27*, 596-602.
9. Chenault, H. K.; Whitesides, G. M. *Appl. Biochem. Biotechnol.* **1987**, *14*, 147-197.
10. Matsue, T.; Yamada, H.; Chang, H.; Uchida, I.; Nagata, K.; Tomita, K. *Biochim. Biophys. Acta* **1990**, *1038*, 29-38.
11. Jaegfeldt, H. *J. Electroanal. Chem.* **1981**, *128*, 355-370.
12. Schmakel, C. O.; Santhanam, K. S.; Elving, P. J. *J. Am. Chem. Soc.* **1974**, *97*, 5083-5092.
13. Jaegfeldt, H. *Bioelectrochem. Bioenerg.* **1981**, *8*, 355-370.
14. Burnett, R. W.; Underwood, A. L. *Biochemistry* **1968**, *7*, 3328-3333.
15. Jansen, M. A.; Elving, P. J. *Biochim.Biophys. Acta* **1984**, *764*, 310-315.
16. Nakamura, Y.; Itoh, S.; Suye, S. *Enzyme Microb. Technol.* **1994**, *16*, 1026-1030.
17. Osa, T.; Kashiwagi, Y.; Yanagisawa, Y. *Chem. Lett.* **1994**, 367-376.
18. Miki, K.; Ikeda, T.; Todoriki, S.; Senda, M. *Anal. Sci.* **1989**, *5*, 269-274.
19. Dempsey, E.; Wang, J.; Wollenberger U.; Ozsoz, M. *Biosens. Bioelectron.* **1992**, *7*, 323-327.
20. Persson, B.; Lan, H. L.; Gorton, L.; Okamoto, Y.; Hale, P. D.; Boguslavsky, L. I.;. Skotheim, T. *Biosens. Bioelectron.* **1993**, *8*, 81-88.
21. Whillner, I.; Riklin, A. *Anal. Chem.* **1994**, *66*, 1535-1539.
22. Shu, H.-C.; Mattiasson, B.; Persson, B.; Nagy, G.; Gorton, L.; Sahni, S.; Geng, L.; Boguslavsky, L.; Skotheim, T. *Biotechnol. Bioeng.* **1995**, *46*, 270-279.
23. Lobo, M. J.; Miranda A. J.; Tunon, P. *Electroanalysis* **1997**, *9*, 191-202.
24. Katakis,; Dominguez, E. *Mikrochim. Acta* **1997**, *126*, 11-32.
25. Huang, T.; Warsinke, A.; Kuwana, T.; Scheller, F. W. *Anal. Chem.* **1998**, *70*, 991-997.
26. Wykes, J.R.; Dunnill, P.; Lilly, M. D. *Biochim. Biophys. Acta* **1972**, *286*, 260-268.

165

27. Aizawa, M.; Coughlin, R. W.; Charles, M. *Biotechnol. Bioeng.* **1976**, *18*, 209-215.
28. Furukawa, S.; Katayama, N.; Iizuka, T.; Urabe, I.; Okada, H. *FEBS Lett.* **1980**, *121*, 239-242.
29. Yamazaki, Y.; Maeda, H. *Agric. Biol. Chem.* **1981**, *45*, 2277-2288.
30. Riva, S.; Carrea, G.; Veronese, F. M.; Bueckmann, A. F. *Enzyme. Microb. Technol.* **1986**, *9*, 556-560.
31. Nakamura, Y.; Suye, S.; Tera, H.; Kira, J.; Senda, M. *Biochim. Biophys. Acta* **1996**, *1289*, 221-225.
32. Simon, M. S.; Moore, P. T. *J. Polym. Sci. Polym. Chem. Ed.* **1975**, *13*, 1-16.
33. Oyama, N.; Ohsaka, T.; Yamamoto, H.; Kaneko, M. *J. Phys. Chem.* **1986**, *90*, 3850-3856.
34. Endo, T.; Takada, T.; Kameyama, A.; Okawara, M. *J. Polym. Sci. A, Polym. Chem.*, **1991**, *29*, 135-137.
35. Hoogvliet, J. C.; van Os, P. J. H. J.; van der Mark, E. J.; van Bennekom, W. P. *Biosens. Bioelectron.* **1991**, *6*, 413-423.
36. Schuhmann, W. *Biosens. Bioelectron.* **1993**, *8*, 191-196.
37. Nakamura, Y.; Cheng, J.-Y.; Tabata, I.; Suye, S.; Senda, M. *DENKI KAGAKU* **1994**, *62*, 1235-1237.
38. Suye, S.; Yokoyama, S. *Enzyme Microb. Technol.* **1985**, *7*, 418-424.
39. Ellman, G. L. *Arch. Biochem. Biophys.* **1959**, *32*, 70-77.
40. Parellada, J.; Narvaec, A.; Lopez, M. A.; Dominguez, E.; Fernandez, J. J.; Pavlov, V.; Katakis, I. *Anal. Chim. Acta* **1998**, *362*, 47-57.
41. Fennouh, S.; Casimiri, V.; Gelso-Meyer, A.; Burstein, C. *Biosens. Bioelectron.* **1998**, *13*, 903-909.
42. Malone, V. F.; Chastain, A. J.; Ohlsson, J. T.; Poneleit, L. S.; Nemecek-Marshall, M.; Fall, R. *Appl. Environ. Microbiol.* **1999**, *65*, 2622-2630.
43. Brosdus, R. M.; Haddock, J. D. *Arch. Microbiol.* **1998**, *170*, 106-112.

MICROBIAL-BASED SENSORS

Chapter 12

A Panel of Bioluminescent Biosensors for Characterization of Chemically-Induced Bacterial Stress Responses

T. K. Van Dyk[1,2], D. R. Smulski[1], D. A. Elsemore[1], R. A. LaRossa[1], and R. W. Morgan[2]

[1]Central Research and Development Department, DuPont Company, P.O. Box 80173, Willmington, DE 19880-0173
[2]Department of Animal and Food Sciences, University of Delaware, Newark, DE 19717-1303

The pattern of stress responses induced by chemical treatment can characterize types of damage sustained by the cell. Such analyses, which supply insight into the molecular mechanisms of toxicity, may be applied to environmental monitoring. A panel of six whole cell biosensors containing selected stress-responsive *Escherichia coli* promoters fused to the *Photorhabdus luminescens luxCDABE* reporter was assembled. These fusions were: *recA-lux* (in the SOS regulon), *grpE-lux* (in the heat shock regulon), *katG-lux* (in the OxyR regulon), *inaA-lux* (internal acidification responsive and in the Sox and Mar regulons), *yciG-lux* (in the σ^S-dependent stress response regulon) and *o513-lux* (a σ^S-independent, stationary phase inducible fusion). Two versions of this panel were constructed; one set of strains was in a *tolC*$^+$ background and the other harbored a mutation in *tolC* that inactivates an outer membrane channel for efflux pumps. The responses of this collection were found to be biologically appropriate when stressed by five chemicals with well-defined modes of action, H_2O_2, nalidixic acid, ethanol, sodium salicylate, and methyl viologen. Treatment of this panel with several potential animal feed contaminants, 2,4-dichlorophenoxyacetic acid, cadmium chloride, aluminum chloride, and aflatoxin B1 also yielded characteristic stress fingerprints.

Introduction

The responses of an organism to chemical insult can reveal molecular modes of toxicity because these responses typically include repair mechanisms specific for the damage incurred. Many such damage induced stress responses are common throughout biology. For example, the heat shock response, which results in elevated expression of proteins such as molecular chaperones and proteases, is induced in essentially all organisms by stresses that result in the accumulation of non-native proteins in the cell (1). DNA damage induced stress responses are also prevalent in biological systems. In E. coli, the SOS response, which results in increased levels of proteins specific for repair of DNA lesions, is induced by DNA damage (2). Consequently, induction of the SOS response in E. coli serves as a biological indicator of DNA damage caused by chemicals and other treatments. Accordingly, induction of the SOS response in E. coli correlates well with carcinogenicity in mammals (3). Thus, the bacterial heat shock, SOS and other stress responses may be good models to detect and discover modes of chemical toxicity.

Many damage inducible responses of microbial cells are regulated at the transcriptional level. As a result, an approach to rapidly gain insight into the cellular stresses sustained by chemical treatment is genetic fusions of reporter genes to stress-inducible promoters. In bacterial systems, a panel of strains utilizing the lacZ reporter has been demonstrated to yield specific responses to various chemical stresses (4). Another sensitive reporter of changes in transcription is that of bacterial bioluminescence encoded by luxCDABE operons. Use of the five gene lux reporter system allows facile monitoring of gene expression because all components necessary for light production are present in the cell. Panels of E. coli strains containing fusions of stress-responsive promoters to the Vibrio fischeri luxCDABE reporter have been described (5) and applied to characterizing the stresses induced by pollutant molecules (6), wastewater (7,8), antibacterial treatments (9,10), and antibiotics (11-13).

Here, a new panel of six stress-responsive E. coli promoters fused to the Photorhabdus luminescens luxCDABE is described. The P. luminescens lux operon encodes proteins that are more thermostable than those encoded by the V. fischeri lux operon (14). Thus, these biosensor strains were used at optimal temperatures for E. coli growth and, hence, optimal kinetics of stress response detection were possible. The six members of the panel were chosen to represent a range of stress responses such that different chemical treatments were expected to result in different patterns of induced gene expression. The recA gene is both a regulator and member of the SOS regulon (2). Thus, the transcriptional fusion of the recA promoter to the lux reporter responded to DNA damage. Likewise, because grpE is in the heat shock regulon controlled by σ^{32} (15), the grpE-lux fusion responded to stresses that induce this protein-damage responsive regulon. The katG gene is a member of the OxyR regulon (16) thus, the katG-lux fusion responded to oxidative damage mediated by hydrogen peroxide. A role for the stationary phase sigma factor, σ^S, has also been demonstrated in expression of katG in E. coli (17). The inaA gene is induced by treatments that result in internal acidification (18,19) and is controlled by the regulator of the multiple antibiotic stress response circuit, MarA, the regulator of the superoxide anion stress response circuit, SoxS, and by a protein related to both MarA and SoxS, Rob (20,21). Accordingly, the inaA-lux fusion responded to stresses that induced these regulatory circuits. Expression of the *yciG* gene is under control of σ^S (*21*), and thus, the *yciG-lux* fusion reported on the activation of the σ^S-dependent stress response. Finally, expression of a *lux* fusion to an otherwise uncharacterized open reading frame *o513* is induced by treatment with the herbicide sulfometuron methyl (*21*). The expression of this *lux* fusion was also observed to be highly induced as the culture aged, suggesting stationary phase regulation (T. K. Van Dyk, unpublished observations); however, expression is not controlled by σ^S (*21*).

To demonstrate the utility of this panel, the characteristic stress fingerprints generated by treatment with several toxic substances that may be found as contaminants of animal feed were determined. These included the commonly used herbicide, 2,4-dichlorophenoxyacetic acid (2,4-D),which is an EPA-regulated hazardous substance and if present in feed may result in accumulation of this molecule within the animal (*22*). Also tested were two metal salts that have adverse effects on animal performance when at sufficiently high concentrations in the feed; aluminum salts are toxic in the range of 500 to 2000 parts per million (ppm) in poultry feed, and cadmium salts are toxic in the range of 12 to 40 ppm for chickens and turkeys (*23*). Finally, aflatoxin B1, a particularly toxic mycotoxin that can be produced following fungal contamination of animal feed components, was tested. Aflatoxin B1 has adverse effects on poultry production when present at about 1 ppm in feed (*24-26*); the lethal dose for broiler chicks is 70 ppm in the feed (*27*).

Materials and Methods

Bacterial Whole Cell Biosensor Strains

Table I lists two versions of the panel of stress-responsive strains. One set carries a *tolC* mutation, rendering each strain hypersensitive to a variety of organic molecules; the other set is *tolC⁺*. As detailed in the following paragraphs, these strains contain either a chromosomal integrant of the *lux* fusion or a plasmid-borne *lux* fusion. Furthermore, two of the strains with plasmid-borne *lux* fusions carry a chromosomal mutation in the *pcnB* gene resulting in a reduced copy number of the plasmid. These modifications fine-tuned the basal level of bioluminescence from unstressed cells to the same general range.

E. coli strain DPD2227 contains a plasmid harboring the *E. coli katG* promoter fused to the *P. luminescens luxCDABE* in a host strain containing a *pcnB* mutation that resulted in reduced copy number of the plasmid. The plasmid in this strain, pDEW228 was constructed by digesting plasmid pKatGLux2 (*28*) with restriction endonuclease *Sal*I, filling in the ends with the Klenow fragment of DNA polymerase I, isolating the 650 basepair fragment containing the *E. coli katG* promoter region, and ligating it into pDEW201 (*29*) that had been digested with the restriction endonuclease *Sma*I. The orientation of the *katG* promoter driving expression of the *luxCDABE* operon was confirmed by digestion with the restriction endonuclease *Hin*dIII and by the H_2O_2-inducible bioluminescence increases in strains containing pDEW228. *E. coli* strain DPD2200 contained pDEW228 in host strain GC4468 (F-Δ*lac4169 rpsL*) (*30*). The *pcnB* mutation was introduced into this strain by P1*clr100* mediated transduction (*31*) to tetracycline resistance using phage lysates of *E. coli* strain MRi93 (*pcnB80, zad-2084*::Tn*10*, Δ(*argF-lac*)169, *flhD5301, fruA25, relA1, rpsL150* (*strR*), Δ*rbs-7, deoC1*) (*32*). The cotransduction of the *pcnB80* allele, yielding strain DPD2227, was detected by screening for reduced bioluminescence (about 20 fold).. *E. coli* strain DPD2238 was made by transduction of strain DPD2200 to tetracycline resistance with P1*clr100* phage grown on strain DE112 (*strR, galK2, lacΔ74 tolC*::miniTn*10*) (*33*) and a second transduction to kanamycin resistance with P1*clr100* phage grown on strain MM38K26 (Δ*pcnB*::*kan argG6 asnA31* or *asnB32 his-1 leuB6 metB1 pyrE gal-6 lacY1 xyl-7 supE44 bgl fhuA2 gyrA rpsL104 tsx-1 uhp*) (*34*). The presence of the *tolC* mutation in strain DPD2238 was scored by sensitivity to the bile salts in MacConkey agar (*35*); the presence of the *pcnB* mutation was confirmed by the approximately 20-fold reduced bioluminescence.

Table I. Stress-responsive *E. coli* Lux Fusion Strains

Stress Response	Regulatory Circuit	Promoter Fused to lux	tolC allele	Strain Name	Reference
Oxidative damage	OxyR & σ^S	*katG*	+	DPD2227	This work
			-	DPD2238	This work
Internal acidification etc.	Mar/Sox/ Rob	*inaA*	+	DPD2226	This work
			-	DPD2240	This work
DNA damage	SOS	*recA*	+	DPD1710	This work
			-	DPD2222	This work
Protein damage	Heatshock (σ^{32})	*grpE*	+	DPD3084	This work
			-	DPD2234	This work
"Super-stationary phase"	?	*o513*	+	DPD2173	(*21*)
			-	DPD2232	This work
Sigma S stress response	Stationary phase (σ^S)	*yciG*	+	DPD2161	(*21*)
			-	DPD2233	This work

E. coli strain DPD2226, carrying a plasmid-borne fusion of the *E. coli inaA* promoter to the *P. luminescens luxCDABE* in a host strain with a *pcnB* mutation, was constructed by transduction of strain DPD2146 (pDEW218 in host strain GC4468) (*21*) to tetracycline resistance with P1*clr100* phage grown on strain MRi93 and screening for cotransduction of the *pcnB80* mutation, as described above. *E. coli* strain DPD2240 was made by transduction of strain DPD2146 to tetracycline resistance with P1*clr100* phage grown on strain DE112 followed by transduction to kanamycin resistance with P1*clr100* phage grown on strain MM38K26. The *tolC* and *pncB* mutations were confirmed as described above.

E. coli strains DPD1710 and DPD3084 contain chromosomal insertions at the *lacZ* locus of the *E. coli recA* or *grpE* promoters, respectively, fused to the *P. luminescens luxCDABE*. These strain were constructed in the host strain MM28 (F-*galK2* IN(*rrnD-rrnE*)*1 rpsL200*) (*36,37*) by a previously described method (*38*) using plasmids pDEW14 or pDEW107, respectively. Plasmid pDEW14 was constructed by ligating a *PstI-EcoRI* fragment containing a *recA-luxCDABE* fusion from plasmid pRecALxx1 into plasmid pBRINT.Cm (*39*). Previously, plasmid pRecALxx1 had been made by ligating into *Bam*HI and *Sal*I digested pJT205 (*29*) a fragment containing the *recA* promoter region that had been obtaining by PCR amplification using pRecALux3 (*40*) as the template. Plasmid pDEW107 was constructed by ligating the 7.2 kbp *PstI-EcoRI* fragment containing a *grpE-luxCDABE* fusion from plasmid pDEW106 into plasmid pBRINT.Km (*39*). Prior to that, plasmid pDEW106 had been made by ligating a 0.6 kbp *Bam*HI fragment containing the *grpE* promoter region from plasmid pGrpELux5 (*33*) into plasmid pJT205 (*29*). *E. coli* strains DPD2222 and DPD2234 were derived from DPD1710 and DPD3084, respectively, by P1*clr100* transduction using phage grown on strain DE112 as the donor,selection for tetracycline resistance, and confirmation of the *tolC* mutation.

E. coli strains DPD2173 and DPD2161 contain plasmids pDEW223 and pDEW215 in *E. coli* host strain GC4468 (*21*). *E. coli* strains DPD2232 and DPD2233 were derived from these by P1*clr100* transduction using phage grown on strain DE112 as the donor, selection for tetracycline resistance, and screening for lack of growth on MacConkey agar plates.

Media and Growth of Bacterial Cultures

LB medium (*31*) was used for all stress response experiments. Overnight cultures were grown in LB medium containing 100 µg/ml of ampicillin for the plasmid-containing strains, or LB medium for the non-plasmid containing strains. These cultures were diluted the following day into LB medium without ampicillin that had been prewarmed to 37°C. A 1:200 dilution was made for the plasmid-containing strains, while a 1:250 dilution was made for the non-plasmid containing strains. Following incubation with shaking at 37°C for 2 hours and 20 minutes the turbidity of the cultures was measured with a Klett-Summerson colorimeter with a red filter and was typically at or near a reading of 30. These actively growing cells were used immediately for bioluminescence analysis. For analysis of the stress responses induced by aluminum chloride, LB medium adjusted to pH5.2 with hydrochloric acid was used because the solubility and toxicity of aluminum chloride is greater at lower pH (*41*). The cell cultures were grown in the reduced pH medium and used, as described below, such that the untreated control was the culture at low pH, thereby allowing aluminum-specific stress responses to be detected.

Chemicals

A 30% solution of hydrogen peroxide (EM Science) was diluted into LB medium to the desired concentrations. A stock solution of 20 mg/ml nalidixic acid (Sigma Chemical Co.) in 1 M NaOH was further diluted into LB medium. Likewise, stock solutions of 1M sodium salicylate (EM Science) in water, 100 mM cadmium chloride (Johnson Matthey Electronics) in water, and 100 mg/ml methyl viologen (Sigma Chemical Co.) in water were further diluted into LB medium. 2,4-D (Janssen Chimica) was dissolved directly in LB medium. Aluminum chloride (Sigma Chemical Co.) was dissolved directly into LB medium that had previously been adjusted to pH 5.2 with hydrochloric acid. Ethanol (200 proof, Quantum Chemicals) was diluted directly into LB medium. Aflatoxin B1 (Sigma Chemical Company) was dissolved in DMSO prior to metabolic activation.

Bioluminescence Analysis

Sterile white 96-well microplates (Microlite, Dynex) were used. The typical arrangement was that the top row (row A) contained a control chemical for each stress-responsive strain. DPD2227 and DPD2238 (*katG-lux*) were tested with 0.004% H_2O_2. DPD2226 and DPD2240 (*inaA-lux*) were tested with 5 mM sodium salicylate. DPD1710 (*recA-lux* in the *tolC*$^+$ host) was tested with 30 μg/ml nalidixic acid; while DPD2222 (*recA-lux* in the *tolC*$^-$ host) was tested with 5 μg/ml nalidixic acid. DPD3084 and DPD2234 (*grpE-lux*) were tested with 3% ethanol. DPD2173 and DPD2232 (*o513-lux*) and DPD2161 and DPD2233 (*yciG-lux*) were tested with 10 mM sodium salicylate. In each case, these chemicals resulted in increased bioluminescence of the indicated biosensor strains, except for the strains containing the *o513-lux* fusion where sodium salicylate was an inhibitor of bioluminescence.

The microplates were prepared shortly before use. Two-fold or three-fold dilutions of the chemical to be tested were made in rows B through G; row H was reserved as the untreated control. When the cell cultures were ready, they were added to all wells of duplicate rows in the prepared microplates, covered with a thin clear plastic film (Fasson FasRoll®), and placed in a ML3000 luminometer (Dynex), the incubation chamber of which had been prewarmed to 37°C. Luminescence data was collected at 37°C in the cycle mode using the following settings: High gain; 20 cycles; 300 second pause; 20 A/D reads per well; all data; auto gain- ON; mixing-ON. The data were processed using a specially written Excel macro to convert to a form for graphing and the average of the duplicate data was plotted as relative light units (RLU) versus time (*42*). These plots were examined to insure that the control chemicals yielded the expected response. Response ratios, which are the ratio of the RLU of the chemically-treated cells culture divided by the RLU of the untreated culture (*42*), were calculated at the point of maximal positive response, which was at 40 to 45 minutes for each stress-responsive strain, except for the *recA-lux* fusion strain where the final time point of 110 minutes gave the maximal response. The other exception was the *o513-lux* fusion strain for which the response ratios were calculated at the point of maximal decrease of bioluminescence, which was at 110 minutes. The concentration of chemical that gave the largest overall stress induction ratios was chosen for display of the stress fingerprints.

Metabolic Activation with Rat Liver Post-Mitochondrial S9 Extract

Postmitochondrial supernatant (S-9) of Sprague Dawley rat liver induced with Aroclor 1254 was purchased from Molecular Toxicology, Inc., stored at -80°C, and thawed on ice prior to use. A cofactor solution of 78 mM sodium phosphate, pH 7.2, 117 mM NaCl, 7.8 mM glucose-6-phosphate, monosodium salt, 5.9 mM β-NADP, sodium salt , 11.4 mM MgCl$_2$, 47 mM KCl, was made by minor modification of a previous method (43), filter sterilized, stored at -80°C, and thawed on ice prior to use.

A 30% solution of the S-9 extract in cofactor mixture was made by mixing 330 μl of thawed S-9 with 770 μl of the thawed cofactor mix. This remained on ice until used. The chemical to be activated was dissolved in DMSO and 10 μl (or less) was added to 90 μl (or more, to total of 100 μl) of the 30% S-9 mixture. Three controls were also set up: No S-9 extract, which was the chemical in DMSO added to sodium phosphate, NaCl buffer; no chemical, which was DMSO added to the 30% S-9 mixture, and no S-9 and no chemical, which was DMSO added to sodium phosphate, NaCl buffer. These were incubated at 37°C for 10 minutes. The activated mixture or controls were diluted directly into a luminometer microplate by adding 10 μl to 90 μl LB medium. Further dilutions were subsequently made in the luminometer microplate. Actively growing cells were added immediately following the dilutions and luminescence was analyzed as above.

Results and Discussion

Characteristic Fingerprints Induced by Five Chemical Stresses

The ability of the stress-responsive *lux* fusion strains to distinguish between various types of chemical stresses was tested by using five chemicals with well-defined modes of action: hydrogen peroxide, nalidixic acid, ethanol, sodium salicylate, and methyl viologen (also known as paraquat). Similar results were obtained with the set of six *tolC*[+] biosensor strains and the set of six *tolC* biosensor strains, although the concentration of chemical required to induce the maximum stress responses was, in some cases, higher for the *tolC*[+] strains. The results from the *tolC* strains (DPD2238, DPD2240, DPD2222, DPD2234, DPD2232, and DPD2233) are shown in Figure 1; these "stress fingerprints" are the response ratios of the set of biosensors to each chemical. In these bar graphs and those that follow, a line is drawn at a response ratio of 1.0; response ratios of this value indicate that the treatment had no effect on the strain. Response ratios greater than 1.0 ("lights-on") indicate increased expression of the *lux* gene fusion, and therefore induction of the particular stress response. Ratios less than 1.0 indicate bioluminescence loss caused by the chemical treatment. This non-specific "lumitoxicity" or "lights-off" can be assumed to be due to inhibition of cellular metabolism required for production of energy or reducing power (44).

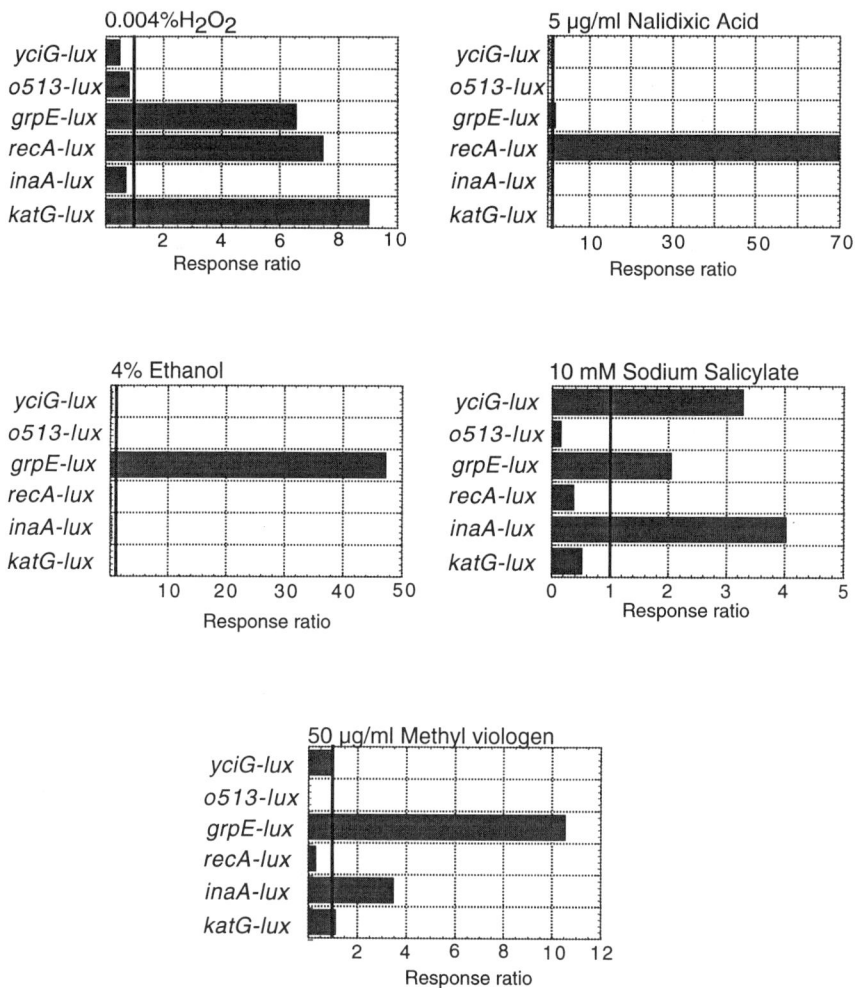

Figure 1. Stress fingerprints induced by treatment of a panel of six biosensor strains with hydrogen peroxide, nalidixic acid, ethanol, sodium salicylate, and methyl viologen.

Hydrogen Peroxide

Hydrogen peroxide induced increased bioluminescence from the *katG-lux* fusion strain in accordance with the known regulation of *katG* expression by the OxyR protein and its activation by hydrogen peroxide *(45)*. Additionally, increased bioluminescence was induced in two other *lux* fusion strains. The increased expression of the *recA-lux* fusion, indicated induction of the DNA damage-sensing SOS response. The increased expression of the *grpE-lux* fusion strain indicated induction of the protein damage sensing heat shock response. These were also in accordance with known effects of the hydoxyl radical, formed from the reaction of hydrogen peroxide with transition metals such as Fe^{2+}, in damaging DNA and proteins *(16)*, and in accordance with previous observations using *V. fischeri luxCDABE* fusions to the *E. coli recA* and *grpE* promoters *(6,46,47)*. Thus, this three pronged fingerprint of hydrogen peroxide was consistent with its biological mode of action of non-specific damage to macromolecules through an oxidative mechanism.

Nalidixic Acid

In contrast to the three pronged fingerprint induced by hydrogen peroxide, nalidixic acid treatment primarily induced expression of the *recA-lux* fusion, indicating that the principal stress was that of DNA damage. Nalidixic acid-mediated inhibition of DNA gyrase results in double stranded breaks in DNA that are processed by the RecBCD nuclease to expose a region of single stranded DNA, which likely acts as a signal for induction of the SOS response *(2)*. Thus, nalidixic acid is a more specific DNA damaging agent than hydrogen peroxide and this was reflected in the stress fingerprint. Note, however, that there was a modest 2-fold induction of bioluminescence in the *grpE-lux* fusion strain, which is in agreement with previous results showing induction of the heat shock response by nalidixic acid *(47,48)*. This result was, therefore, consistent with previous observations that the heat shock response is induced by a large variety of insults to the cell *(48-50)*. Thus, this *grpE* fusion to the *P. luminescens luxCDABE* responded to a variety of stresses, as has also been observed with a *grpE* fusion to the *Vibrio fischeri luxCDABE* operon *(6,33,47,51)*. The bioluminescence of the remaining four fusion strains, each of which contains a plasmid-borne *lux* fusion, was not increased by nalidixic acid treatment. These results suggested that the reported increased plasmid copy number upon induction of the SOS response *(52)* did not result in false positive report of stress induction within the time frame of these assays.

Ethanol

Ethanol is known to be a potent inducer of the protein damage sensing heat shock response *(33,50)*. Of the six stress-responsive fusions represented in this panel, only the expression of *grpE-lux*, which reports on the heat shock response, was induced by ethanol treatment. The absence of ethanol-induced bioluminescence increases from the *katG-lux* strain was in contrast to results with a fusion of the *E. coli katG* promoter region to the *V. fischeri lux* operon *(28)*. This suggests that the effect of ethanol on the *V. fischeri lux* fusion may be related to a non-specific solvent effect *(42,53)* such as stabilization of the *V. fischeri lux* polypeptides, and furthermore suggests increased specificity as an advantage of the *P. luminescens lux* reporter.

Sodium Salicylate

The weak acid sodium salicylate causes cytoplasmic acidification by diffusing through the membrane in a protonated form, followed by release of a hydrogen ion inside the cell (*18,54*). Treatment with this compound induced a different, three pronged stress fingerprint. Expression of two of these fusions, *inaA-lux* and *yciG-lux*, had previously been shown to be induced by this treatment, with induction of the *inaA-lux* expression being under control of *marA* and *rob* (*21*). Increased expression of the *grpE-lux* fusion demonstrated, again, the general stress-responsive nature of the heat shock response.

Methyl Viologen

A two pronged fingerprint was induced by treatment with the redox cycling agent methyl viologen. Once again, induction of expression of the *grpE-lux* fusion was observed. Increased bioluminescence from the *inaA-lux* fusion strain was also seen. In contrast to induction of *inaA* expression by sodium salicylate, the induction by methyl viologen is under control of the SoxR and SoxS regulon (*20*). This regulon responds to oxidative damage caused by superoxide anion radical that is generated by one-electron reduction of molecular oxygen mediated by methyl viologen (*16*). Lack of induction of the *katG-lux* fusion in response to methyl viologen contrasts to the moderate induction observed using a *katG* fusion to the *V. fischeri luxCDABE* reporter (*6*). Similarly to the case with ethanol, this lack of response suggests that fusions to the *P. luminescens luxCDABE* reporter confer increased specificity of responses.

Overall, each of these five chemicals with defined modes of action yielded a characteristic stress response fingerprint that was reflective of the biological effects caused by these compounds. Thus, these results validated the responsiveness of this panel to characteristic insults.

Stress Fingerprint Induced by a Herbicide, 2,4-D, and two toxic metal salts

Chemicals with modes of action that are not as well-defined were addressed next. Stress fingerprints induced by a herbicide with a mode of action thought to mimic plant hormones and two toxic metal salts are shown in Figure 2. The four pronged fingerprint induced by 2,4-D treatment indicates that it causes multiple stresses on the bacterial cell. This fingerprint was most like that induced by sodium salicylate because the expression of the *yciG*, *grpE* and *inaA* fusions were induced. Thus, 2,4-D may, like sodium salicylate, be functioning to bring hydrogen ions into the cytoplasm and thereby lowering internal pH. Interestingly, the slight induction of the *katG* fusion indicates that the cell may also sense an oxidative stress; the mechanism of this effect is not known.

Both $CdCl_2$ and $AlCl_3$ induced expression of the *grpE-lux* fusion, although to varying degrees. This again indicated that the heat shock response is generally inducible by numerous stresses to the cell. Treatment with cadmium chloride has been shown to induce synthesis of the GrpE protein (*48*); however, a similar fusion of the *grpE* promoter to the *V. fischeri luxCDABE* reporter does not respond to cadmium chloride unless ethanol or other inducers of the heat shock response are

177

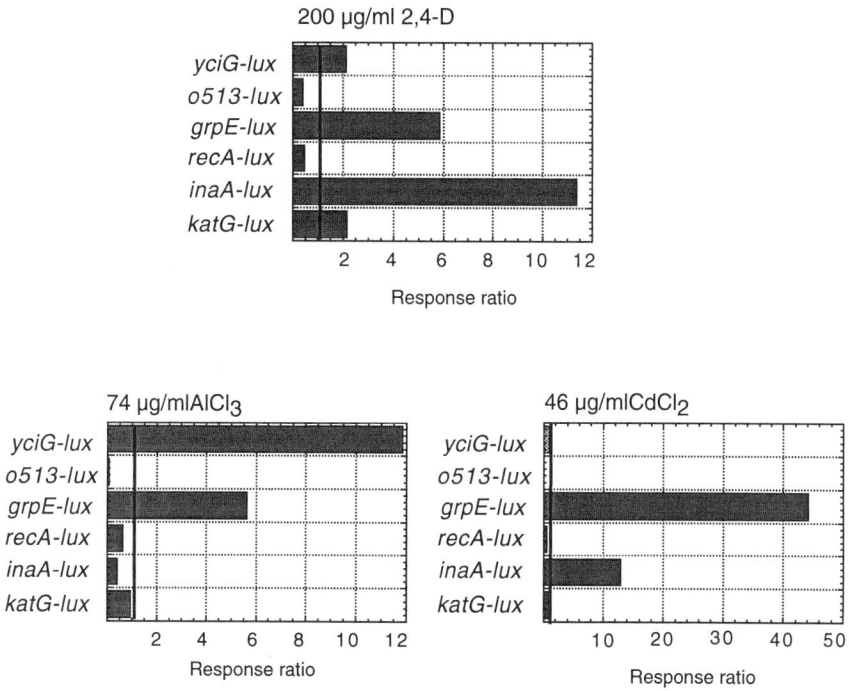

Figure 2. Stress fingerprints induced by treatment of a panel of six biosensor strains with 2,4-dichlorophenoxyacetic acid, aluminum chloride and cadmium chloride.

simultaneously present (51). Thus, an advantage of the *P. luminescens luxCDABE* reporter was demonstrated because it can directly report on this stress. Cadmium chloride did not induce expression of the *recA-lux* fusion. A similar negative results for induction of the SOS response is also observed in *E. coli* strains containing an SOS responsive promoter fusion to the *lacZ* reporter and correlates with the negative result in the Ames test for mutagenesis (3). However, results obtained by two dimensional gel electrophoresis suggest that the amount of the RecA protein is transiently increased following a cadmium chloride stress (55). This study and an earlier two dimensional gel study (48) did not identify other members of the SOS regulon as induced by cadmium chloride; thus, expression of the RecA protein may be regulated by posttranscriptional mechanisms following cadmium stress.

The remainder of the patterns of gene expression induced by these two metal salts differed. Aluminum chloride induced the *yciG-lux* fusion, while cadmium chloride did not. On the other hand, cadmium chloride induced the *inaA-lux* fusion, while aluminum chloride did not. Thus, the biological effects of these two inorganic metal salts were distinguished by their stress fingerprints. It worth noting the induction of some but not all of these fusions indicated that aluminum chloride was not a non-specific activator of the *P. luminescens luxCDABE*. This result, which contrasts with observations from transcriptional fusions using *V. harveyi luxAB* in *E. coli* (56), points to another advantage of the *P. luminescens luxCDABE* reporter. It is also noteworthy that the stress pattern induced by cadmium chloride was like that induced by methyl viologen. Although these two compounds are structurally unrelated, they both function as oxidizing agents (16,48); this similar mode of action was reflected in the similar stress fingerprints induced.

Stress Fingerprints Induced by Aflatoxin B1 With and Without Metabolic Activation

Aflatoxin B1, a mycotoxin that is a carcinogen in many animal species (57) and a risk factor for human hepatocellular carcinoma (58), requires metabolic activation to the mutagenic epoxide form (59). Because bacteria lack the enzymes for this and other transformations of promutagens to mutagens, many bacterial tests have incorporated an *in vitro* activation step (6,60-64). Thus, stress fingerprints for this important environmental contaminant were done with and without pretreatment with a rat liver microsomal S9 extract. These fingerprints are shown in Figure 3. The unactivated aflatoxin B1 induced modest stress responses reported by the SOS-controlled *recA-lux* and the multiply regulated *inaA-lux* fusions. Interesting, even without activation by a mammalian extract, the DNA damage response was observed. This modest induction of the *recA-lux* fusion without S9 treatment required the presence of the *tolC* mutation in the host strain (data not shown). Also interesting was the induction of *inaA-lux* expression by untreated aflatoxin B1. One of the regulators controlling the expression of this gene fusion is the multiple antibiotic resistance response (Mar). It may be that the Mar response is induced by mycotoxins. Activation of aflatoxin B1 with the rat liver microsomal extract resulted in an altered stress fingerprint. The response of the DNA damage-responsive *recA-lux* fusion was increased, as expected for an increased amount of the mutagenic form of aflatoxin B1, and was the major stress induced. In contrast, the induction of the *inaA-lux* fusion was decreased, although at lower concentration of activated aflatoxin B1 it was still present (data not shown). Use of this panel of six stress responsive strains, thus, provided a measure of the DNA damage response caused by aflaxtoxin B1, as do other bacterial tests (*64-67*). However, additional information about the biological effects of aflatoxin B1 in its activated and unactivated forms was also obtained. Accordingly, the stress fingerprint of the two forms of aflatoxin B1 are distinguished from each other (Figure 3), as well as from that of nalidixic acid (Figure 1), a DNA damaging agent with a different mode of action.

Detection Limits for Aflatoxin B1

The limits of aflatoxin B1 detection by the *recA-lux* and *inaA-lux* fusion strains were determined by using various concentrations of untreated and activated compound. The bioluminescence of the culture with chemical added was tested to determine if it was greater than that of the control without chemical. The unpaired t test was used and $P<0.005$ values in a one-tailed test were considered significant. With this criterion, Table II shows the lowest concentrations of aflatoxin B1 detected.

Table II. Detection Limits for Aflatoxin B1

Sensor	With Metabolic Activation	Without Metabolic Activation
recA-lux	11 μg/ml (P<0.0001)	11 μg/ml (P<0.0001)
inaA-lux	3.7 μg/ml (P<0.0036)	1.2 μg/ml (P<0.0005)

The best detection limit was at 1.2 μg/ml (or 1.2 ppm) unactivated aflatoxin B1 using the *inaA-lux* fusion strain. This detection at the ppm level does not compare favorably with alternative detection methods such as ELISA, TLC, or HPLC where detection levels are in the parts per billion (ppb) range or lower (*68-70*). Likewise, the detection levels here are not useful as analytical tests for food and feed safety where the typical action levels are in the range of 2-40 ppb (*71*). However, the levels of aflatoxin B1 detected by these bioluminescent sensors were in the range where biological effects are typically observed in poultry (*24-26*), and were well below the LD50 for broiler chicks of 70 ppm in the feed (*27*). Of course, these bioluminescent sensors were not specifically induced by aflatoxin B1 and also responded to other substances that produce a similar stress on the cell. This non-specificity of the sensors may be an advantage in that mycotoxins other than aflatoxin B1 that induce the DNA damage response or the *inaA* response may also be detected. Thus, these tests do not rely on prior knowledge of the nature of the possible chemical contaminants and will, for example, report on the presence of any substance that induces the SOS response in *E. coli*.

Possible improvements could be made to the panel of strains to lower the detection levels for compounds such as aflatoxin B1. Additional mutations affecting the outer membrane could result in increased sensitivity. The combination of a *tolC*

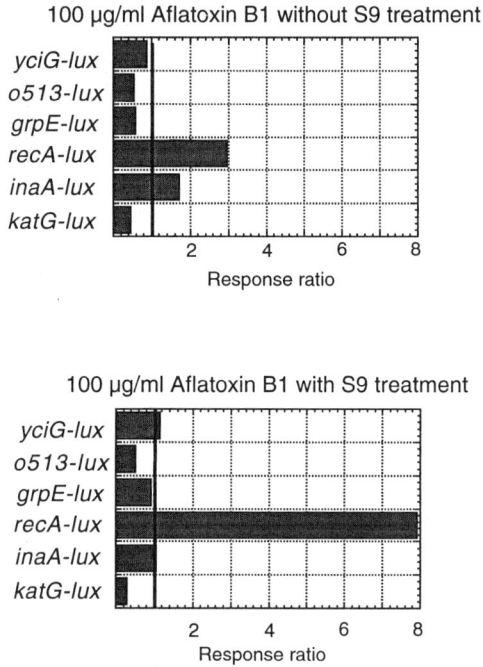

Figure 3. Stress fingerprints induced by treatment of a panel of six biosensor strains with aflatoxin B1 without and with prior metabolic activation.

mutation that prevents pumping out of substances with an *rfa* mutation that alters the permeability of the outer membrane results in increased sensitivity to hydrophobic compounds as compared with strains carrying either mutation alone (*72*). Furthermore, if detection of a particular stress response is desired, the genes encoding repair mechanisms for that response could be destroyed by mutation. For example, a mutation that destroys an excision repair pathway has been used to increase sensitivity of bacterial tester strains to DNA damaging agents (*73,74*). Also, altering the growth and testing medium may alter sensitivity to various compounds.

Conclusions

Overall, this six member panel reported on various stresses to the cell and gave characteristic stress fingerprints that were related to the biological mode of action of the chemical stressor. The least characterized genetic fusion in the group, *o513-lux*, although previously found to be induced by a herbicide, sulfometuron methyl (*21*), was not observed to respond with increased bioluminescence to any of the chemical stresses tested here. Nevertheless, this strain was useful as a general indicator of toxicity through a "lights off" response; all chemical treatments resulted in a response ratio of less than 1.0 for the *o513-lux* fusion. In contrast, bioluminescence of the *grpE-lux* fusion strain was induced to varying degrees by all but one of the chemicals tested. Thus, this fusion reported on general stress to the cell, as does a fusion of the *V. fischeri luxCDABE* to the *E. coli grpE* promoter (*6,33,47,51*). The remaining four fusions were differentially induced by chemical treatments and resulted in the characteristic stress fingerprints observed. Thus, the use this panel of stress responsive gene fusions for characterizing the toxic effects of single chemicals was demonstrated; this and other such molecular toxicology approaches will be useful in defining the modes of toxicity of pure chemicals. On the other hand, environmental samples are likely to contain a mixture of chemicals; in this circumstance, use a panel of stress responsive strains will give a stress fingerprint of the predominant modes of toxicity found associated with such samples, whether these toxicities are due to one or more chemicals. Furthermore, interactions of chemicals to yield new modes of toxicity may be detected.

Literature Cited

1. Morimoto, R. I.; Tissières, A.; Georgopoulos, C. In *The Biology of Heat Shock Proteins and Molecular Chaperones*; Morimoto, R. I., Tissières, A., Georgopoulos, C., Eds.; Cold Spring Harbor Laboratory Press: Cold Spring Harbor, NY, 1994; pp 1-30.
2. Walker, G. C. In *Escherichia coli and Salmonella: Cellular and Molecular Biology*; Neidhardt, F. C., Ed.; ASM Press: Washington, DC, 1996; pp 1400-1416.
3. Quillardet, P.; Hofnung, M. *Mutatation Res.* **1993**, *297*, 235-279.
4. Orser, C. S.; Foong, F. C. F.; Capaldi, S. R.; Nalezny, J.; MacKay, W.; Benjamin, M.; Farr, S. B. *In Vitro Toxicology* **1995**, *8*, 71-85.

5. Van Dyk, T. K.; Belkin, S.; Vollmer, A. C.; Smulski, D. R.; Reed, T. R.; LaRossa, R. A. In *Bioluminescence and Chemiluminescence: Fundamentals and Applied Aspects*; Cambell, A. K., Kricka, L. J., Stanley, P. E., Eds.; John Wiley & Sons: Chichester, England, 1994; pp 147-150.
6. Belkin, S.; Smulski, D. R.; Dadon, S.; Vollmer, A. C.; Van Dyk, T. K.; LaRossa, R. A. *Wat. Res.* **1997**, *31*, 3009-3016.
7. Belkin, S. In *Methods in Molecular Biology: Bioluminescence Methods and Protocols*; LaRossa, R. A., Ed.; Humana Press Inc.: Totowa, NJ, 1998; Vol. 102; pp 247-258.
8. Belkin, S.; Van Dyk, T. K.; Vollmer, A. C.; Smulski, D. R.; LaRossa, R. A. *Environ. Toxicol. Water Qual.* **1996**, *11*, 179-185.
9. Pedahzur, R.; Shuval, H. I.; Ulitzur, S. *Wat. Sci. Tech.* **1997**, *35*, 87-93.
10. Dukan, S.; Dadon, S.; Smulski, D. R.; Belkin, S. *Appl. Environ. Microbiol.* **1996**, *62*, 4003-4008.
11. Oh, J.-T.; Van Dyk, T. K.; Cajal, Y.; Dhurjati, P. S.; Sasser, M.; Jain, M. K. *Biochem. Biophys. Res. Comm.* **1998**, *246*, 619-623.
12. Oh, J.-T.; Cajal, Y.; Skowronska, E. M.; Belkin, S.; Chen, J.; Van Dyk, T. K.; Sasser, M.; Jain, M. K. *Unpublished* .
13. Oh, J.-T.; Cajal, Y.; Dhurjati, P. S.; Van Dyk, T. K.; Jain, M. K. *Biochim. Biophys. Acta* **1998**, *1415*, 235-245.
14. Szittner, R.; Meighen, E. *J. Biol. Chem.* **1990**, *265*, 16581-16587.
15. Ang, D.; Georgopoulos, C. *J. Bacteriol.* **1989**, *171*, 2748-2755.
16. Demple, B. *Annu. Rev. Genet.* **1991**, *25*, 315-337.
17. Ivanova, A.; Miller, C.; Glinsky, G.; Eisenstarck, A. *Mol. Microbiol.* **1994**, *12*, 571-578.
18. Slonczewski, J. L.; Gonzalez, T. N.; Bartholomew, F. M.; Holt, N. J. *J. Bacteriol.* **1987**, *169*, 3001-3006.
19. White, S.; Tuttle, F. E.; Blankenhorn, D.; Dosch, D. C.; Slonczewski, J. *J. Bacteriol.* **1992**, *174*, 1537-1543.
20. Rosner, J. L.; Slonczewski, J. L. *J. Bacteriol.* **1994**, *176*, 6262-6269.
21. Van Dyk, T. K.; Ayers, B. L.; Morgan, R. W.; LaRossa, R. A. *J. Bacteriol.* **1998**, *180*, 785-792.
22. Volchonok, M. G. In *Gig. Aspekty Okhr. Zdorov'ya Naseleniya*, 1977; pp 184-185.
23. *Nutrient Requirements of Poultry*; National Academy Press: Washington, DC, 1984.
24. Aletor, V. A.; Kasali, O. B.; Fetuga, B. L. *Zentrabl. Veterinaermed.* **1981**, *28*, 774-781.
25. Smith, E. E.; Kubena, L. F.; Braithwaite, C. E.; Harvey, R. B.; Phillips, T. D.; Reine, A. H. *Poultry Sci.* **1992**, *71*, 1136-1144.
26. Thaxton, J. P.; Tung, H. T.; Hamilton, P. B. *Poultry Sci.* **1974**, *53*, 721-725.
27. Wyatt, R. D. *Egg Industry* **1991**, 12-16.
28. Belkin, S.; Smulski, D. R.; Vollmer, A. C.; Van Dyk, T. K.; LaRossa, R. A. *Appl. Environ. Microbiol.* **1996**, *62*, 2252-2256.
29. Van Dyk, T. K.; Rosson, R. A. In *Methods in Molecular Biology: Bioluminescence Methods and Protocols*; LaRossa, R. A., Ed.; Humana Press Inc.: Towowa, NJ, 1998; Vol. 102; pp 85-95.
30. Greenberg, J. T.; Chou, J. H.; Monach, P. A.; Demple, B. *J. Bacteriol.* **1991**, *173*, 4433-4439.

31. Miller, J. H. *Experiments in molecular genetics*; Cold Spring Harbor Laboratory Press: Cold Spring Harbor, NY, 1972.
32. Lopilato, J.; Bortner, S.; Beckwith, J. *Mol. Gen. Genet.* **1986**, *205*, 285-290.
33. Van Dyk, T. K.; Majarian, W. R.; Konstantinov, K. B.; Young, R. M.; Dhurjati, P. S.; LaRossa, R. A. *Appl. Environ. Microbiol.* **1994**, *60*, 1414-1420.
34. Masters, M.; Colloms, M. D.; Oliver, I. R.; He, L.; MacNaughton, E. J.; Charters, Y. *J. Bacteriol.* **1993**, *175*, 4405-4413.
35. Davis, R. W.; Botstein, D.; Roth, J. R. *Advanced bacterial genetics*; Cold Spring Harbor Laboratory: Cold Spring Harbor, NY, 1980.
36. Gottesman, M. E.; Yarmolinski, M. B. *J. Mol. Biol.* **1968**, *31*, 487-505.
37. Pratt, D.; Erdahl, W. S. *J. Mol. Biol.* **1968**, *37*, 181-200.
38. Elsemore, D. A. In *Methods in Molecular Biology: Bioluminescencent Protocols.*; LaRossa, R. A., Ed.; Humana Press Inc.: Totowa, NJ, 1998; Vol. 102; pp 97-104.
39. Balbás, P.; Alexeyev, M.; Shokolenko, I.; Bolivar, F.; Valle, F. *Gene* **1996**, *172*, 65-69.
40. Vollmer, A. C.; Belkin, S.; Smulski, D. R.; Van Dyk, T. K.; LaRossa, R. A. *Appl. Environ. Microbiol.* **1997**, *63*, 2566-2571.
41. Wood, J. M. In *Metal speciation: theory, analysis and application*; Kramer, J. R., Allen, H. E., Eds.; Lewis Publishers: Chelsea, MI, 1988; pp 309-312.
42. Van Dyk, T. K. In *Methods in Molecular Biology: Bioluminescence Methods and Protocols*; LaRossa, R. A., Ed.; Humana Press Inc.: Totowa, NJ, 1998; Vol. 102; pp 153-160.
43. Clare, C. In *Methods Molecular Biology: In Vitro Toxicity Testing Protocols*; O'Hare, S., Atterwill, C. K., Eds.; Humana Press Inc.: Totowa, NJ, 1995; Vol. 43; pp 297-306.
44. Chatterjee, J.; Meighen, E. A. *Photochemistry and Photobiology* **1995**, *62*, 641-650.
45. Rosner, J. L.; Storz, G. *Curr. Top. Cell. Regul.* **1997**, *35*, 163-177.
46. Belkin, S.; LaRossa, R. A. In *Progress in Microbial Ecology*; Martins, M. T., Sato, M. I. Z., Tiedje, J. M., Hagler, L. C. S. N., Döbereiner, J., Sanchez, P. s., Eds.; Brazilian Society for Microbiology: São Paulo, Brazil, 1997; pp 565-570.
47. Van Dyk, T. K.; Smulski, D. R.; Reed, T. R.; Belkin, S.; Vollmer, A. C.; LaRossa, R. A. *Appl. Environ. Microbiol.* **1995**, *61*, 4124-4127.
48. VanBogelen, R. A.; Kelley, P. M.; Neidhardt, F. C. *J. Bacteriol.* **1987**, *169*, 26-32.
49. Blom, A.; Harder, W.; Matin, A. *Appl. Environ. Micro.* **1992**, *58*, 331-334.
50. Welch, W. In *Stress proteins in biology and medicine*; Morimoto, R. I., Tissières, A., Georgopoulos, C., Eds.; Cold Spring Harbor Laboratory Press: Cold Spring Harbor, NY, 1990; pp 223-278.
51. Van Dyk, T. K.; Reed, T. R.; Vollmer, A. C.; LaRossa, R. A. *J. Bacteriol.* **1995**, *177*, 6001-6004.
52. Bertrand-Burggraf, E.; Oertel, P.; Schnarr, M.; Daune, M.; Granger-Schnarr, M. *Plasmid* **1989**, *22*, 163-168.
53. Heitzer, A.; Applegate, B.; Kehrmeyer, S.; Pinkart, H.; Webb, O. F.; Phelps, T. J.; White, D. C.; Sayler, G. S. *J. Microbiol. Methods* **1998**, 45-57.
54. Padan, E.; Silberstein, D.; Schuldiner, S. *Biochim. Biophys. Acta* **1981**, *650*, 151-166.
55. Ferianc, P.; Farewell, A.; Nyström, T. *Microbiology* **1998**, *144*, 1045-1050.

56. Guzzo, J.; Guzzo, A.; DuBow, M. S. *Toxicol Lett* **1992**, *64-65*, 687-93.
57. In *Foodborne Pathogenic Microorganisms and Natural Toxins Handbook*; U. S. Food & Drug Administration; Center for Food Safety & Applied Nutrition, 1998.
58. Qian, G. S.; Ross, R. K.; Yu, M. C.; Yuan, J. M.; Gao, Y. T.; Henderson, B. E.; Wogan, G. N.; Groopman, J. D. *Cancer Epidemiol Biomarkers Prev* **1994**, *3*, 3-10.
59. McLean, M.; Dutton, M. *Pharmac. Ther.* **1995**, *65*, 163-192.
60. Maron, D. M.; Ames, B. N. *Mutation Res.* **1983**, *148*, 25-34.
61. Quillardet, P.; Huisman, O.; D'Ari, R.; Hofnung, M. *Proc. Natl. Acad. Sci. U S A* **1982**, *79*, 5971-5975.
62. Shimada, T.; Oda, Y.; Yamazaki, H.; Mimura, M.; Guengerich, F. P. *Methods Mol. Genet.* **1994**, *5(Gene and Chromosome Analysis, Pt. C)*, 342-355.
63. van der Lelie, D.; Regniers, L.; Borremans, B.; Provoost, A.; Verschaeve, L. *Mutat. Res.* **1997**, *389*, 279-290.
64. Sun, T. S. C.; Stahr, H. M. *J. AOAC Int.* **1993**, 893-898.
65. Quillardet, P.; de Bellecombe, C.; Hofnung, M. *Mutation Res.* **1985**, *147*, 79-95.
66. Reisenfeld, G.; Kirsch, I.; Weissman, S. H. *Food Add. Contam.* **1985**, *2*, 253-257.
67. McCann, J.; Choi, E.; Yamasaki, E.; Ames, B. N. *Proc. Natl. Acad. Sci. USA* **1975**, *72*, 5135-5139.
68. Park, D. L.; Miller, B. M.; Hart, L. P.; Yang, G.; McVey, J.; Page, S. W.; Pestka, J.; Brown, L. H. *J. - Assoc. Off. Anal. Chem.* **1989**, *72*, 326-332.
69. Scott, P. M.; Lawrence, G. A. *J. AOAC Int.* **1997**, *80*, 1229-1234.
70. Vicente, E.; Soares, L. M. V. *Cienc. Tecnol. Aliment.* **1995**, *15*, 201-205.
71. Chu, F. S. *Mutat. Res.* **1991**, *259*, 291-306.
72. Fralick, J. E.; Burns-Keliher, L. L. *J. Bacteriol.* **1994**, *176*, 6404-6406.
73. Ames, B. N.; McCann, J.; Yamasaki, E. *Mutation Res.* **1975**, *31*, 347-364.
74. Quillardet, P.; Hofnung, M. *Mutation Res.* **1985**, *147*, 65-78.

Chapter 13

Continuous Monitoring of Protein Damaging Toxicity Using a Recombinant Bioluminescent *Escherichia coli*

Man Bock Gu, Robert J. Mitchell, and Joong Hyun Kim

Department of Environmental Science and Engineering, Kwangju Institute of Science and Technology (K-JIST), 1 Oryong-dong, Puk-gu, Kwangju, 500–712, South Korea (email: mbgu@kjist.ac.kr)

Through the use of the recombinant bacterium TV1061, containing a *grpE-luxCDABE* fusion, and a previously developed continuous toxicity monitoring system we were able to detect protein damaging toxicity in aqueous systems. This bacterial strain is responsive to the toxicity of protein-damaging agents and produces a bioluminescent response in their presence. The system uses the two-stage mini-bioreactors previously developed in our lab which allows stability even under the conditions that the toxicity greatly exceeded the sub-lethal level. The response of the strain under these conditions shows a unique signature when compared to previously tested strains.

INTRODUCTION

Significant efforts have been directed toward the development of a sensitive biosensor that can be used in the monitoring of effluents from wastewater treatment plants and natural water systems. The primary characteristics of the biosensor are that it should be a rapid detector of toxicity, produce a measurable and quantifiable signal when under toxic conditions and show stability and reproducibility.

Commercialized Systems Using Live Biosensors

The most commonly used toxicity monitoring systems are fish and/or daphnia as test organisms. The toxicity of effluent water is determined through differences in the behaviour of the animals, including death, from which the water may be declared unfit for consumption and use without further treatment.

Also in use are tests that utilize protozoa. Spirotox® uses the protozoan *Spirostomum ambiguum* as a test organism to determine the quality of water (19). This ciliate is sensitive to many compounds, especially heavy metals, that may be found in wastewater. Other eucaryotes have been used in toxicity monitoring, including algae (5,20), but they are not as widely used as fish or daphnia.

Bioluminescent Bacteria

As in any monitoring system, the most important factors to consider are reliability, reproducibility, time, ease of use and cost. The idea of using bioluminescent bacteria grew out of this ideology. Bioluminescent bacteria express *luxAB, luxCDABE*, the fire-fly luciferase gene or the click beetle luciferase gene and require the addition of decanal, no substrate, or the substrate luciferin, respectively. In all species of naturally bioluminescent bacteria, these five genes, *luxCDABE*, are expressed and are necessary for the production of light. Since bacteria have a faster growth rate than higher organisms, such as daphnia and fish, and require less maintenance, they are easier to handle and cheaper to grow. Also, the time between the induction and the detection of the signal is greatly reduced resulting in a biosensor that is quick and that has an easily quantifiable signal.

This idea of using bacterial bioluminescence in the monitoring of the environment was further developed with the Microtox® test (4). This test uses the bacteria *Vibrio fischeri* which produces bioluminescence naturally after reaching a certain cell density. In their Microtox® Acute™ test, a sample of test water is added and the bacteria were tested for reduction in their light output. In the Microtox® Chronic Toxicity™ test, the test sample is serially diluted and added to the bacterial strain, which is then grown in the mixture of media and sample. The inhibitory effect is then measured compared to the control and is the result of the toxicity on the growth, reproduction and luciferase expression (18). Many researchers have evaluated the Microtox® system, including the testing of more than 1300 pure compounds (4,11,18), and have compared the sensitivity of the bacteria to that of fish and daphnia (30). The sensitivity of the bacteria compared to higher organisms varies depending on the chemical being tested (18).

Recombinant Bacteria

The use of bioluminescence has resulted in many different tests, including the use of naturally bioluminescent bacteria (27) aside from Microtox®. The genes from the bacterium *Vibrio fischeri*, and other bioluminescent bacteria, have been cloned into bacteriophage (15) and into plasmids that are functional in a wide range of bacterial hosts, including *E. coli* and strains of *Pseudomonas* (3,21,22,29). Many different bacterial strains have been constructed with these plasmids and produce the lux or luc (fire-fly) genes, and, therefore, light, either under constitutive expression or by some method of induction. These bioluminescent bacterial strains have been used in various experiments, including viability tests (16,34), antibiotic studies (1,7,8), and toxicity monitoring (14,17,19,29,32).

Some of the plasmids that have been constructed are known as promoter probe plasmids, and have a multiple cloning site (MCS) upstream from the *lux* genes. Into this MCS a promoter of interest can be inserted and the strength and characteristics of this promoter can be determined by measurement of the bioluminescence, which corresponds well with the quantity of luciferase protein in the cell. The use of these promoter probe plasmids has allowed the construction of luciferase biosensors for a wide variety of specific substances, including toluene (25), naphthalene (12), tetracyclines (13), alkanes (28), arsenite (31), antimonite (31), and mercury (26). Some plasmids have also been constructed using promoters that are induced during different stress responses in *E. coli*. Some of the promoters inserted into these include *katG* (2), *grpE* (2,9,23,33), *fabA* (6) and *recA* (2,10,17,35).

Principles of Toxicity Sensing Using Recombinant Bacteria

E. coli, like all organisms, has the ability to adjust to varying environmental situations and toxicity. When the bacterium experiences an environmental stress, a signal transduction cascade occurs in which certain promoters are induced and their proteins are expressed. Some of the more elucidated responses include the SOS response(24), which is concerned with damage to the cellular DNA, the oxidative-damage response and the response to protein damage.

Using the pUCD615 (22) promoter-probe vector, several plasmids were constructed that respond to these different stress responses. The strain used in this study is TV1061 (*grpE::luxCDABE*) with the host being RFM443 (33). TV1061 is a strain that detects protein damage (2,9,23). The *grpE* promoter is one of about 20 promoters that are transcribed from under conditions that the cells experience protein damage, due to the addition of chemicals such as ethanol or phenol or by an increase in temperature. When this happens, the intracellular level of a special sigma factor, σ^{32}, elevates. It is this factor that recognizes these promoters specifically, resulting in transcription from the *grpE* promoter and the induction of light.

Mini-bioreactor System

Gu *et al* (9) developed a continuous single-stage mini-bioreactor that was used in toxicity monitoring. This mini-bioreactor works as a turbidostat as media is added and the grown cells and spent media are passed out of the bioreactor. Through the addition of a fiber-optic probe connection, it became possible to measure the bioluminescence as the reaction proceeded, allowing on-line monitoring of waste effluent. This mini-bioreactor, however, also faced the problem that when the toxicity is greater than sub-lethal, system failure might occur resulting in the washing out of all the bacteria. Also, the daughter cells may not be as responsive to toxicity as the parent cells due to changes in the cellular DNA brought on by the presence of mutagens.

188

For these reasons, our lab has previously developed a two-stage mini-bioreactor for the detection of toxicity (10). The bacterial cells are grown in the first mini-bioreactor and then are passed into the second mini-bioreactor, and it is in this reactor where they are mixed with the effluent or test stream. This allows the separation of the growth stage from that of the test stage, therefore, keeping the bacterial population consistent over time, and since there will always be bacteria being pumped into the test reactor, it lends the system stability even with a highly toxic insult. Therefore, in the situations that the toxicity of the effluent is greater than the sub-lethal level, the mini-bioreactor should show stability as fresh bacteria cells are introduced and the second mini-bioreactor system stabilizes.

A Model for a Continuous Two-Stage Toxicity Monitoring System

The schematic diagram of the continuous two-stage mini-bioreactor developed in this study is shown in Figure 1. Sterile media is supplied into the first stage with a working volume of V_1 at a flow rate F_1. The recombinant bioluminescent *E. coli* cells are pumped into the second stage which contains the toxic chemical at a concentration of C_{t21}, at a flow rate of F_1. The cells are then pumped out from the second stage at a flow rate of F_2. The material balance equations for the toxicity-sensing cells and toxic chemical in the first and second stage can be described as follows:

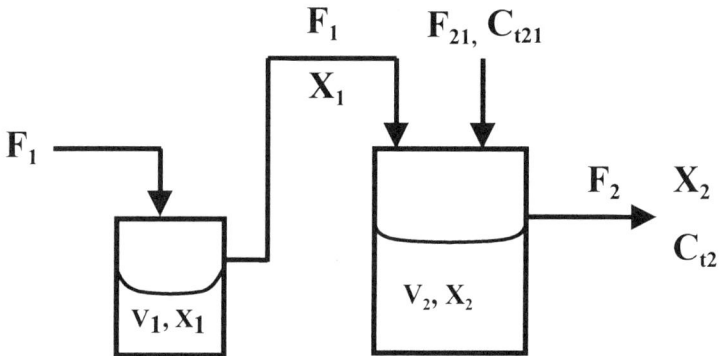

Figure 1. Schematic diagram of the continuous two-stage mini-bioreactor

For the first stage,

$$V_1\frac{dX_1}{dt}=\mu_1 X_1 V_1 - F_1 X_1 \qquad (1)$$

where X_1 is the cell mass and μ_1 is the specific growth rate of cells, both in the first stage. At steady state

$$\mu_1 = D_1 \qquad (2)$$

where D_1 is the dilution rate in the first stage, or F_1 divided by V_1. From equation (2), the specific growth rate is controlled by D_1.

For the second stage,

$$V_2 \frac{dX_2}{dt} = F_1 X_1 + \mu_2 X_2 V_2 - F_2 X_2 \qquad (3)$$

where X_2 and μ_2 are the cell mass and specific growth rate in the second stage, respectively.

At steady state

$$\mu_2 = D_2 - D_{12} \frac{X_1}{X_2} \qquad (4)$$

where D_2 is the dilution rate in the second stage, or F_2 divided by V_2, and D_{12} is dilution rate, or F_1 divided by V_2. Two mass balance equations for the toxic chemical are formulated as follows:

For pulse injection of a toxic chemical into the second stage

$$V_2 \frac{dC_{t2}}{dt} = -F_2 C_{t2} \qquad (5)$$

$$C_{t2} = C_{t20} \exp(-D_2 t) \qquad (6)$$

where C_{t20} is initial concentration of toxic chemical in the second stage.

For the step injection of toxic chemical at a flow rate F_{21}.

$$V_2 \frac{dC_{t2}}{dt} = F_{21} C_{t21} - F_2 C_{t2} \qquad (7)$$

$$C_{t2} = C_{t21} \frac{F_{21}}{F_2} [1 - \exp(-D_2 t)] \qquad (8)$$

where C_{t21} is the concentration of toxic chemical which is injected into the second stage.

MATERIALS AND METHODS

The *E. coli* strain used was TV1061(33). Seed cultures were grown in 100 ml LB media containing kanamycin monosulfate (25 μg/ml) in 250 ml flasks overnight. The temperature was set at 30 ℃ and the cultures were agitated on a rotary shaker at 250 rpm. All mini-bioreactor equipment was sterilized by autoclaving. Media had kanamycin monosulfate added and was stored on ice and in the dark the duration of the experiment. The system was inoculated with 2 ml of the overnight culture into the first mini-bioreactor. The system was allowed to stabilize, determined through the bioluminescence and the cell density, before addition of the test chemicals. Kanamycin monosulfate, Mitomycin C and phenol were all purchased from Sigma Chemical Co. Ethanol was purchased from Merck.

Concentrations of chemicals were determined by dividing the stock solution concentration by the final volume, assuming instantaneous mixing. Bioluminescence within the mini-bioreactor was measured using a fiber-optic connection to the luminometer, which was factory adjusted so that a single photon tritium light source gave a value of 5 AU (arbitrary units). The use of this system allowed on-line monitoring, which was shown to correspond well with off line measurements. (Data not shown) As discussed earlier for the single-stage mini-bioreactor, this system was run as a turbidostat for all experiments

RESULTS AND DISCUSSION

Pulse Injection

In many situations, there is a sudden increase in chemical concentration for only a short time. To mimic these situations, we implemented a pulse injection experiment where a chemical is added all at once and the response of the strain is monitored.

TV1061, as stated earlier, responds to chemicals that cause protein damage. In the two-stage mini-bioreactor system, the response kinetics of TV1061 to various chemicals were measured. Figure 2 shows the response of TV1061 to a 1.5 % ethanol pulse injection. As shown, shortly after induction the bioluminescence increases sharply after an initial upset. Then after reaching a maximum value for the relative bioluminescence, (RBL = BL of the induced cells / BL of the control) of around 4, the bioluminescence decreases following a washing out pattern, similar to that expected for a dye, and comes back down to the basal level before induction.

After the system reached stability, different concentrations of phenol were added by pulse injection into the second mini-bioreactor and the bioluminescence(BL) was measured and recorded. (Figure 3) As the concentration of phenol increased, the BL also increased until the concentration of the phenol exceeded 300ppm. After this, there was a very apparent initial drop in the bioluminescence due to the toxicity of the

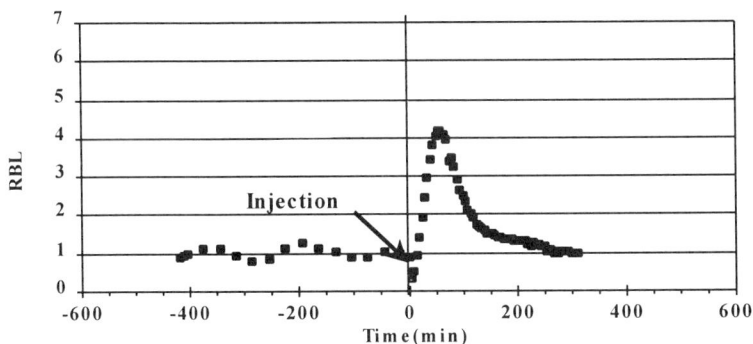

Figure 2. Response kinetic of TV1061 to a pulse-injection of 1.5% ethanol
$(V_1=10mL, V_2=20mL, D_1=0.83/hr, D_2=1.14/hr)$

phenol. This is in contrast to the results of DPD2794, which shows a plateau signature when exposed to high concentrations of chemical in the two-stage minibioreactor system. (10) At 1400 ppm, the BL dropped to nearly zero due to the toxicity of the phenol. Typically, after phenol addition the BL follows the washing out profile of a dye. This result is due to the washing out of the phenol and the induced cells as they are being replaced with the media and the fresh cells growing in the first mini-bioreactor. For the 1400ppm induction, however, no washing out profile is apparent but the system demonstrated its ability to stabilize even after a highly toxic insult.

Step Injection

Another possibility is the situation where a toxic chemical is present in the effluent over an extended period of time. TV1061's response to this type of insult was determined using a step injection, the maintaining of the chemical concentration for the duration of the experiment. Using this method, it is possible to measure the response of the bacteria to the continuous presence of toxic material. TV1061 was tested with 100ppm of phenol over a period of about 11 hrs. The response ratio, or RBL, is plotted in Figure 4. After the initial addition, the system reaches equilibrium around an RBL value of 2.5 and maintains this level for the rest of the experiment. After stopping the phenol injection, the system demonstrates the washing-out profile previously discussed and then levels off near the original BL it had before induction.

The dotted line represents the concentration of the phenol calculated mathematically using the species balance equations listed earlier. The washing out effect due to dilution is shown and correlates well with the measured bioluminescence wash out pattern. Figure 5 shows the results of TV1061 induced with ethanol using the step injection method. Again the bioluminescence leveled off, here around a value of 3.5, and follows the wash out pattern as well. These results demonstrate the

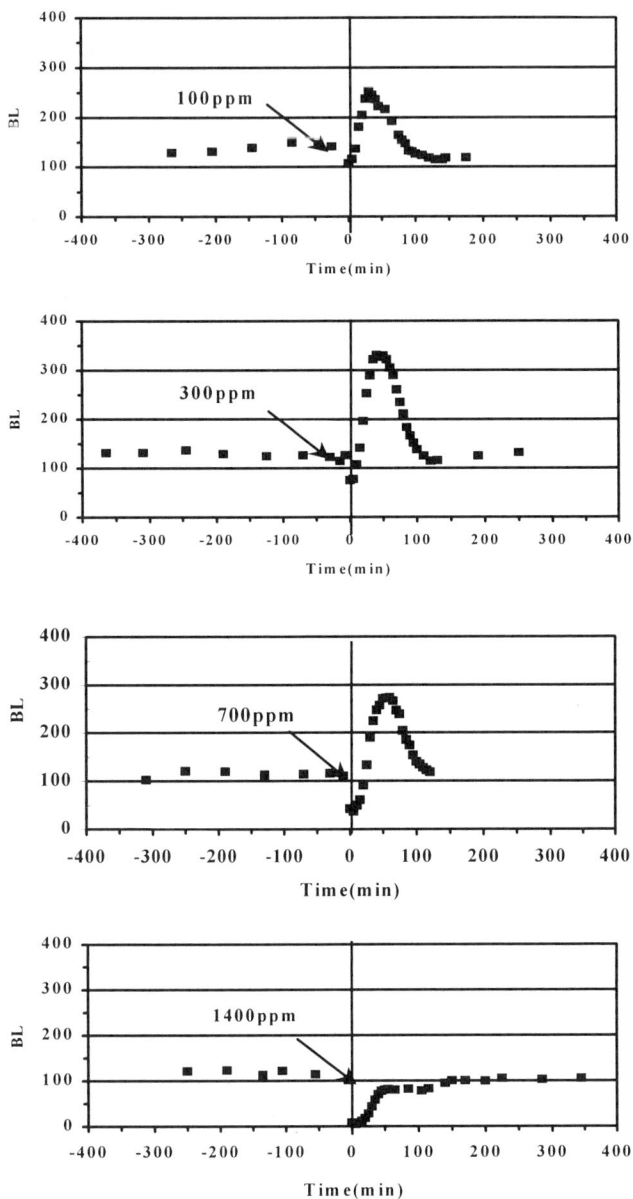

Figure 3. Response kinetics of TV1061 to a pulse injection of different concentrations of phenol (V₁=10mL,V₂=20mL,D₁=D₂=0.83/hr)

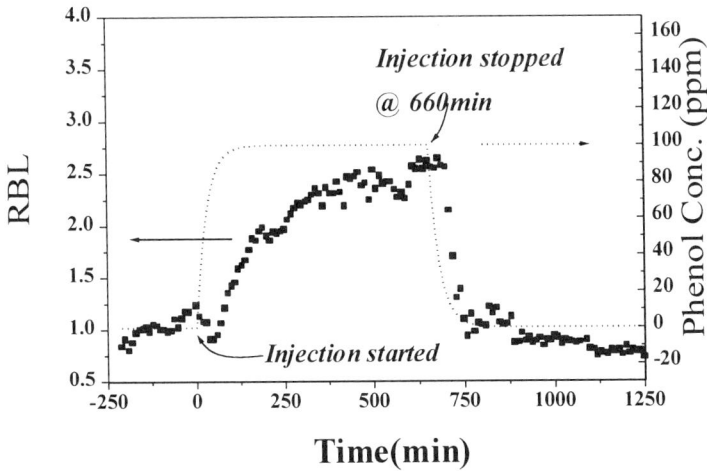

Figure 4. Response ratio of TV1061 to a step-injection of 100 ppm phenol.
(V_1=10ml, V_2=20, D_1=0.8/hr, D_2=2.1/hr)

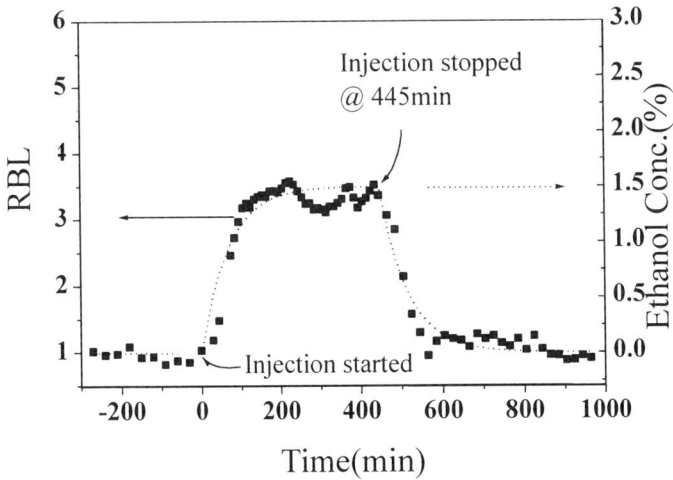

Figure 5. Response of TV1061 to a step injection of 1.5% ethanol
(V_1=10mL, V_2=20mL, D_1=0.87/hr, D_2=2.1/hr)

response of the strain when exposed for to a chemical for a short period of time or for a long period of time with the final BL level being very near that of the basal level.

Figure 6 shows the results of a step injection where the concentration of phenol exceeds the sub-lethal concentration. As with the pulse injection, there is a sudden decrease in the bioluminescence, shown by the decrease in the RBL values. As soon as the injection was stopped, the RBL value again rose to a value near that of the pre-induced cells. These results demonstrate the effectiveness of this system to deal with either a sudden increase for a short period of time or for a long period of time with the final level near that of the basal level before induction. It also demonstrates the differences inherent in the bacteria. When exposed to high concentrations for a prolonged period, DPD2794 had a depressed plateau formation with increases on either side (10 – Figure). TV1061, however, displays no significant increase but had a large decrease in bioluminescence shortly after induction and remained low until the injection of chemical was stopped.

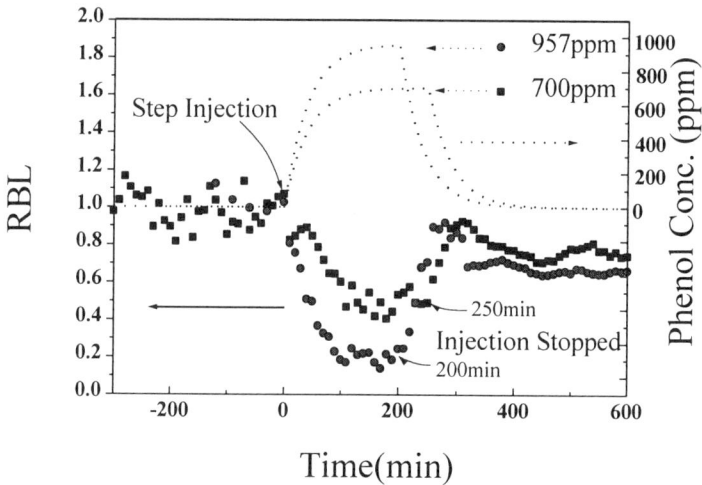

Figure 6. Responses of TV1061 to the phenol over sub-lethal concentration
(V1=10mL, V2=20mL, D1=0.87, D2=1.5/hr)

CONCLUSION

The responses of TV1061 demonstrate its usefullness as a biosensor for the detection of protein damaging toxicity. This strain combined with the minibioreactors provided a system which was consistent and reliable. Theoretically, as long as the plasmid is stable and the bacteria retain their ability to sense toxic upsets, this system may be operated continuously.

Through the use of recombinant techniques, bacteria biosensors that are both specific and sensitive to a series of different toxic avenues, dependent on the promoters cloned, and produce a measurable signal under such conditions are possible. Though they have many similarities, the strains have their own "personality" and respond to different situations. Their response is even unique, as shown by the response kinetics of TV1061, with its initial drop in bioluminescence as well as the prolonged decrease when exposed to an insult over an extended time. By using these bacteria, it is possible to not only detect the presence of toxicity but to also "identify" it by the mode that it works. Also, the signal produced when there are chemicals present is quick, around 120 minutes, and definite as the BL remains fairly constant prior to induction. This difference may be used as an indicator of toxicity and, thus, provide a warning to the presence of hazardous substances.

As for the system, its potential use in environmental monitoring is apparent, but the results of these experiments show that the use of any strain requires its previous characterization within the two-stage minibioreactors before implementation. Based upon previous results one would have thought the system failed initially due to the reduction in light, something previously unseen. Therefore further strain characterization needs to be done using this system.

ACKNOWLEDGEMENTS

This research was funded by a KJIST independant project (1997-1999) and in part by the Korea Science and Engineering Foundation (KOSEF) through the ADEMRC at the Kwangju Institute of Science and Technology (K-JIST). We would like to thank Dr. LaRossa at DuPont Co. for kindly providing the strain TV1061 to be used in our study.

REFERENCES

1. Arain, T.M.; Resconi, A.E.; Hickey, M.J.; Stover, C.K.; *Anti. Agents Chemo.*, 1996, 40, 1536
2. Belkin, S.; Smulski, D.R.; Vollmer, A.C.; Van Dyk, T.K.; LaRossa, R.A.; *Appl. Env. Microbiology*, 1996, 42
3. Boyd, E.M.; Killham, K.; Wright, J.; Rumford, S.; Hetheridge, M.; Cumming, R.; Meharg, A.A.; *Chemosphere*, 1997, 35, 1967
4. Bulich, A.A.; Greene, M.W.; Isenberg, D.L.; In D.R. Branson and K.L. Dickson(eds.), Aquatic Toxicology and Hazard Assessment: Fourth Conference, ASTM STP 737, Philadelphia, PA
5. Chen, C.Y.; Lin, K,C.; Yang, D.T.; *Chemosphere*, 1997, 9, 1959
6. Choi, S.H.; Gu, M.B.; Not published yet
7. Cooksey, C.C.; Morlock, G.P.; Beggs, M.; Crawford, J.T.; *Anti. Agents Chemo.*, 1995, 39, 754
8. Fabricant, J.D.; Chalmers, J.H.; Bradbury, M.W; *Bull. Environ. Contam. Toxicology*, 1995, 54, 90

196

9. Gu, M.B.; Dhurjati, P.S.; Van Dyk, T.K.; LaRossa, R.A.; *Biotechnol. Prog.*, 1996, 12, 393

10. Gu, M.B.; Gil, G.C.; Kim, J.H.; *Biosensors and Bioelectronics*, 1999, 14, 355

11. Gustavson, K.E.; Svenson, A.; Harkin, J.M.; *Env. Toxic. Chem.*, 1998, 17, 1917

12. King, J.M.H.; DiGrazia, P.M., Applegate, B.; Burlage, R.; Sanseverino, J.; Dunbar, P.; Larimer, F.; Salyer, G.S.; *Science*, 1990, 249, 778

13. Korpela, M.T.; Kurittu, J.S.; Karvinen, J.T., Karp, M.T.; *Anal. Chem.*, 1998, 70, 4457

14. Lampinen, J.; Virta, M.; Karp, M.; *Appl. Env. Microbiology*, 1995, 61, 2981

15. Maillard, K.; Benedik, M.J., Willson, R.C.; *Environ. Sci. Technol.*,1996, 30, 2478

16. Matrubutham, U.; Thonnard, J.E.; Salyer, G.S.; *Appl. Microbial. Biotechnol.*, 1997, 47, 604

17. Min, J.; Lee, C.W.; LaRossa, R.A.; Gu, M.B.: Submitted for Publication (*Environ. Rad. Biol.*)

18. Munkrittick, K.R.; Power, E.A.; *Environ. Toxicol. Water Qual.*, 1991, 6, 35

19. Nalecz-Jawecki, G.; Rudz, B.; Sawicki, J.; *J. Biomed. Mat. Res.*, 1997, 35, 101

20. Pandard, P.; Vasseur, P.; Rawson, D.M.; *Water Research*,1993, 3, 427

21. Ramanathan, S.; Ensor, M.; Daunert, S.; *Trends in Biotechnology*, 1997, 15, 500

22. Rogowsky, P.M.; Close, T.J.; Chimera, J.A.; Shaw, J.J.; Kado, C.I.; *J. Bacter.*, 1987, 169, 5101

23. Rupani, S.P.; Gu, M.B.; Konstantinov, K.B.; Dhurjati, P.S.; Van Dyk, T.K.; LaRossa, R.A.; *Biotechnol. Prog.*, 1996, 12, 387

24. Sassanfar, M; Roberts, J.W.; *J. Mol. Biol.*, 1990, 212, 79

25. Simpson, M.L.; Salyer, G.S.; Applegate, B.M.; Ripp, S., Nivens; D.E., Paulus, M.J.; Jellison, G.E; *Trends in Biotechnology*, 1998, 16, 332

26. Smith, T.; Pitts, K.; McGarvey, J.A.; Summers, A.O.; *Appl. Env. Microbiology*, 1998, 64, 1328

27. Steevens, J.A.; Vansal, S.S.; Kallies, K.W.; Knight, S.S.; Cooper, C.M.; Benson, W.H.; *Chemosphere*, 1998, 36, 3167

28. Sticher, P.; Jaspers, M.C.M.; Stemmler, K.; Harms, H.Zehnder; A.J.B., Van Der Meer, J.R; *Appl. Env. Microbiology*, 1997, 63, 4053

29. Sousa, S.; Duffy, C.; Weitz, H.; Glover, L.A.; Bar, E.; Henkler, R.; Killham, K.; *Env. Toxic. Chem.*, 1998, 17, 1039

30. Sweet, L.I.; Travers, D.F.; Meier, P.G.; *Env. Toxic. Chem.*, 1997, 16, 2187

30. Tauriainen, S.; Karp, M.; Chang, W.; Virta, M.; *Appl. Env. Microbiology*, 1997, 63, 4456

32. Thomulka, K.W.; Abbas, C.G.; Young, D.A.; Lange, J.H. *Bull. Environ. Contam. Toxicology*, 1996, 56, 446

33. Van Dyk, T.K.; Majarian, W.R.; Konstantinov, K.B.; Young, R.M.; Dhurjati, P.S.; LaRossa, R.A. *Appl. Env. Microbiology*, 1994, 60, 1414-1420

34. Virta, M.; Lineri, S.; Kankaanpaa, P.; Karp, M.; Peltonene, K.; Nuutila, J.; Lilius, E.; *Appl. Env. Microbiology*, 1998, 64, 515

35. Vollmer, A.C; Belkin, S.; Smulski, D.R; Van Dyk, T.K.;LaRossa, R.A.; *Appl. Env. Microbiology*, 1997, 63, 2566

Chapter 14

Whole-Cell Environmental Monitoring Devices: Bioluminescent Bioreporter Integrated Circuits

Steven Ripp[1], Bruce M. Applegate[1], David E. Nivens[1], Michael J. Paulus[2], George E. Jellison[2], Michael L. Simpson[2], and Gary S. Sayler[1,3]

[1]Center for Environmental Biotechnology, University of Tennessee, Knoxville, TN 37996
[2]Oak Ridge National Laboratory, P.O. Box 2008, MS 6006, Oak Ridge, TN 37831

Bioluminescent bioreporter integrated circuits (BBICs) are novel biosensor devices that utilize light emitting microorganisms as biological sensors for chemical contaminants. The microorganisms are coupled to a full-custom integrated circuit containing photodetector units and low-noise signal processing circuitry for measuring and communicating chemically induced cellular light responses. Two prototype BBICs have been designed and tested using naphthalene and toluene sensitive bioluminescent bioreporters. Measured sensitivities have approached the low parts-per-billion range.

Whole-Cell Biosensors

Chemical sensing methods involving microorganisms have been widely applied for the detection and measurement of industrial pollutants and toxicants. These 'whole-cell' biosensors consist of living microbial cells that respond in some manner to a target chemical, coupled to an instrument capable of detecting this response. The use of intact microbes as chemical sensors allows for detection and monitoring of chemical bioavailability rather than just chemical presence. This is in stark contrast to analytical techniques that simply determine whether a chemical is present and at what concentration, without providing requisite information as to the biological effect of the chemical.

[3]Corresponding author.

197

198

In their earliest forms, whole-cell biosensors consisted of electrochemical transducers that monitored basic changes in cell growth activity (i.e., oxygen utilization or carbon dioxide production) *(1)*. The microbial cells and sensing elements have since evolved into more complex systems. Genetic engineering techniques have produced microorganisms that exhibit enhanced responses to specific biological or physical agents in their environment, and technological advances have produced highly sensitive transducers capable of detecting electrochemical, acoustic, or optical signals. Consequently, whole-cell biosensors are now being developed for the detection of a wide range of chemicals for environmental as well as medical, agricultural, defense, and food applications *(2)*.

Bioluminescent Bioreporter Integrated Circuits

An innovative whole-cell biosensor currently undergoing testing is the bioluminescent bioreporter integrated circuit (BBIC) *(3)*. The biological component of the BBIC consists of bacterial cells that emit light in response to a specific chemical, suite of chemicals, or physical agent within the cell's environment. The sensor is a photodetector unit capable of detecting the bacterial light signal. The bioluminescent bacteria, photodetectors, and all circuitry necessary for processing the light signals and communicating the results are situated on a 4 mm^2 integrated circuit (Fig. 1).

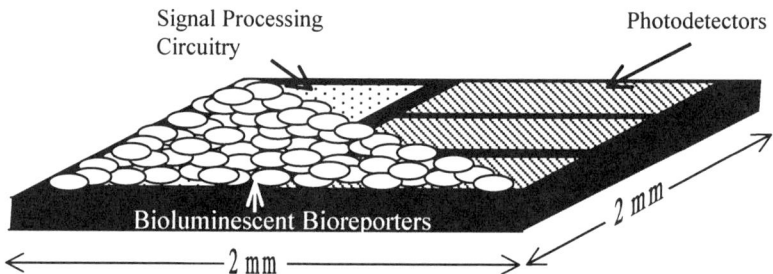

Figure 1. Basic embodiment of a bioluminescent bioreporter integrated circuit.

Bioluminescent Bioreporters

The bioluminescent bacteria adhered to the integrated circuit are genetically engineered cells capable of generating light in response to chemical exposure. Light production is dependent on a genetic operon designated *lux*, derived from luminescent marine bacteria like *Vibrio fischeri*, *Vibrio harveyi*, or *Photobacterium leiognathi (4)*. The *lux* operon consists of five genes, *luxC*, *D*, *A*, *B*, and *E*. *luxA* and *B* together encode for the enzyme luciferase which is responsible for generating the light response. Luciferase converts long-chain aldehydes into fatty acids with a concomitant production of photons, visible as blue-green light at 490 nm. The fatty acids are recycled back into the aldehyde substrate by a multienzyme fatty acid complex consisting of three proteins, a reductase, transferase, and synthetase encoded by the *luxC*, *D*, and *E* genes, respectively.

Using genetic engineering techniques, the *lux* operon can be placed under the control of a promoter element that is induced in the presence of a specific chemical. Activation of the promoter results in transcription of the *lux* genes and light production, thereby forming a direct correlation between chemical presence and bioluminescence. *lux* fusions can consist of just the *luxAB* complex or the complete *luxCDABE* cassette. If only the *luxAB* genes are used, the cells must be supplied with an aldehyde substrate, usually decanal, before light production is possible. Utilization of the entire *luxCDABE* cassette, however, allows for a completely self-contained bioluminescent response, negating the requirement for any exogenous substrate additions. Consequently, these bioreporters can be used for continuous on-line, near real-time measurements.

Bioluminescent bioreporters can also be engineered using a eukaryotic genetic system designated *luc (5, 6)*. *luc* genes are derived from the firefly (*Photinus pyralis* or *Luciola mingrelica*) or click beetle (*Pyrophorus plagiothalamus*) and have been incorporated into *Escherichia coli* cells for biosensing applications. However, similar to the *luxAB* constructs described above, *luc*-based bioreporters must be supplied with an exogenous substrate, luciferin, before bioluminescence can occur. Cells must also typically be destructively lysed. Thus, as currently designed, *luc* bioreporters cannot be continuous fully independent biosensors.

Pseudomonas fluorescens HK44

A well-studied example of a *lux*-based bioluminescent bioreporter is the microorganism *Pseudomonas fluorescens* HK44, which is capable of sensing naphthalene, a common constituent of polyaromatic hydrocarbons (PAHs). *P. fluorescens* HK44 contains the complete *luxCDABE* cassette, and is therefore capable of intrinsic bioluminescence *(7)*. The original parental *P. fluorescens* strain from which HK44 was derived was isolated from a manufactured gas plant facility heavily contaminated with PAHs. The strain was specifically isolated due to its ability to degrade naphthalene, which was experimentally determined to occur via a two pathway catabolic process situated on a plasmid designated pKA1 (Fig. 2). The upper pathway converts naphthalene to salicylate, which is then further converted to catechol through the lower pathway. In strain HK44, genes within the lower pathway were replaced with the *luxCDABE* genes, producing a cell that continued to degrade naphthalene to salicylate, but rather than further converting salicylate to catechol, salicylate now served as the inducer for the artificial *lux* operon. Therefore, *P.*

fluorescens HK44 became a bioluminescent bioreporter for naphthalene (or salicylate), and has been applied in several experimental assessments to measure naphthalene bioavailability under environmental conditions *(8-11)*. Similar bioluminescent bioreporters incorporating the complete *luxCDABE* cassette have since been engineered to detect a number of biological and physical agents (Table I).

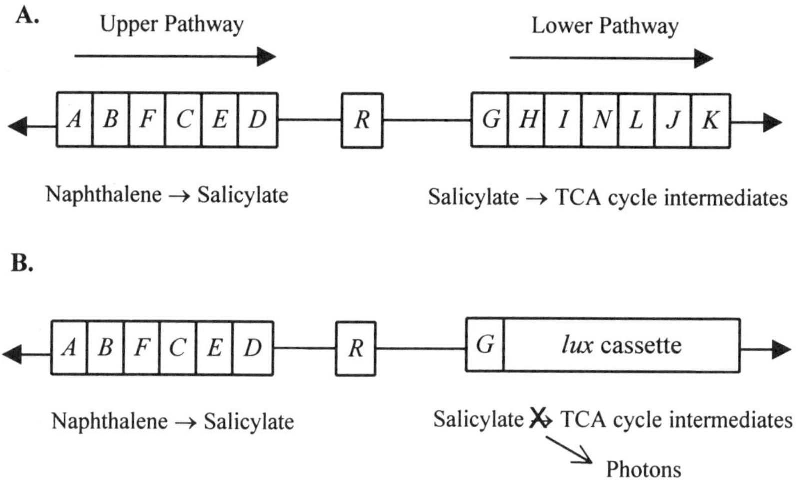

A.

Upper Pathway   Lower Pathway

◄─ [A | B | F | C | E | D]──[R]───[G | H | I | N | L | J | K]─►

Naphthalene → Salicylate   Salicylate → TCA cycle intermediates

B.

◄─ [A | B | F | C | E | D]──[R]───[G | *lux* cassette]─►

Naphthalene → Salicylate   Salicylate ✗ TCA cycle intermediates
                                                    ↘ Photons

Figure 2. A) Genetic organization of the pKA1 catabolic plasmid. Genes of the upper naphthalene regulatory system encode for proteins that mediate the conversion of naphthalene to salicylate. Salicylate is then further degraded to TCA cycle intermediates. B) In Pseudomonas fluorescens *HK44, genes within the lower pathway were replaced with genes of the* lux *cassette to produce a bioluminescent bioreporter sensitive to naphthalene and salicylate.*

Integrated Circuit

The light response generated by the bioluminescent bioreporters is typically measured with optical transducers such as photomultiplier tubes, photodiodes, microchannel plates, charge-coupled devices, or photographic films. Additionally required is usually some means of transferring the bioluminescent signal to the transducer, which necessitates the need for fiber optic cables, lenses, or liquid light guides. What results is a large, bulky instrument anchored to power and optic cables. In contrast, integrated circuit technology positions all the required instrumentation onto a 2 mm by 2 mm silicon chip, and, by adhering the bioluminescent bioreporters directly to the chip, the need for external signal transfer cables can be bypassed (Fig. 3). In essence, the integrated circuit consists of only two main components;

Table I. Chemical and physical agents capable of detection by *luxCDABE*
bioluminescent bioreporters

Xylene	2,4-D
Phenol	Dinitrotoluene
Benzene	Ethylbenzene
Zinc	Lead
Cadmium	Cobalt
Chromium	Thallium
Copper	Polychlorinated Biphenyls (PCBs)
Mercury	Nitrate
Nickel	Naphthalene
Toluene	Arsenic
Trichloroethylene	DNA damaging agents (ultraviolet light)
4-chlorobenzoic acid	Isopropylbenzene
Oxidative stressors (peroxides, superoxide radicals)	Environmental stressors (heat shock, alginate production)

photodetectors for capturing the on-chip bioluminescent bioreporter signals and signal processors for managing and storing information derived from bioluminescence. If required, remote frequency (RF) transmitters can also be incorporated into the overall integrated circuit design for wireless data relay, which can approach levels of satellite transmission coupled to global position sensing to create intelligent distributed biosensing networks. These networks will ultimately allow BBICs to communicate with each other, referencing time and location along with other physical parameters, to provide vector or transport information to pinpoint locations of potential biohazardous concern.

The integrated circuit design is based on an industry standard complementary–metal-oxide-semiconductor (CMOS) process *(12)*. CMOS possesses several characteristics desirable for BBIC applications, including low power drain, high noise immunity, the ability to perform digital, analog, and electro-optical signal processing, and highly reliable performance. Furthermore, the nature of CMOS design allows for the incorporation of accessory devices such as RF communication and position-sensing circuitry. CMOS is also amenable to optical application-specific integrated circuit (OASIC) fabrication. These 'personalized' integrated circuits can be produced quickly and at relatively low cost to custom fit specific applications. Therefore, BBICs can be crafted to fit a large number of user-defined functions.

Adhering bioreporters to integrated circuits

BBICs will require methods for encapsulating and immobilizing bioreporter cells directly to integrated circuits. A large number of natural and synthetic polymers have been developed for cell encapsulation, all of which are potentially applicable in the BBIC format *(13, 14)*. It must be demonstrated, however, that the matrix itself does

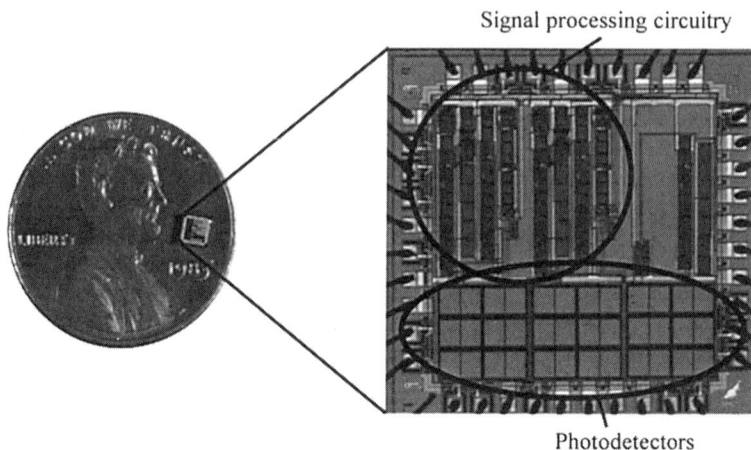

Signal processing circuitry

Photodetectors

Figure 3. Size of prototype integrated circuit compared to a U.S. penny. The magnified view displays the photodetector units and signal processing circuitry required for detecting and monitoring bioluminescence.

not induce bioluminescence nor optically interfere with bioluminescent signals. To best fulfill BBIC requirements, it is desirable to have a liquid matrix that can be combined with the cells, directly applied to the integrated circuit surface as a thin film, and allowed to polymerize. Alternatively, bioreporter cells can be encapsulated within blocks or sheets of polymer, from which thin slices can be dissected and attached to the chip. The encapsulation matrix additionally needs to provide an environment in which cell survivability can be maintained for extended periods. Thus, nutrients and co-factors, as well as a means of sustaining proper hydration, must be incorporated within the final matrix. An encapsulation matrix functional in both aqueous and vapor phases is also desirable. Overcoming all of these obstacles within a single matrix is a difficult task, but experiments utilizing the family of hydrogel and xerogel polymers are showing great promise. These polymers, hydrophilic in nature and optically clear, are amenable to bioreporter encapsulation and have been shown in our own studies to not interfere with bioluminescent signals. Cell survivability, however, remains problematic, with bioreporters remaining fully functional for periods approaching one to two weeks. Current attempts are aimed at incorporating lyophilized bioreporter cells within the encapsulation matrix for long-term storage intervals. Bioreporters could then be resuscitated directly on the integrated circuit for use when needed.

Prototype integrated circuits

We have developed two prototype integrated circuits for initial BBIC testing purposes. The first prototype was fabricated in a 1.2 μm CMOS process employing a photodiode that has a strong response to the 490 nm bioluminescent signal *(15)*. Leakage and noise of the photodiode was shown to be minimal *(3)*. The second prototype was fabricated in a newer 0.5 μm CMOS process to allow the inclusion of an on-chip remote frequency transmitter. Again, this photodiode generated very low noise (~120 e⁻/sec with an integration time of 13 minutes).

The front-end signal conditioning circuit for both chips was a current-to-frequency converter (CFC). When compared to conventional electrometer circuits, this signal conditioning circuit provided lower noise, faster recovery from overloads, and larger dynamic range. A digital signal proportional to the sum of the leakage and photo-current was generated by counting pulses from the CFC for a specified time (i.e., the integration time). To determine background sensitivity, multiple measurements were taken in the absence of bioluminescence with the integration time set to one minute. Leakage currents consistently produced a signal of approximately 6 ± 0.22 counts/minute.

Since the integrated circuits will be in constant contact with encapsulated bioreporter cells, it was necessary that they be overlaid with a protective coating. Amorphous silicon dioxide is commonly used for this purpose, but is unsuitable in BBIC applications due to its susceptibility to attack by various biological and chemical agents. Rather, a silicon nitride film, which has been shown to be much more resistant to chemical and biological degradation, will be deposited on the integrated circuits *(16)*. Encapsulated bioreporter cells can then safely be placed on the integrated circuit without harm to the circuit or the bioreporters. Conveniently, silicon nitride also possesses optical properties that will actually improve the response of the photodetectors to the 490 nm bioluminescent signal *(17)*.

Prototype BBICs

The integrated circuits described above were initially tested in the BBIC format using the bioluminescent bioreporters *Pseudomonas fluorescens* 5RL and *Pseudomonas putida* TVA8. *P. fluorescens* 5RL is a naphthalene bioreporter containing the same *nah-lux* fusion as described previously for the *P. fluorescens* HK44 bioreporter, but is incapable of metabolizing salicylate *(18)*. *P. putida* TVA8 is a bioreporter for BTEX (benzene, toluene, ethylbenzene, and xylene), representing water soluble components of petroleum fuels *(19)*. Toluene was chosen as the representative chemical inducer for strain TVA8. Since only two integrated circuits were available for initial testing, bioreporter cells could not be directly affixed to the circuits. Rather, a flow cell was developed to confine the bioreporters directly above the integrated circuit within a light-tight environment. As more integrated circuits become available, actual adhesion of bioreporters to the circuits will be performed to produce a fully functional BBIC as shown in Figure 1. Bioluminescence was induced in the bioreporter cells and a control sample of cells by exposure to naphthalene or toluene vapor. Bioluminescent data was transferred from the BBIC directly to a computer interface. The minimum detectable concentrations achieved by *P. fluorescens* 5RL and *P. putida* TVA8 on the BBIC both approached approximately 10 ppb (Fig. 4).

204

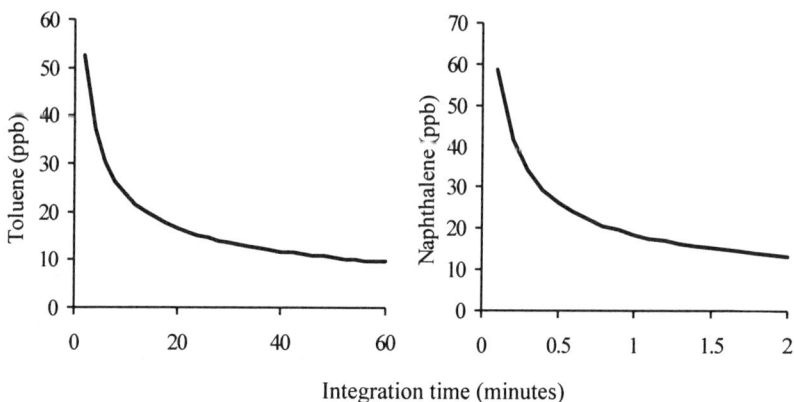

Figure 4. Minimum detectable concentrations of toluene and naphthalene as a function of integration time for the prototype BBIC containing either the Pseudomonas putida *TVA8 (toluene) or* Pseudomonas fluorescens *5RL (naphthalene) bioluminescent bioreporters.*

Towards better BBICs

Improving BBIC technology relies on refining integrated circuit design as well as enhancing bioreporter attributes. Integrated circuit complexity has approximately doubled every year since its inception, with corresponding increases in reliability and decreases in costs. Therefore, improvements in chip design are likely unmitigated. New chips with more sensitive photodetectors and faster processing units are currently undergoing testing, and will undoubtedly lead to BBIC chemical detection limits below the current 10 ppb threshold. These integrated circuits can also carry sensors to monitor environmental parameters such as temperature, pH, or oxygen concentration, thereby producing a more comprehensive 'laboratory-on-a-chip' device.

Bioluminescent bioreporters are also evolving into organisms suitable for a wide range of BBIC applications. Genetic engineering modifications are now underway to create mammalian *lux*-based bioreporters that will produce bioluminescence at 37°C without cell lysis or exogenous reagent requirements. Such bioreporters, when placed on integrated circuits, may have huge potential in the continuous, on-line monitoring of medically important diagnostic agents or in the rapid assessment of novel chemicals for combinatorial chemistry purposes.

Acknowledgments

Funding support provided by The Perkin-Elmer Corporation, NASA, the Department of Energy, and the National Institutes of Health.

Literature Cited

1. Buerk, D. G. *Biosensors: Theory and Applications*; Technomic Publishing: Lancaster, Pennsylvania, 1993.
2. Rogers, K. R.; Gerlach, C. L. *Environ. Sci. Technol.* 1996, *30*, 486-491.
3. Simpson, M. L.; Sayler, G. S.; Ripp, S.; Nivens, D. E.; Applegate, B. M.; Paulus, M. J.; Jellison, G. E. *Trends Biotech.* 1998, *16*, 332-338.
4. Meighen, E. A. *Annu. Rev. Genet.* 1994, *28*, 117-139.
5. Cebolla, A.; Vazquez, M. E.; Palomares, A. J. *Appl. Environ. Microbiol.* 1995, *61*, 660-668.
6. Hastings, J. W. *Gene* 1996, *173*, 5-11.
7. King, J. M. H.; DiGrazia, P. M.; Applegate, B.; Burlage, R.; Sanseverino, J.; Dunbar, P.; Larimer, F.; Sayler, G. S. *Science* 1990, *249*, 778-781.
8. Heitzer, A.; Webb, O. F.; Thonnard, J. E.; Sayler, G. S. *Appl. Environ. Microbiol.* 1992, *58*, 1839-1846.
9. Heitzer, A.; Malachowsky, K.; Thonnard, J. E.; Bienkowski, P. R.; White, D. C.; Sayler, G. S. *Appl. Environ. Microbiol.* 1994, *60*, 1487-1494.
10. Sayler, G. S.; Cox, C. D.; Burlage, R.; Ripp, S.; Nivens, D. E.; Werner, C.; Ahn, Y.; Matrubutham, U. In *Novel Approaches for Bioremediation of Organic Pollution*; Fass, R., Flashner, Y., Reuveny, S., Eds.; Kluwer Academic/Plenum Publishers: New York, 1999.
11. Webb, O. F.; Bienkowski, P. R.; Matrubutham, U.; Evans, F. A.; Heitzer, A.; Sayler, G. S. *Biotech. Bioeng.* 1997, *54*, 491-502.
12. Frederiksen, T. M. *Intuitive CMOS Electronics*; McGraw-Hill: New York, 1989.
13. Armon, R.; Dosoretz, C.; Starosvetsky, J.; Orshansky, F.; Saadi, I. *J. Biotech.* 1996, *51*, 279-285.
14. Cassidy, M. B.; Lee, H.; Trevors, J. T. *J. Ind. Microbiol.* 1996, *16*, 79-101.
15. Simpson, M. L.; Dress, W. B.; Ericson, M. N.; Jellison, G. E.; Sitter, D. N.; Wintenberg, A. L.; French, D. F. *Rev. Sci. Instr.* 1998, *69*, 377-383.
16. Jellison, G. E.; Modine, F. A.; Doshi, P.; Rohatgi, A. *Thin Solid Films* 1998, *313*, 193-197.
17. Doshi, P.; Jellison, G. E.; Rohatgi, A. *Appl. Opt.* 1997, *36*, 7826-7837.
18. Johnston, W. H. , University of Tennessee, Knoxville, 1996.
19. Applegate, B. M.; Kehrmeyer, S. R.; Sayler, G. S. *Appl. Environ. Microbiol.* 1998, *64*, 2730-2735.

AFFINITY-BASED SENSORS

Chapter 15

Affinity Biosensors for Characterization of Environmental Endocrine Disruptors

Hongwu Xu, Miriam M. Masila, and Omowunmi A. Sadik[1]

Department of Chemistry, State University of New York at Binghamton, P.O. Box 6016, Binghamton, NY 13902–6016

Bioaffinity sensors for the identification and quantitation of environmental endocrine disruptors and metabolites are described. The procedure relies on the recognition of an artificial template having a predefined affinity for the endocrine disruptor of interest. When exposed to the analyte, the template competes with, and displaces a receptor generating a signal. The optimal binding activity was determined from the standard curves generated from the quantity of the labeled molecule remaining after the system reaches equilibrium. Detection limits in the range low ppb range was obtained for PCBs, triazines and metal ions. A rational strategy for assessing the additive and synergetic effects of several structurally similar analogs of PCBs is presented.

Introduction

There are accumulating evidences indicating that certain pesticides, environmental pollutants, industrial chemicals and naturally occurring phytoestrogens can dramatically alter the normal physiological functioning of the endocrine system (1-3). These chemicals are commonly reffered to as "endocrine disrupting chemicals" (EDCs). EDCs can alter the endocrine system by affecting hormone synthesis or the degradation, transport, receptor binding, and gene transcription and elimination of natural-borne hormones responsible for the regulation of homeostasis, reproduction, and developmental processes [1,2]. The desire to understand the cellular and molecular basis for these actions, and the need for taking appropriate regulatory steps are the major reasons for the establishment of the endocrine disrupting working group (3). Obtaining information on which chemicals in the environment should be labeled

[1]Corresponding author (fax: 607–777–4478; email: osadik@binghamton.edu).

as EDCs is very critical and this poses a signifivant challenge to existing analytical techniques.

Currently, there are certain problems making the monitoring and surveillance studies of EDCs difficult. These include the cost and time involved in screening a wide variety of synthetic EDCs and their metabolites. According to a recent EPA advisory committee on endocrine disruptors, several commercial chemicals, including nearly 87,000 mixtures, should be screened to assess and determine their effects on the endocrine system [3]. The hormonal activities of man-made or naturally occurring EDCs interacting with estrogen receptors have so far been evaluated using bioassays (1). Other techniques include DNA-binding assays, receptor gene assays, fluorimetry, and chromatographic methods. Although, simple in operation, receptor-binding assays are limited by the inability to differentiate the binding of agonist or antagonist EDCs. Consequently, researchers are seeking alternative ways of narrowing their current focus by first learning about the molecular processes contributing to endocrine disrupting characteristics. Some experiments are being designed to help identify and locate sites and actions of endocrine disrupting molecules in laboratory animals and wildlife (4-10). These include interference with normal molecular pathways, e.g. antibody inactivation of natural agonists and the creation of molecular models based on structure–activity relationships.

Bioaffinity sensors are capable of separating individual or selected range of components from complex mixtures of biomolecules based on their chemical structures and/or biological functions. This is possible by designing a selective interface and containing a biological molecule having a high affinity for the EDC of interest. The biologically selective layer can then be immobilized on the surface of a transducer to generate unique signals for the analyte. In addition, through the design of sensor arrays for multianalyte detection, bioaffinity sensors can be used for validating endocrine targets whose actions remain largely unknown.

The minimum affinity allowed for EDC-receptor interactions depends on the design of the assay and the concentration of the ligand to be measured. Since most suspected EDCs exist in the low parts-per-billion (ppb) levels, the rate constant can therefore be the same order of magnitude as the molarity of the EDC. An ideal screening system is defined by the molecular or biological characteristics that the system should recognize. The ability to recognize certain molecular characteristics is also determined partly by the receptor system produced, and partly by the bioaffinity sensor approach used. We have previously reported immunosensors for some known and suspected EDCs such as polychlorinated biphenyls (PCBs), and 2,4,5 TCP, and (11-13). Experiments performed using PCB antibodies resulted in a detection limit of 0.1 ng/ml for selected Aroclors, and with a total analysis time of about 20 minutes (11,12). This paper provides recent data obtained for bioaffinity characterization of some potential EDCs such as PCBs, triazines and heavy metals.

Experimental

The following instruments were used for the experiments described in this paper: Hewlett-Packard Diode-array UV/Vis spectrophotometer was used for the characterization of all protein conjugates. ELX 800 UV Plate Reader (from Bio-Tek Instruments) was used for all of the enzyme-linked immunosorbent assay (ELISA) experiments. EG&G PAR potentiostat/ galvanostat Model 263A and EG&G 270 software were employed for the electrochemical experiments using silver/silver chloride reference electrode, platinum wire counter electrode and gold (A = 0.2 cm^2) as working electrode. Quartz crystal microbalance (QCM) measurements were carried out using EG&G quartz crystal analyzer (Model QCA917). A 9MHz EG&G At-cut quartz crystal was sandwiched between the two gold electrodes (A = 0.186 cm^2).

Design of artificial interfaces with predefined functions

Biosensor sensing elements were designed by first synthesizing an analog of the desired EDC. In this work, different analogs of PCBs, triazines and toxic metals were prepared. The analogs were then covalently immobilized on gold substrates or gold quartz crystal electrodes (Figure 1). A monolayer of thin film was coated to serve as the matrix. Due to co-adsorption, a mixed monolayer molecular imprinting of the template and the matrix is formed with a well-defined cavity. The monolayer cavity were shaped like that of the template and had high affinity for the analyte. Upon exposure of this surface to the EDC-analyte, the analog was displaced due to its significantly higher affinity for the substrate. This generated an analytical signal corresponding to the concentration of the EDC.

Production of Templates and Antibody

PCB Analogs:
Anti-PCB antiserum were produced as follows under EPA Contract No. 68-03-3511 (14) and the procedures employed are summarized: Rabbit polyclonal antibodies were produced against 4-hydroxy analogues of 2,2',4,5,5'-pentachloro biphenyl, namely 6-2,2',4',5,5'-pentachloro-4-biphenylol)-hexanoic acid. These analogues were covalently-linked to keyhole limpet hemocyanin. The coating antigen was prepared from 4-(2,4,5-trichlorophenoxy)-butyric acid and bovine serum albumin using a modified procedure described by Langorne et.al (15).

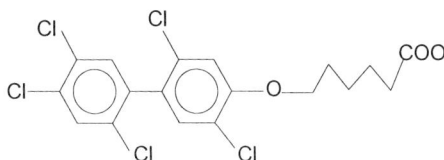

PCB immunization hapten

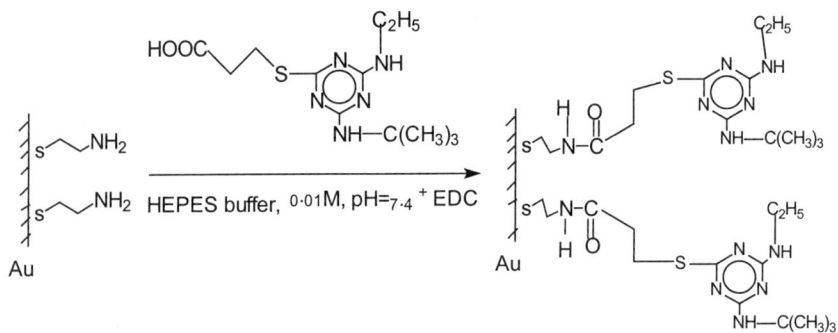

Figure 1. Design of templates for the identification and detection of EDC.

Using this modified procedure, 1-ethyl-3-(3-dimethylaminopropyl) carbodiimide was substituted for N,N'dicyclohexylcarbodiimide. The antibody was purified using Protein A affinity Column which was desalted over an Excellulose desalting column (Pierce). Protein concentration was monitored by measuring the absorbance at 280 nm. Antibodies were divided into aliquots and stored in deep freezer (-18°C) for later use.

Triazine Analogs and Antibody:
Triazine analogs were (Dr. Fleeker) prepared from active esters of the carboxylic acid analogs of triazine haptens using N-hydroxysuccinamide. The triazines were coupled to a high molecular weight carrier, bovine serum albumin (BSA), or keyhole limpet hemocyanin (KLH) used for the production of the antibodies. The antibodies were purified by gel filtration and protein-A immunoaffinity columns, and subsequently characterized using ELISA and nuclear magnetic resonance (NMR) techniques. By using these antibodies, sensors for *s*-triazine were developed based on antibody inhibition of the current generated by the ferricyanide mediator on antigen-immobilized gold electrodes.

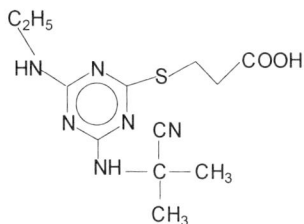

Cyanazine Hapten   Cyanazine BSA receptor

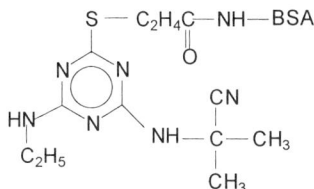

Sensor Preparation and Characterization

Electrochemical Immobilization:
Au electrode was pretreated in the same way as the Quartz immobilized technique before being modified with atrazine hapten using EDAC as the coupling reagent. The modified electrode was used for electrochemical analysis, first without soaking in an antibody solution. Later the electrode was incubated in an anti-cyanazine antibody solution at 35°C using a thermostated water-bath. All cyclic voltammetry experiments were conducted at the same temperature. Other electrochemical immobilization procedures were as recently reported (12,13).

Polymer Synthesis:
Various pyrrole derivatives were polymerized through electrochemical oxidation to enable the conducting polymer films to be used for conductivity, electrochemical, and

mass measurements. Some selective electrodes for phenols, PCBs and *s*-triazines were prepared by electropolymerization of pyrrole onto platinum electrodes in the presence of tetrabutyl ammonium perchlorate. The selectivities obtained were comparable to a range of structurally similar organic compounds, including 2,3,5,6-tetra chloroanisole, 2,3,4-trichloroanisole, 2-chloroanisole, 2,4,5-trichlorophenol, simazine, cyanazine, and substituted benzenes.

Characterization of Sensing Elements

Antibody membrane electrodes were characterized using cyclic voltammetry (CV), chronoamperometry (CA) and ELISA protocols. The electrolyte solution used for the CV consisted of 0.1 M phosphate buffer saline (PBS) at pH 7.4, 0.1 M NaCl and 0.1 M $NaHCO_3$. The ELISA test was conducted to assess the bioactivity of the antibody protein incorporated into the polymers. This was performed directly on the polymers deposited on platinum strips, which were coated on polyester and prepared as described above. A section of the polymer was removed from the bulk using a hole puncher and was placed in the bottom of microtitre plates for ELISA analysis. The plates were read at 405 nm every 10-minute interval.

RESULTS & DISCUSSIONS

The ability of EDCs to interfere with the hormone system was a major factor used to develop a detection method. Most studies have focused on estrogens using bioassay techniques. Bioassay involves the quantitation of biological response that follows the application of a stimulus to a living organism. The applied stimulus is represented by standard (or test) samples containing biologically active substances or analytes. The stimulus can be applied to a biological system such as a whole animal or plant, isolated organs or tissues, whole cells or biologically active macromolecules (e.g. antibodies, enzymes, DNA or receptor proteins). Quantitative response can be observed in some aspects of the biological system, thus resulting in a positive or negative signal such as an increase in activity, or negative response (inhibition), or even death to the biological system.

In this study, we have utilized the response of the bioaffinity sensors to determine the biological activity normally attributed to the analytes. The achitecture of the sensors was similar to those commonly used for bioassays. In this case, the endocrine disruptor of interest was permitted to compete with, and displace a receptor label (e.g. enzyme label). The quantity of endocrine disruptor was determined using the standard curve obtained from the quantity of the labeled-molecule remaining after the system reaches equilibrium. The attainment of equilibrium required only 1-3 minutes. The sensor was first characterized using ELISA in order to determine the optimal binding by plotting the absorbance vs. the concentration of coating antigen for the various concentrations of antibody (Figure 2). The optimal antibody dilution was obtained by choosing the most diluted concentration that provides a strong signal for each antigen concentration.

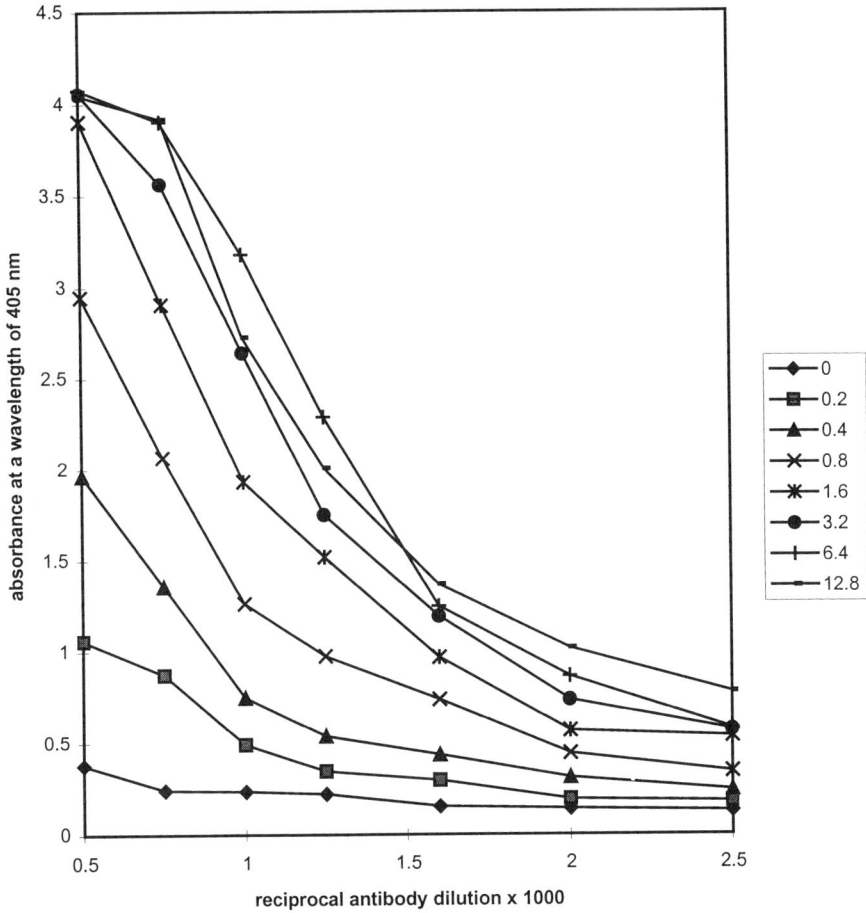

Figure 2. ELISA checkerboard titration of PCB coating antigen and PCB antibody. Conditions: Varying concentrations of PCB antisera diluted in carbonate/bicarbonate buffer (pH 9.5), plates were washed with phosphate buffer pH 7.2, PCB antibody was diluted in PBS Tween 20, seconday antibody was IgG goat anti-rabbit with alkaline phosphatase label.

For the PCB antigen, this concentration was determined to be about 1/1000. The antigen concentration was obtained in a similar manner by plotting the absorbance vs. the reciprocal dilution factor for each antigen concentration. The optimal antigen concentration is the one that yields the steepest slope in the linear region of the curve. The optimal antigen concentration of 1.6 μg/ml obtained was used to fabricate the electrochemical sensor using gold as the working electrode that was pretreated as described in the Experimental Section. Using the PCB sensor developed, the method detection limit was in the low ppb range. In addition it was possible to determine the apparent binding constants for several of the EDCs detected. On-going efforts include the determination of endocrine disrupting index using the relationship between detection limit, receptor concentration, apparent binding constants and molecular weight.

Differentiating the Binding of Interfering Compounds

There are concerns that the effects of different EDCs may be additive, or even synergistic, such that regulating individual compounds may be inadequate. In a particular yeast-based assay, the binary mixture of weak estrogenic pesticides, including dieldrin, chlordane, toxaphene, and endosulfan, exhibited greater than 90-fold relative to the analyte concentration [16]. A typical range is between 160 to 1600-fold of synergistic endocrine receptor binding, which includes the induction of reporter gene activity compared to that observed for the compounds alone. In addition, the process of enabling estrogenic activities in some tissues while completely blocking the same activities in others could result in an increase in estrogenic activities of certain EDCs found in some tissues. This phenomenon is known as "agonistic and antagonistic" effect.

We have tested the synergistic effects of several structurally similar analogs of PCBs under different concentrations of the target compound. Receptors or antibody systems displaying high structural specificities were selected based on a statistical cut-off range of 0.1%. For individual compounds, a particular source of biasing is the presence of cross-reactivity substances, which simulate the analyte to react with an antibody or a receptor. The extents to which these substances disrupt the assay results can be measured partly from their relative binding constants. For example, in the simple case of a labeled receptor binding with PCB having an affinity represented by K_A, and a cross reactant having an affinity represented by K_{CR}, the relative potency (R_P) of the analyte and the cross reactant is given by:

$$R_P = (f K_A / K_{CR}) + b \qquad \text{(Equation 1)}$$

where f and b are free and bound labelled fractions respectively. Equation 1 implies that the R_P of the cross reactant varies as a function of assay response. So, if the analyte binds with an affinity which is 100 times greater than that of the cross reactant, then the analyte will display a 50.5 times potency when 50% of the labeled material is receptor bound. This will increase to a 100-fold in the potency when 0% is bound. This implies a 2-fold change in R_P over the entire range of the dose-response curve assuming this extends from 50% binding of the label at zero dose to 0% at infinite dose. The change in R_P for two different EDCs competing for receptor-binding site occupancy depends on the relative concentrations of the receptor and the EDC in the system. If R_P is unity (i.e. when $b = 1$), this will be considered equipotent, assuming K (the binding constant) is 1. Similarly, the relative potency of two EDCs in the assay reacting with the same (labeled), receptor-binding site is given by:

$$R'_p = ([f_{Rep}] + 1/K_u)([f_{Rep}] + 1/K_s]$$ (Equation 2)

where $[f_{Rep}]$ is the free (labeled) receptor concentration; K_u and K_s are the affinity constants for the EDCs U and S respectively. This step will be used to eliminate chemicals that do not exhibit low binding constants or endocrine characteristics. Low binding constant is defined as $1/e$ value of the detection limit of the particular EDC. Figure 3 is a summary of the relative potency obtained for Aroclor 1248 and different structurally related compounds including substituted anisoles, benzenes, phenols, dichlorodiphenyl trichloroethanes. Results show that the PCB binds with about 2-fold change in the relative potency over the concentration range tested relative to the substituted benzenes, phenols, anisoles, DDT and DDEs. On the other hand, pentachlorophenol, exhibited about 1.5 fold change in potency.

Triazine Sensors

Several pesticides and herbicides are routinely used to improve crop harvesting and pest-control. Due to the growing concern about health effects, several investigations have been conducted in order to understand how pesticides and herbicides degrade in the environment. Consequently, some of these compounds have been implicated as potential endocrine disruptors (2,3). Current methods of monitoring pesticides include liquid chromatography (LC) as well as gas chromatography with mass spectrometry (GC-MS). The high costs and labor involved in chromatographic methods have led to the search for low-cost alternatives, which are capable of providing rapid analysis. We hereby report the development of immunosensors for atrazine, cyanazine, simazine and their metabolites.

Hapten monolayer electrode sensor assembly was used to detect triazine in a flow injection analysis mode. The interaction of the electrode with different antibody concentrations resulted in the formation of an antibody-antigen (Ab-Ag) complex which insulated the electrode towards the $[Fe(CN)_6]^{4-}/Fe(CN)_6]^{3-}$ redox probe and this in turn resulted in no charge transfer. The extent of insulation depends on the antibody concentration and the time of exposure to the antibody solution. The decrease in amperometric response of the antigenic monolayer to corresponding antibody solution for a fixed time produces a quantitative measurement of the antibody concentration. Typical responses obtained for cyanazine-hapten monolayer electrode to different antibody concentrations is shown in Figure 4. The lowest detection limit achieved for cyanazine sensor was 4.0 μg/ml at a response time of few minutes and a less-than 2% cross-reactivity to atrazine, simazine and other metabolites.

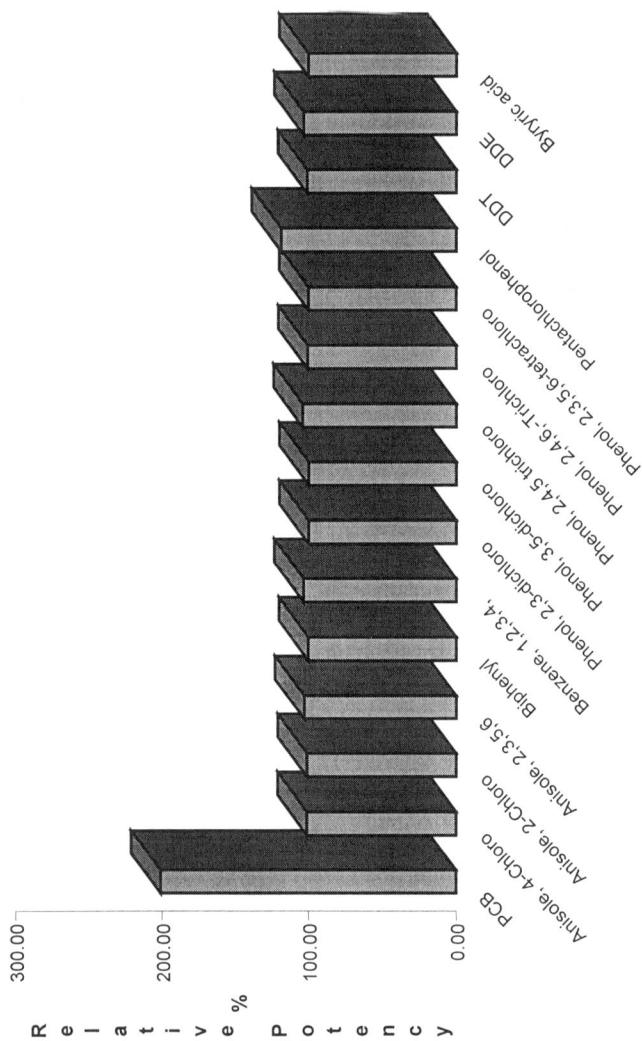

Figure 3. Relating Potency[a] obtained for PCBs and other structurally related compounds. Signal generation obtained using pulsed amperometric detection. Condition: injection volume 100µL, flow rate = 0.5 ml/min, initial potential = 0.6V, final potential = -0.6 V, initial time = 60ms, final time = 480 ms. ((a) given by Equation 1, $R_P = (f K_A/K_{CR}) + b$)

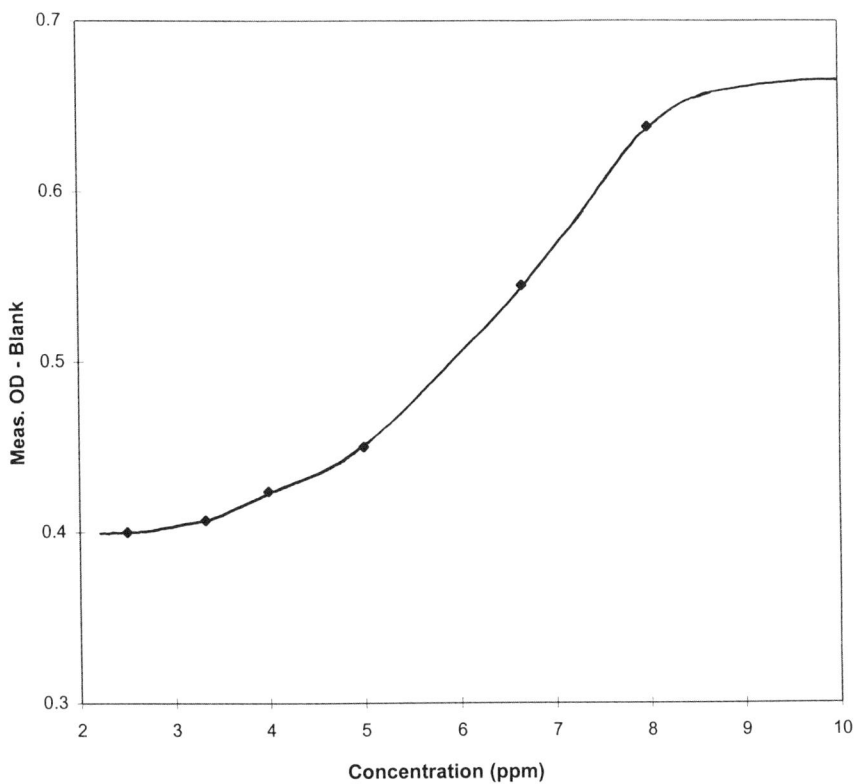

Figure 4. Calibration curve obtained with cyanazine hapten monolayer using different concentrations of anti-cyanazine antibody

Artificial Immunosensors for the Detection of Metal Ions

We have utilized the high affinity chelation of 2-pyridylazo ligands and simple ELISA techniques to develop a highly sensitive and selective analysis of gallium (III) (17). Compounds of 2-pyridylazo (PAR) and their derivatives are attractive because their selectivity can be controlled through pH adjustment and the use of masking agent (18). Selectivity is important in potential application for detoxification of heavy metals, inhibiting cell growth and as labels in clinical diagnosis. Due to its useful spectroscopic and luminescence properties, the use of PAR as complexing agents was used in the design of enzyme-like receptors. The design of PAR-labeled with Ga(III) has to take account of the fact that Ga(III) forms very stable chelates with PAR(K $\cong 10^{20}$ l/mol).

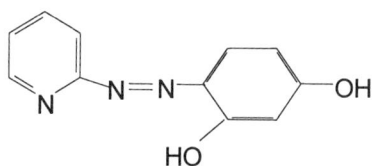

4-(2-Pyridylazo)resorcinol (PAR)   1-(2-Pyridylazo)-2-naphthol (PAN

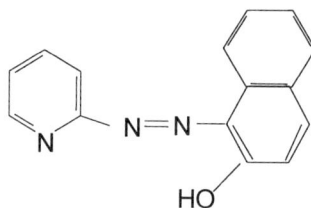

We synthesized PAR–protein conjugates and used the conjugates for the determination of gallium in a solid-phase immunoassay methodology. The principle of metal detection using PAR chelator conjugated to proteins is shown in Figure 5. The PAR was conjugated to ovalbumin to form PAR-OVA conjugate. PAR conjugated to alkaline phosphatase (PAR-AP) was used as detection chelator in the assay format. This detection format uses no antibody unlike in conventional immunoassay. The conjugates were synthesized using water-soluble carbodiimide and N-hydroxy succinamide coupling chemistries. Experimental results obtained from UV/Vis spectrometry, Fourier-transform infrared techniques and mass spectrometry confirmed that a new class of protein conjugates had been synthesized. The characterization showed that the metal ions bind to PAR conjugated to the proteins and not directly on the protein (Figure 6).

The conjugates were employed in a sandwich enzyme-linked immunosorbent assay (ELISA) format for the determination of different metal ions including lead, mercury, cadmium and gallium. The study showed that gallium ions could be conveniently detected with IC_{50} value of 10^{-6}M, a linear concentration range of 5 x 10^{-7}M to 1 x 10^{-4}M and a correlation coefficient of 0.9980. The assay detection limit which is defined as three standard deviations above the zero standard, was recorded to be 5 x 10^{-8}M. Moreover, the method showed a remarkable selectivity for Gallium relative to several other metals investigated. The selectivity was modulated by three

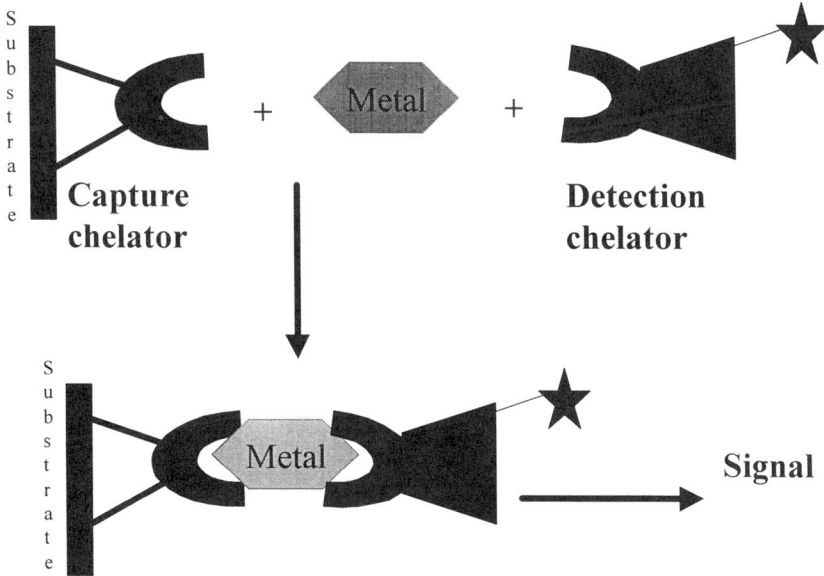

Figure 5. Principle of metal detection using PAR-bioconjugate. The PAR conjugate binds to the solid phase or the transducer for metal sensing. Analyte recognition occurs through the metal sandwiched between another PAR-enzyme-label. The metal shows a co-ordination number of six and the binding is through phenolic oxygen, nitrogen of the pyridine group and nitrogen of the azo group present on the PAR chelator.

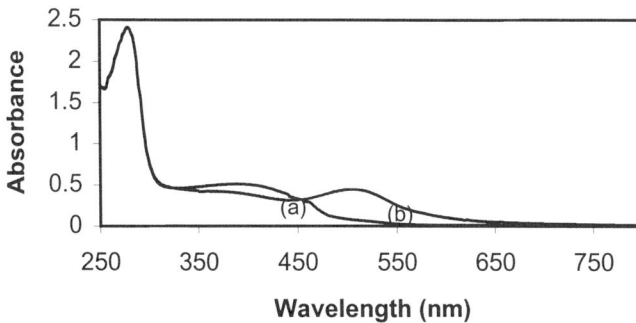

Figure 6. UV/Vis spectra recorded for (a) 3 ml of PAR chelate conjugated to ovalbumin (PAR-OVA) and (b) PAR-OVA in the presence of 0.01 ml $10^{-2}M$ Ga (III). Phosphate buffer pH 7.4 was used as the eluent.

parameters; the binding constants, the optimal pH for the ionization of the chelate and the size-to-charge ratios of the metal to PAR. Figure 7 shows that for metal ions showing no similarity in stoichiometry, the size-to-charge ratio, and optimal pH, no interference were recorded. The successful replacement of an antibody by a well-known chelator conjugated to protein, as a recognition molecule (19), represents a new approach for designing immunoassays that will have a wide range of future applications, especially for detecting chemicals that elicit poor immunological response.

CONCLUSIONS

Bioaffinity sensors are suitable for the characterization of environmental estrogens. The ability to prepare bioaffinity surfaces with predefined functions provides accurate and highly specific identification and quantitation of EDCs in complex environments. It is envisaged that this approach will provide a viable alternative to the characterization of potential environmental estrogens in lieu of the existing long-term animal studies.

ACKNOWLEDGMENT

This work was supported by the US-Environmental Protection Agency Office of Research & Development, NCERQ Advance Monitoring Program under contract # R825323010.

Literature Cited

1. Colborn T. Dumanoski D., Myers J. P., *"Our Stolen Future,"* New York: Dutton (1996).
2. Bette, Hilleman; Endocrine Testing Scheme Proposed for Most Chemicals, *C&EN* February 9, 1998 , pp 7-8.
3. Keith, L. H., *Environmental Endocrine Disruptors: A Handbook of Property Data.* J. Wiley & Sons, NY, 1997.
4. Oosterkamp, A.J., Hock, B., Seifert, M. and Irth, H.; Novel Monitoring for Xenoestrogens. *Trends in Analytical Chem.* **1997**, 16, 544-553
5. *Endocrine Disruptors – Effects on Male and Female Reproductive Systems* Radish K. Naz, Editor; CRC Press Boca Raton, 1999.
6. Kurtz, F. W., Wood, P.H., and Bottimore, D. P. *Rev, Environ. Contamin. Toxicol.* **1992**., 120, 231.
7. Rappe C., *Chemosphere* **1993**, 27, 211.
8. Falck, F., Ricci A., Wlf M. S., Godbold, Deckers P. *Arch. Environ. Health* **1992**, 47, 143.
9. Safe S. *Environ. Health Perspectives* **1997**, 105 (Suppl.3), 675.

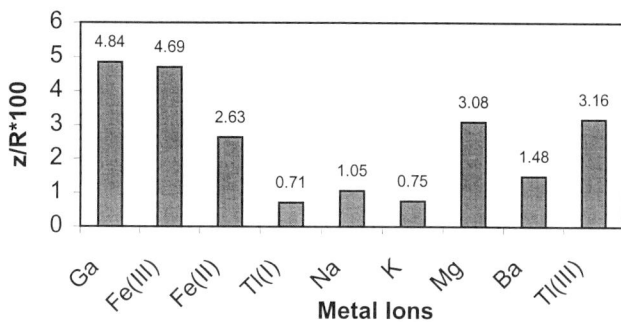

Figure 7. Results obtained for the two-site assay showing the relevance of charge-to-size (z/R) ratios in metal detection. Metal ions having dissimilar z/R values did not interfere in the assay. For the detection of gallium III, the interference recorded for Fe(III) was removed with hydroxylamine.

10. Sadik, O. A., Diane Witt, *Environmental Science & Technology* **1999**, 33(17): A368-75.
11. Sadik, O. A., Van Emon, J. M.; *ChemTech.* **1997**, Vol 27 No.6 38-46.
12. Bender S., Sadik O. A. *Environmental Science & Technology* **1998**, 32, 6, 788-797.
13. Sargent, A., and Sadik, O. A.; *Anal. Chim. Acta* **1998**, 376, 125-131.
14. Johnson, J., Van Emon, J. M., Development and Evaluation of a Quantitative Enzyme-Linked Immunosorbent Assay (ELISA) for Polychlorinated Biphenyls; EPA/600R-94/112; U.S. *Environmental Protection Agency:* Washington, DC, 1994.
15. Langone, J. L., Vunakis, H. V., *Res. Comm. Chem. Path. Pharm.* **1975**, 10, 163.
16. Arnold S., Klotz D., Collins B., Vonier P., Guilette L., Jr., and McLachlan J. *Science* **1996**, 272, 1489-1492.
17. Xu H., Lee E., R. Bakthiar, C. Hendrickson and O. Sadik, *Analytical Chemistry* **1999** (In Press).
18. Flaschka, H. A., and Barnard, A. J., *Chelates in Analytical Chemistry*, Vol. 4, Marcell Dekker, NY 127, 1972.
19. Bekheit, H., Lucas, A. Szurdoki F., Gee S., Hammock B., *J. Agric Food Chem.* **1993**, 41, 2220.

Chapter 16

Use of Charge Coupled Devices for the Simultaneous Detection of Multiple Pesticides by Imaging ELISA Techniques

Ioana Surugiu, Anatoli Dzgoev, Kumaran Ramanathan, and
Bengt Danielsson[1]

Department of Pure and Applied Biochemistry, Box 124 Center for
Chemistry and Chemical Engineering, Lund University, S-221 00 Lund,
Sweden

The chemiluminescent reaction between Horse Radish Peroxidase (HRP)/ Alkaline Phosphatase (AP) and the luminol/CSPD/hydrogen peroxide substrate is used in a multianalytical ELISA approach to simultaneous analysis of different pesticides. The pesticides included in the present study were 2,4-D, Atrazine and Simazine. A novel variant of peroxidase (from transgenic tobacco, TOP) has also been investigated.

The microformat ELISA previously described was employed using thick film hydrophobic pattern on glass plates with flat wells of 2 μL capacity. In addition, sol-gel modified glass capillaries were also employed. As detection system for the chemiluminescent reaction we used a PhotoMultiplier Tube (PMT) or a Charge Coupled Device (CCD) camera. For the PMT/CCD camera based assay the monoclonal antibodies (mAbs) were diluted 1:1000 and bound to the surface during an over night incubation at 4 °C. Non bound antibodies were removed by washing with PBST buffer and the free space was blocked with 2.7 mg mL^{-1} of the gelatin-based blocking reagent. For 2,4-D a detection range of 0.1-100 ng mL^{-1} was obtained. Work with real samples and with mixtures of pesticides is under way.

Pesticides is a term used in a broad sense for chemicals, synthetic or natural, which are used for the control of insects, fungi, bacteria, weeds, nematodes, rodents, and other pests (1,2). The use of pesticides must be regulated in such a manner that

---

[1]Corresponding author.

the intake of a pesticide residue does not exceed the acceptable daily intake. Thereby, the monitoring of pesticides in food, soil, water etc is one of the most important aspects of minimizing potential hazards to human health (3). Several different forms of pesticides have been described in the literature. See Table 1.

Table 1: Examples for various categories of Pesticides

Pesticides	Examples
Organochlorine Insecticides	----------
Organo phosphorus Insecticides and carbamate	Paraoxon,Aldicarb, carbofuran, dimethoate, bromophos,ethylparathion,chlorofenvinphs, pyrimicarb,methylparathio,diethoxy phosphoryl cyanide, pyrethroid insecticides, methyl phosphonic acid, p-aminophenyl 1,2,2,trimethyl-propyl diester (MATP).
Ammonium ion based fertilizers	Propazine, Atrazine, Simazine.
Herbicides	Glyphosate,sulfonylurea,2,4-dichloro phenoxy acetic acid (2,4-D), chlortoluron.
Thiocarbamate insecticides	Dimethoate,triallate,fenitrothion,methiocar, fenthion,methomyl, thiocarbamate, aldicarb sulfone,aldicarbflfoxe, oxamyl, mathrocarb, eptam, triasulfuron.

Organophosphorus compounds are powerful inhibitors of the enzymes involved in the nerve function. Such compounds can form stable complexes with acetylcholinesterase thus preventing, by phosphorylation its function. One of the most used herbicide is 2,4-dichlorophenoxyacetic acid (2,4-D), a pesticide with a wide range of biological effects and very toxic to mammals. The human oral lethal dose is 0.1 ppm (4). This is also the basis for the most common approach to detect cholinesterase inhibitors (5). Numerous electro- chemical assays have also been proposed. The acetylcholine-esterase activity can also be assayed by measuring the choline with choline oxidase resulting in sensitivities in the 10 ppm range (6). Similar experiments can be carried out with thermometric sensing. The most sensitive assay realised with immobilized choline oxidase/catalase column was used in the enzyme thermistor (ET) and the acetylcholine esterase (AchE) was recently immobilized in a precolumn with ConA-Sepharose (7).

The advantages of immunoassays as a selective, sensitive, and cost-effective method for assaying/screening clinical and environmental samples are now widely acknowledged (8). Despite the success of the microtitre plate based enzyme-linked immunosorbent assay (ELISA) (9), its extensive use is limited due to the requirement of trained personnel and long assay time. These drawbacks have initiated the development of a variety of other formats such as; surface plasmon resonance (*10*), optical fibres (*11*), planar (*12*)/ surface acoustic waveguides (*13*) and piezoelectric devices (*14*).

Another class of immunosensors utilizes enzyme conjugates with antigens at the surface of antibody coated capillaries or thick film patterned glass plates (*15*). This report describes a PMT and CCD camera based immunosensor for the herbicide (2,4-D) and fertilizers (Atrazine and Simazine). The detection principle improves the sensitivity of the sensor by atleast three orders of magnitude compared to a microtitre plate assay.

Capillary based chemiluminescent assays offer the advantage of high surface to volume interaction and efficient signal to noise ratio, compared to microtitre plate assays. Although capillary based chemiluminescent assays have proven to be very successful with clinical analytes (*16*) there are no reports of them being used in immunosensors for pesticide detection. Several recent reports, however, do describe other types of chemiluminescence based sensors for pesticides. Ayyagari et al (*17*) reported on the development of chemiluminescence assays for organophosphorus compounds based on the inhibition of alkaline phosphatase. Gao et al (*18*) investigated the effect of different tapered fibre tip construction on the chemiluminescent intensity. Heineman et al (*19*) reported on the use of capillary enzyme immunoassay for electrochemical detection for the determination of atrazine in water.

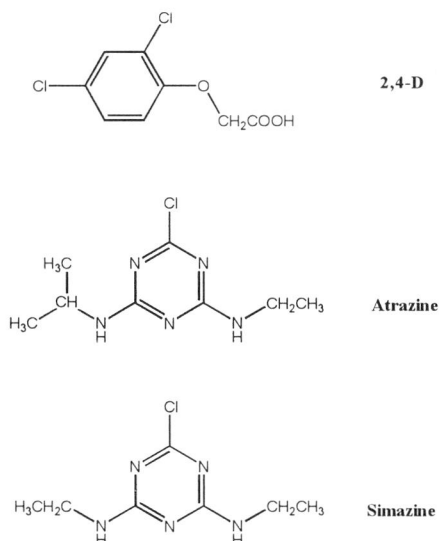

Figure 1: Structures of the pesticides: 2,4-D, Atrazine and Simazine.

We have chosen chemiluminescence based detection of 2,4-D, Atrazine and Simazine (See Figure 1) using the capillary based immunosensor since a) there is a need for a fast, sensitive, reliable, and cost-effective means to detect pesticides for environmental monitoring and the food industry, and b) we have previously investigated the CCD camera based immunosensor schemes using microformat and "patterned thick film" based immunoassay formats. The analytical figures of merit for various chemiluminescence based sensors previously reported in the literature for pesticides are summarized in Table II. It is clear from Table II that there is a trade off between detection limit and kind of assay format chosen. In contrast, the PMT and CCD based immunosensors offer high sensitivity as well as faster analysis time. Limits of detection (S/N>3) using the CCD and PMT based immunosensors for 2,4-D are about 6 pg. The total assay time (measuring time excluding the preparation) is less than 120 sec using both the approaches.

Experimental

Materials

Chemicals, Immunoreagents and Standards:

The 2,4-dichlorophenoxyacetic acid (2,4-D), N-hydroxysuccinimide (NHS), N,N'-dicyclohexylcarbodiimide (DCC), Tween 20 and bovine serum albumin (BSA) were purchased from Sigma chemical Co. (St. Louis, MO). The diethanolamine was obtained from Merck-Schuchardt (Hannover, Germany). The 1,4-dioxan was bought from Riedel-De Haen AG (Hannover, Germany). The disodium 3-(4-methoxy spiro[1,2- dioxetane- 3,2'- (5'-chloro) tricyclo[3.3.1]decan]-4-yl) phenyl phosphate (CSPD), 11.6 mg mL^{-1} stock solution and Emerald II enhancer, 10 mg mL^{-1} stock solution were purchased from Tropix Inc. (Bedford, MA). Calf intestinal alkaline phosphatase (EIA grade) and 4-nitrophenyl phosphate (4-NPP) were obtained from Boehringer Mannheim (Mannheim, Germany). The anti-2,4-D/Atrazine/Simazine mAbs (clone 1/F6/C10) were raised at the Veterinary Research Institute (Brno, Czech Republic) and kindly provided by Dr. Milan Franek and Dr. Sergei Eremin, Moscow State University, Russia. Rabbit anti-mouse IgG-AP was purchased from DAKO (Copenhagen, Denmark). The HRP, AP and TOP were obtained from Boehringer Mannheim (Germany).

Buffers and standards were prepared using distilled and deionized water. Phosphate-buffered saline (PBS), pH 7.4, contained 0.13 M NaCl, 2.6 mM KCl, 4.0 mM $Na_2HPO_4.7H_2O$, and 1.0 mM KH_2PO_4. Washing buffer solution (PBST) contained PBS with 0.1% Tween-20. Blocking solution contained PBST with 27 mg/ml blocking reagent for ELISA (Boehringer Manheim). Sodium carbonate buffer was used for coating and contained 13 mM Na_2CO_3 and 85 mM $NaHCO_3$, pH 9.6. A stock solution of 2,4-D (10 mg mL^{-1}) was prepared in methanol. For calibration, a serial dilution of the stock solution with PBS was prepared from 0.001 to 1000 ng mL^{-1}.

The CCD camera employed in this study was a Photometrix 200 (Photometrix, Tucson, AZ). The camera was thermoelectrically cooled to -45 °C, equipped with a Thompson TH 7895 chip which was 512 x 512 pixels, dark current 0.3 electrons s^{-1}

Table II:Application of various chemiluminescent assays for pesticide analysis

Pesticide	Method of detection	Detection limit	Exposure time	Reference
Paraoxon	AchE inhibition [#1]	0.75 µg L^{-1}	----- [#4]	Roda et al (20)
Aldicarb	"	4 µg L^{-1}	----- [#4]	Roda et al (20)
Pyricarb (carbamate)	"	ng L^{-1}	5 min	Moris et al (21)
Paraoxon	AP inhibition	50 ppb	30 sec	Ayyagari et al (22)
Methyl parathion	"	500-700 ppb	30 sec	Pande et al (23)
NH4$^+$ ions in fertilizers (glyphosate)	N-compounds using CL [#2].	0.03 µg mL^{-1}	----- [#4]	Halvatzis et al (24)
Thiocarbamate	S-compounds using CL [#3].	4 pg	----- [#4]	Karen Chang et al (25)
Dichlorprop methyl ester (DME)	Agrochemical products by immunoassay.	0.11 ng mL^{-1}	600 sec	Navaz Diaz et al (26)
2,4-dichlorophenoxy acetic acid (2,4-D)	"	6 pg	90 sec	Dzgoev et al (27)
Methyl phosphonic acid (MATP)	"	10-6 µmol	5 min	Erhard et al (28)
Chlortoluron	"	0.1 µg L^{-1}	40 sec	Fawaz et al (29)
Triasulfuron	"	0.02 µg L^{-1}	7.5 min	Schlaeppi et al (30)

[#1]
$$\text{Acetylcholine} \xrightarrow{AchE} \text{acetate + choline} \xrightarrow{CO} \text{betaine} + H_2O_2$$
$$\text{Luminol} + 2H_2O_2 + OH^- \xrightarrow{HRP} \text{aminophthaleine anion} + N_2 + 3 H_2O + \text{light.}$$
AchE=Acetylcholinesterase; CO=Choline oxidase.

[#2] N-compound + O_2 ----- NO* + other products
NO* + O_3 ----- NO_2 + O_2 ----- NO_2 + O_2 + light.
*excited state

[#3] $SO_{(g)} + O_{3(g)}$ ----- $SO_2^*_{(g)} + O_{2(g)}$
$SO_2^*_{(g)}$ ----- $SO_{2(g)}$ + light
*excited state.

[#4] Based on a chemiluminescent (CL) flow analysis.

pixel^{-1}, a well capacity of 3.66x10^5 electrons, a 14 bit AD converter, and a quantum efficiency of 0.36 at 542 nm.

Microscope slides patterned with a proprietary hydrophobic layer using thick-film technology were purchased from Cel-Line (Newfield, NJ). The rectangular pattern consisted of 4 x 20, 2 mm^2 squares/1.5 diameter circular pads (hydrophilic), separated by 1 mm lines (hydrophobic). The film was 100 μm thick. The 2 x 96 well plates were purchased from Nunclon, Denmark.The 32 x 1 mm disposable glass capillaries (microcaps) were purchased from Drummond Scientific (Broomall, PA, USA). The polished silicon wafers (75 mm i.d.) were purchasd from Wacker Chemie (Munich, Germany). The gold coated silicon wafers were made by sequentially depositing a 30 μm layer of Cr (to improve adherence) and then a 2000 μm layer of Au. The commercial 2,4-D tube-test kit EnviroGard was from Millipore (Bedford, MA, USA).

The PMT (photomultiplier) Sensor Module HC 135-01 with embedded microcontrol system for chemiluminescence intensity detection was purchased from Hamamatsu (Hamamatsu Corporation, Bridgewater, NJ, USA), and mounted in a light-tight holder for capillary made in black Delrin ^{TM}. The standard LKB Wallac 1250 luminometer (EC & G Wallac, Turku, Finland) was used as a reference detector. Multiskan MCC/340 (Helsinki, Finland) was used as the microtiter plate reader.

Methods

Synthesis of 2,4-D-BSA, 2,4-D-TOP and 2,4-D HRP conjugates:
The synthesis of 2,4-D-BSA conjugate was carried out as previously described (*31*). The 2,4-D-TOP and 2,4-D HRP conjugates were synthesized based on the periodate oxidation method (*32*) with minor modifications as described elsewhere (*33*). Briefly, a) to 1 mL of 10 mg mL^{-1} TOP (*34*) solution in 1 mM Na-acetate buffer, pH 4.5, 0.5 mL of 35 mg mL^{-1} NaIO$_4$ solution prepared in the same buffer was added. The reaction mixture was stirred for 30 min at room temperature (RT), and thereafter purified with gel filtration through a G-25 column using 1 mM Na-acetate buffer as eluent. Next, 1 mL of 36 mM diaminopropane in 0.1 M Na-carbonate buffer, pH 9.2, was added to the enzyme solution, and the reaction mixture was incubated overnight at RT. Modified enzyme solution was purified with gel filtration b) to activate the antigen molecule, 17 μmoles of NHS and 48 μmoles of DCC was added to 9 μmoles of 2,4-D in 1.2 mL of DMF and the mixture was stirred for 1.5 h at RT. During this time, approximately 40 μmoles of DCC were added to the mixture c) The activated 2,4-D was added to the activated TOP solution, and the mixture was incubated under stirring for 2 h at RT, and then left overnight at 4 °C. The conjugate obtained was purified twice by gel-filtration through Sephadex G-25 packed in a 40 cm column.

The 2,4-D conjugate was synthesised in the same way, except that 40 mg of enzyme was used for the coupling reaction, and diaminohexane was used for the enzyme modification. The 2,4-D enzyme molar ratios of the conjugates obtained were determined by comparing the number of surface amino groups in the native and conjugated enzymes using the 2,4,6-trinitrobenzene sulfonic acid method. The Simazine and Atrazine conjugates were a gift from Dr. Milan Franek and Dr. Sergei Eremin, Moscow State university, Russia.

Assay and Imaging Procedures

The CCD Camera-Based Chemiluminescence System, luminometer based Chemiluminescence ELISA and microtiter plate based Colorimetric ELISA were performed as described earlier (27).

Apparent binding constants

Anti-2,4-D IgG were serially diluted in PBS-T2 and applied to a microtiter plate coated with anti-mouse IgG, and incubated for 1 h at 37 °C. The plate was washed with PBS-T1 and unoccupied sites were blocked with the blocking solution for 30 min. The plate was washed, and conjugates diluted to appropriate concentrations in PBS-T2 were added. The plate was incubated for 2 h at RT, and thereafter washed. ABTS substrate solution was added, and the absorption at 405 nm was measured after 45 min incubation at RT. The absorbance vs IgG concentration profile was plotted, and the apparent binding constants were calculated using the values representing 50% of the bound IgG fraction.

Sol-gel coating procedure:

The protocol as described elsewhere (35) was used to increase the adsorption properties of the surface of glass capillaries. Cleaned glass capillaries were filled with the sol solution and incubated for 1 h on a shaker to facilitate uniform coating. The excess sol was removed and the coated capillaries were left overnight at RT to obtain the dry gel. Later, the capillaries were washed with buffer followed by copious amounts of water, and then dried at RT.

Assay procedures

Direct competitive ELISA

200 μL well^{-1} of 10 μg mL^{-1} anti-2,4-D IgG in PBS-T2 was added in duplicates to microtiter plate wells coated with anti-mouse IgG and incubated 1 h at 37 °C. The plate was washed four times with PBS-T1, and unoccupied sites were blocked with the blocking solution. The plate was washed with PBS-T1, and 50 μL of 2,4-D standard solution or analyte was added to each well and preincubated (in case of 2,4-D-TOP) for 15 min and thereafter, 150 μL of 25 nM conjugate in PBS-T2 was added to each well. The plate was incubated for 2 h at RT and then washed. Next, 200 μL well^{-1} of ABTS substrate solution was added and after 45 min incubation at RT, the absorbance at 405 nm was measured.

Standard chemiluminescent ELISA

500 μL tube^{-1} of 10 μg mL^{-1} rabbit anti-mouse IgG in 50 mM sodium carbonate buffer, pH 9.6 was incubated overnight at 4 °C. The tubes were washed with PBS-T1 three times, and unoccupied sites were blocked with the blocking solution. 500 μL of 10 μg mL^{-1} anti-2,4-D IgG in PBS-T2 were added to each tube, and incubated for 1 h at 37 °C. The tubes were washed, and 100 μL tube^{-1} of the 2,4-D standards or analyte in 10 mM PBS was added and preincubated for 15 min. Thereafter, 400 μL of 30 pM 2,4-D-TOP conjugate was added to each tube, and after 2 h incubation at RT, the

unbound fraction was washed out with PBS-T1. 500 µL of luminol substrate solution was added to each tube and the maximum chemiluminescent intensity was measured with a standard LKB luminometer.

Capillary chemiluminescent ELISA

Initially, 50 µL capillary[-1] of 10 µg mL[-1] rabbit anti-mouse IgG in 50 mM sodium carbonate buffer solution, pH 9.6 was incubated overnight at 4 °C. The capillaries were washed three times with PBS-T1, and then twice with water. Thereafter, 50 µL of 10 µg mL[-1] anti-2,4-D IgG in PBS-T2 was added to each capillary, and incubated for 1 h at 37 °C. The capillaries were washed with washing buffer and water. Thereafter, 50 µL mixture of 2,4D-TOP and standards or samples were added to each capillary, and the competitive reaction was allowed to proceed by incubation for 2 h at RT. To determine nonspecific binding, four capillaries received non specific IgG and no pesticide, while ten capillaries received specific IgG but no pesticide, in order to determine the maximum chemiluminescent intensity for the bound fraction. Finally, the capillaries were washed and 50 µL capillary[-1] of luminol substrate solution was added, and the maximum chemiluminescent intensity was measured with the modified Hamamatsu PMT.

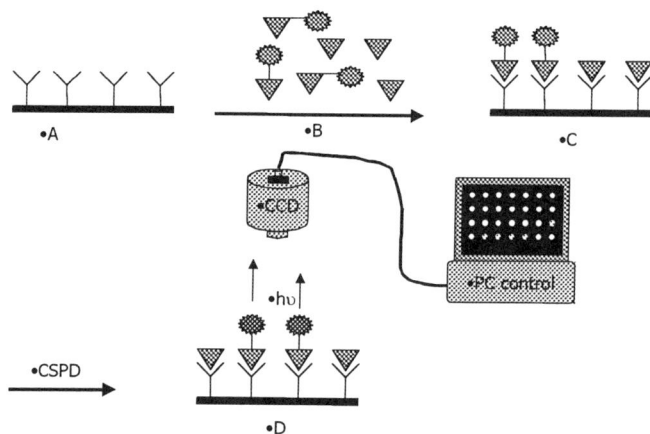

Figure 2. Schematic of the set-up based on the CCD camera approach

Results & discussion

A schematic of the assay procedure is shown. See Figure 2. During the CCD camera based assay the standard or the sample is pre-incubated with excess of the 2,4-D specific antibody in the solution phase. This mixture is transferred to a 2,4-D-BSA conjugate coated well, where the unreacted antibodies bind. After washing, the AP-labeled rabbit anti-mouse IgG is added followed by the chemiluminescent substrate (CSPD). The emitted light is detected by the CCD and quantified using a Photometrix interface software.

Microformat Chemiluminescent ELISA

This is a single sample detection approach. The assay conditions such as mixing, substrate and enhancer concentrations, buffer composition, assay time, CCD camera positioning, exposure settings and cross talk between adjacent spots were optimised. The absolute sensitivity of the assay for detection of AP using CSPD was determined by analyzing a serial dilution of AP ranging between 0.4×10^{-9} and 0.4×10^{-15} M. A range between 2×10^{-9} and 2×10^{-12} M with a detection limit of 8×10^{-13} and 1.6×10^{-19} mol was obtained.

The kinetics and stability of the chemiluminescent emission from the AP/CSPD reaction were optimised and the signal increased rapidly and remained stable for over 20 min. The substrate had a limiting effect only at very high AP concentrations. During the assay the exposure time was set at 90 s and the sensitivity could be improved upon longer exposures.

The CCD based assay was also combined with a competitive immunoassay using mAb against 2,4-D. The calibration curve was linear between 10 and 0.01 ng mL^{-1} and the pecision varied between 3 and 14% (CV%). The detection limit was 96 pg mL^{-1} or 2.7×10^{-11} M corresponding to 5.4×10^{-17} mol of 2,4-D with a CV of 12.5%. Both the within-assay and day-to-day assay CVs were less than 10% for samples containing more than 1 ng mL^{-1} of 2,4-D, while for samples containing less than 1 ng mL^{-1} the CV was more than 10%.

A comparison of the colorimetric and the chemiluminometric assay indicated that the sensitivity of the immunoassay was highly dependent on the antibody used. A preliminary comparison revealed that the detection limit of the tube based chemiluminescent assay was 0.1 ng mL^{-1} 2,4-D while the detection limit of the colorimetric assay was 22 ng mL^{-1}. This indicated the microformat imaging ELISA to be 20 times more sensitive than the luminometric assay and 4×10^{3} times more sensitive than the colorimetric ELISA. The precision of the luminometric assay was 6% with a workable range between 0.1 and 10^{2} ng mL^{-1}. Similarly for the colorimetric assay the precision was 33% with a workable range between 10 and 10^{3} ng mL^{-1}.

A comparison of the samples from southern Sweden using the microformat chemiluminescent assay and a commercially available 2,4-D assay kit from Millipore revealed that six of the samples were found to have 2,4-D concentrations in the range 0.06 and 0.1 ng mL^{-1} and five of the six samples were also positive with the Millipore kit. Cross talk effects between two samples placed adjacent to each other were also observed, although their effect on the measurements was minimal with possibilities to make corrections.

The thick-film patterned microformat chemiluminescent imaging ELISA

Following the optimization of the CCD based chemiluminescent single sample assay, an attempt was made for simultaneous detection of samples. In brief, 2,4-D-BSA conjugates were adsorbed to the patterned glass plates and incubated for 1 hour at RT. The standard and unknown samples was pre-incubated with 2,4-D specific antibodies. Aliquots of this mixture were transferred to the microwells where the unreacted antibodies bound to the 2,4-D-BSA conjugates. This was followed by preincubation with AP-labeled rabbit anti-mouse antibody and the substrate plus CSPD were added. The emitted light from the reaction was imaged on a CCD camera and quantitated using PMIS. To minimise timing errors a maximum of 24 samples were simultaneously analyzed.

A calibration graph constructed from the mean peak light emission values provided a linear range between 0.1 and 10^2 ng mL^{-1} (see Figure 3). The precision of this range varies between 8 and 18%. The detection limit was found to be 4.3 x 10^{-10} M or 96 pg mL^{-1} with a CV of 12.5% at S/N of 3.8. The recovery studies performed by adding of 2,4-D between 0.5 and 50 ng mL^{-1} to a sample having 0.5 to 100 ng mL^{-1} gave a recovery of 79 to 145%.

The reliability of the assay was evaluated by spiking five water samples taken from agricultural area in southern Sweden with four different concentrations of 2,4-D (0.01, 10, 20 and 100 ng mL^{-1}) and analysed using the CCD imaging ELISA. Similar samples were compared with the commercial tube-test kit from Millipore

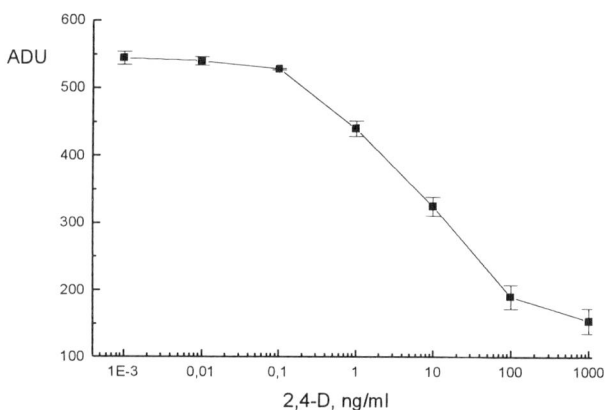

Figure 3. The calibration curve used for the simultaneous detection of 2,4-D samples

"EnviroGard". The imaging ELISA values correlated well with the values obtained with the commercial kit. The within-assay CV were between 6.4 and 11.4% while the day-to-day assay results were between 9.6 and 14.4%. A similar approach could be extended for simultaneous detection of multi-pesticide samples such as Atrazine, Simazine and 2,4-D employing the respective conjugate. The initial optimization of such assays are currently in progress (see Figure 4).

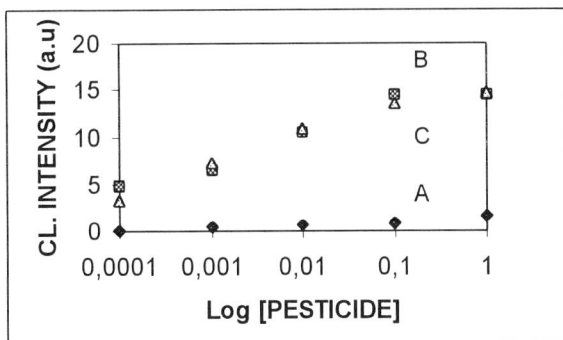

Figure 4. The calibration graph for the detection of 2,4-D (A), Atrazine (B) and Simazine (C) using the CCD camera based Chemiluminescent assay.

Comparison of the thick film patterned surface with capillary based and gold coated support:

Following the success with the thick-film patterned assays, other formats of immobilization supports have also been examined. Initially the supports were treated with three different AP concentrations (1.3×10^{-10}, 1.3×10^{-11} and 1.3×10^{-12} M) along with the substrate and CSPD. The choice of the support affected the sensitivity dramatically. The thick film patterned surface provided a larger linear range and lower blank intensity compared to the glass capillary and gold plate. In the case of the capillary the light was spread over a larger area with the ends of the capillary being the brightest. The possibility of the capillary acting as wave guides to increase the collection efficiency is currently being investigated. Results also indicate that the BSA conjugate saturation is faster in the thick-film patterned glass, gold and glass capillary surfaces. However, the adsorptive capacity of the gold and glass capillary surfaces were 20% higher than the thick-film supports.

The stability of the gold-plate and thick-film patterned microplate were approximately the same. The linear range was between 1 and 10^3 ng mL^{-1} with S/N at 1 ng mL^{-1} being 2.2 and 2.5 for gold plate and the patterned microplate respectively.

The conjugation efficiency of TOP (tobacco peroxidase)

In order to further improve the sensitivity and cost of assays, a recently reported variant (*34*) of HRP from transgenic tobacco plants (TOP) was chosen. The TOP demonstrates higher enzyme activities in addition to providing higher chemiluminescence in the absence of enhancers (p-iodophenol) and using a PMT detector. Based on the periodate method the quality of the conjugate was controlled by the periodic measurement of the peroxidase activity and the immunological activity. The changes in the activity of TOP were minimal after conjugation. The equilibrium binding constant was found to be 0.9×10^8 M^{-1}, which was higher compared to HRP-2,4-D conjugate i.e. 1.25×10^7 M^{-1}. Based on the trinitrobenzenesulfonic acid method the hapten to enzyme ratio was 8:1 and 12:1 for TOP 2,4-D and HRP 2,4-D conjugates.

Based on the calibration plots in the range between 0.1 and 10^4 ng mL^{-1} the TOP-conjugate was 3 times more ative than HRP-conjugate with 2,4-D. Also the TOP exhibited a higher catalytic activity compared to native HRP.

The superiority of the TOP was reflected in the chemiluminescent assays. While TOP emits a response in the 10 pg to 5 ng mL^{-1} range of 2,4-D the native HRP conjugate showed no response. These assays carried out in capillaries using a PMT detector demonstrated a better sensitivity and lower detection limit. Also in the presence of sol-gel as the cladding layer the calibration curve showed a linearity between 30 pg and 500 ng mL^{-1}.

Conclusions

A sensitive chemiluminescent assay based on CCD camera and PMT as a detector for the pesticides 2,4-dichlorophenoxyacetic acid (2,4-D), Atrazine and Simazine are described. The assays employing microformat thick-film patterned glass slide, glass capillaries and gold-coated glass plates are found to be sensitive in addition to possibilities of high speed detection and simultaneous analysis. The results also indicate the flexibility in scaling the assays to various dimensions in addition to employment of both AP and HRP as labels for the chemiluminescent reaction. A novel form of HRP i.e. TOP has also contributed to increasing the sensitivity of the assays and cost reduction, in combination with the wave guiding nature of sol-gel coated capillaries. These approaches would eventually enable the design of high throughput and multianalytical immunoassays.

Literature Cited

1. Mulchandani, A.; Pan, S.T.; Chen, W. *Biotech. Prog.* 1999, 15, 130.
2. Rogers, K.R.; Mascini, M. *Field Anal. Chem. & Tech.* 1998, 2, 317.
3. Rathore, H.S.; Saxena, S.K.; Tahira, B. In *Handbook of Food Analysis*; Nolet, M.L., Ed.; Marcel Dekker Inc: New York, NY, 1996; Vol. II, pp 1381.
4. United States Environmental Protection Agency (1975) Fed. Regist. Dec 16, *Chem. Abst.* 1976, **84**, 79548a.
5. Kröger, S.; Persson, A.; Jonsson, R.; Saityeva, N.; Danielsson, B. (submitted).
6. Mattiasson, B.; Rieke, E.; Munneke, D.; Mosbach, K. *J. Solid-Phase Biochem.* 1979, 4, 263.

7. Mattiasson, B.; Borrebäck, C. *FEBS lett.* 1978, *85*, 119.
8. Kröger, S.; Setford, S.J.; Turner, A.P.F. *Anal. Chem.* 1998, *70*, 5047.
9. Crowthers, J.R. *ELISA Theory & Practice*; Humana Press, Totowa, NJ. 1995.
10. Jorgenson, R.C.; Yee, S.S. *Sens. Actuators B* 1993, *12*, 213.
11. Wolfbeis OS. In. *Fiber Optic Chemical and Biosensors*; Wolfbeis OS Ed: CRC Press, Boca Raton, FL 1991; **Vol. I**, pp 1.
12. Plowman, T.E.; Reichert, W.M.; Peters, C.R.; Wang, H.K.; Christensen, D.A.; Aleuron, J.N. *Biosens. Bioelectron.* 1996, *11*, 149.
13. Tom-Moy M. *Anal. Chem.* 1995, *67*, 1510.
14. Guilbault, G.G.; Schmidt, R.D. *Electrochemical, Peizoelectric and Fiber-optic biosensors.* Jai Press Ltd. 1991; Vol. I, pp 257.
15. Dzgoev, A.; Mecklenburg, M.; Xie, B.; Miyabayashi, A.; Larsson, P.O.; Danielsson, B. *Anal. Chim. Acta.* 1997, *347*, 87.
16. Momeni, N.; Ramanathan, K.; Larsson, P.O.; Danielsson, B.; Bengmark, S.; Khayyami, M. *Anal. Chim. Acta.* 1999, *387*, 21.
17. Ayyagiri, M.S.; Kamtekar, S.; Pande, R.; Marx, K.A.; Kumar, J.; Tripathy, S.K.; Akkara, J.; Kaplan, D.L. *Biotech. Prog.*1995, *11*, 699.
18. Gao, H.H.; Chen, Z.; Kumar, J.; Tripathy, S.K.; Kaplan, D.L. *Optical. Engg.* 1995, *34*, 3465.
19. Jiang, T.J.; Halsall, H.B.; Heineman, W.R. *J. Agric. Food. Chem.* 1995, *43*, 1098.
20. Roda, A.; Rauch, P.; Ferri, E.; Girotti, S.; Ghini, S.; Carrea, G.; Bovara, R. *Anal. Chim. Acta.* 1994, *294*, 35.
21. Moris, P.; Alexandre, Y.; Roger, M.; Remacle, J. *Anal. Chim. Acta.* 1995, *302*, 53.
22. Ayyagiri, M.S.; Kamtekar, S.; Pande, R.; Marx, K.A.; Kumar, J.; Tripathy, S.K.; Kaplan, D.L. *Mat. Sci. Engg.* 1995,*C2*, 191.
23. Pande, R.; Kamtekar, S.; Ayyagiri, M.S.; Kamath, M.; Marx, K.A.; Kumar, J.; Tripathy, S.K.; Kaplan, D.L. *Bioconjugate Chem.* 1996, *7*, 159.
24. Halvatzis, S.A.; Timotheou-Potamia, M.M. *Talanta* 1993, *40*, 1245.
25. Karen-Chang, H.C.; Taylor, L.T. *Anal. Chem.* 1991, *63*, 486.
26. Navas, D.A.; Garcia, S.F.; Lovillo, J.; Gonzalez, G.J.A. *Anal. Chim. Acta.* 1996, *321*, 219.
27. Dzgoev, A.; Mecklenburg, M.; Larsson, P-O.; Danielsson, B. *Anal. Chem.* 1996, *68*, 3364.
28. Erhard, M.H.; Jungling, S.B.; Kellner, J. *J. Immunoassay* 1992, *13*, 273.
29. Fawaz, K.M.; Wynne, A.G.; Stevenson, D. *Analyst* 1996, *121*, 329.
30. Schlaeppi, J.M.A.; Kessler, A.; Förry, W. *J. Agric. Food Chem.* 1994, *42*, 1914.
31. Fleeker, J. *J. Assoc. Off. Anal. Chem.* 1987, *70*, 874.
32. Nakane, P.K.; Kawaoi, A. *J. Histochem. Cytochem.* 1974, *22*, 1084.
33. Dzantiev, B.B.; Zherdev, A.V.; Romanenko, O.G.; Sapegova, L.A. *Intern. J. Environ. Anal. Chem.* 1996, *65*, 95.
34. Gazaryan, I.G.; Lagrimini, L.M. *Phytochemistry* 1996, *41*, 1029.
35. Narang, U.; Bright, F.V.; Prasad, P.N.; Ramanathan, K.; Kumar, N.D.; Kamalasanan, M.N.; Malhotra, B.D.; Chandra, S. *Anal. Chem.* 1994, *66*, 3139.

Chapter 17

Immunosensor for Fast Detection of Bacterial Contamination

Ihab Abdel-Hamid, Plamen Atanasov, Dmitri Ivnitski, and Ebtisam Wilkins

Department of Chemical and Nuclear Engineering, University of New Mexico, Albuquerque, NM 87131

A flow-through immunosensor has been developed and integrated into a portable assay system which includes a pump, liquid reagents and associated electronics. The immunosensor employs an ELIFA format with a disposable porous membrane as a carrier for primary antibodies against bacteria. The 'sandwich' scheme of assay has been used and iodine, formed as a result of the enzymatic oxidation of iodide by the peroxidase-label of the secondary antibody, has been detected amperometrically. The overall analysis time is 40 min. The immunoassay system allows direct (without pre-enrichment) determination of pathogenic bacteria with a low detection limit of 50 cells/mL. The approach has been tested in assays of *E. coli* (total coliform quantitation and *E. coli* O157:H7 individual strain detection) and *Salmonella*. The prototype is a compact (20 x 8 x 25 cm), self-standing device and has the potential for complete automation allowing its use in laboratory settings as well as in field conditions.

Environmental bacterial contamination and food processing quality control, especially methods of detection and enumeration bacterial pathogens are high national priorities. Contamination of drinking water and foods by bacterial pathogens (such as *E. coli, Salmonella, Campylobacter, Legionella, Staphylococcus aureus, Streptococci*, etc.) results in numerous diseases *(1,2)*. Worldwide, infectious diseases account for nearly 40% of the total 50 million annual estimated deaths. Microbial diseases constitute the major cause of death in many developing countries of the world. From a military point of view, there are a number of pathogenic bacteria which can be considered possible biological warfare agents *(3,4)*. These microorganisms are resistant to environmental conditions, most of the human population is completely susceptible, and the diseases they cause are severe with a high fatality rate. A large quantity of organisms could easily be grown and preserved for several years. A growing number of bacterial pathogens have been identified as important food- and waterborne pathogens *(5-8)*. Estimates of the yearly incidence of foodborne illness vary widely from million cases to 81 million cases in the USA, with bacterial foodborne outbreaks accounting for 91% of the total outbreaks *(2, 9)*. In fact, the incidence of human diseases caused by foodborne and waterborne pathogens, such as *Salmonella sp., Escherichia coli, Staphylococcus aureus, Campylobacter jejuni, Campylobacter coli* and *Bacillus cereus* has not decreased since 1994. Examples include the 1997 Hudson ground beef recall due to *E. coli* O157:H7 contamination and more than 9,000 illnesses and 313 deaths in 1996 due to *E. coli* O157:H7 *(6)*.

Current practices for preventing microbial diseases rely upon careful control of various kinds of pathogenic bacteria in clinical medicine, food safety and environmental monitoring. Approximately 5 million analytical tests - for *Salmonella* only - are performed annually in the United States *(10-12)*. Conventional bacterial identification methods usually include a morphological evaluation of the microorganism as well as tests for the organism's ability to grow in various media under a variety of conditions. Although standard microbiological techniques allow the detection of single bacteria, amplification of the signal is required through growth of a single cell into a colony. The methods generally have 4 distinct phases *(12, 13)*: (1) preenrichment, to allow growth of all organisms; (2) selective enrichment, to allow growth of the organism under investigation, and to increase bacterial population to a detectable level; (3) isolation, by using selective agar plates; and (4) conformation, serological and biochemical tests to confirm the identification of a particular pathogenic organism. The main disadvantages of conventional methods of analysis are: the multistep assay, the process is time consuming; completing all phases needs at least 16 h and can take as long as 48 h. The detection limit is 10^5-10^6 cells/ml without pre-enrichment. Thus conventional techniques are not suitable for direct and fast analyses of bacteria without pre-enrichment.

To meet expectations of users, analytical devices for bacterial detection must have the specificity to distinguish between bacteria, the adaptability to detect different analytes, the sensitivity to detect bacteria directly in real samples, on-line and without preenrichment. Time and sensitivity of analysis are the most important limitations related to the usefulness of microbiological testing. Effective screening of environmental and food samples requires a rapid, extremely selective and very sensitive analytical technologies since the presence of even a single pathogenic organism in the body or food may be an infectious dose. At the same time, the device must have simple and inexpensive configurations. Most of these properties may be achieved by using biosensor technology.

Electrochemical biosensors have some advantages over other analytical systems in that they can operate in turbid media, offer comparable instrumental sensitivity, and are more amenable to miniaturization. Modern electroanalytical techniques have very low detection limits (typically 10^{-9} M) that can be achieved using small volumes (1-20 µl) of samples. Furthermore, the continuous response of an electrode system allows for on-line control and the equipment required for electrochemical analysis is simple and cheap compared to most other analytical technique. Amperometric immunosensors aimed at microbial analysis have recently been reported *(14-20)* including some immunomagnetic bead technology *(19-21)*.

Most of the immunoassay methods for bacterial detection are currently performed in solid-phase ELISA format conducted on micro-titer plates. Flow immunofiltration assay, can be an excellent alternative for detection of bacterial pathogens because it not only overrides the effects of diffusional limitations in conventional ELISA, but also allows the concentration of bacteria on the membrane by filtering a large volume of the sample. Heterogeneous flow immunofiltration assays offer extremely accelerated binding kinetics. First, there is a high surface area to volume ratio in the immunosorbent. Also, the flowing stream actively brings the sample in contact with the solid-phase antibody. This factor results in a greatly enhanced antigen-antibody encounter rate and in nearly quantitative immunobinding during the short immunoreaction time(22-24).

We are developing a novel flow-injection immunofiltration amperometric system (25, 26) for direct and rapid detection of low concentrations of bacterial pathogens in solutions. *E. coli* O157:H7, total *E. coli* and *Salmonella* were selected as model organisms. *Escherichia coli* is a typical inhabitant of the human intestinal tract and can also be a causative agent of intestinal and extra-intestinal infections. *Escherichia coli* O157:H7 is considered as one of the most dangerous foodborne pathogens. This O157:H7 strain produces large quantities of a potent toxin in the lining of the intestine and causes severe damage resulting in hemorrhagic colitis or hemolytic uremic syndrome which may lead to death, especially in children. *Salmonella* is a typical foodborne pathogen and its presence is a major concern in microbial food safety and quality control.

Immunosensor and Immunofiltration Assay System

The immunosensor design is based on a combination of a flow-through immunofiltration technique and a down-stream detection electrode assembly. The design concept, previously described in (27, 28), has been modified and adapted for the immunofiltration assay of bacteria. The technical layout of the immunosensor and the immunofiltration assay system developed and used in this study are presented in Fig. 1. The immunocolumn (see Fig. 1.a), which is the *disposable sensing element*, consists of a plastic body with the *immunofiltration membrane* , laying on a *support* and covered by a *flow distributor*. The *electrochemical detection module*, in which the disposable immunocolumn fits (see Fig. 1.a) contains the detecting electrode (graphite rod) and the supporting electrode assembly: a graphite counter electrode and a Ag/AgCl reference electrode that are exposed into a channel which serves as an outlet for the flow-through immunosensor. The *locking inlet adaptor* (see Fig. 1.a) contains a capillary tube which serves as the inlet to the immunosensor while the adaptor body fits inside the disposable sensing element to achieve air- and liquid-tight locking.

The schematic of the flow-injection immunofiltration assay system is shown in Fig. 1.b. The system has the following main components: the *immunosensor* assembly (locked components as shown in Fig. 1.a), a peristaltic pump, a five-way valve with flow *injector* and a mixing chamber, reagents and waste vessels, analyte supplier and an *electrochemical interface* which consists of a dedicated calibrated potentiostat, an LC display and analog data output. The electrochemical interface is used to provide the working potential for the immunosensor and to process the output signal. A manual jack is used for moving the holder to lock with the immunocolumn and the adapter (motion shown by arrows in Fig. 1.a).

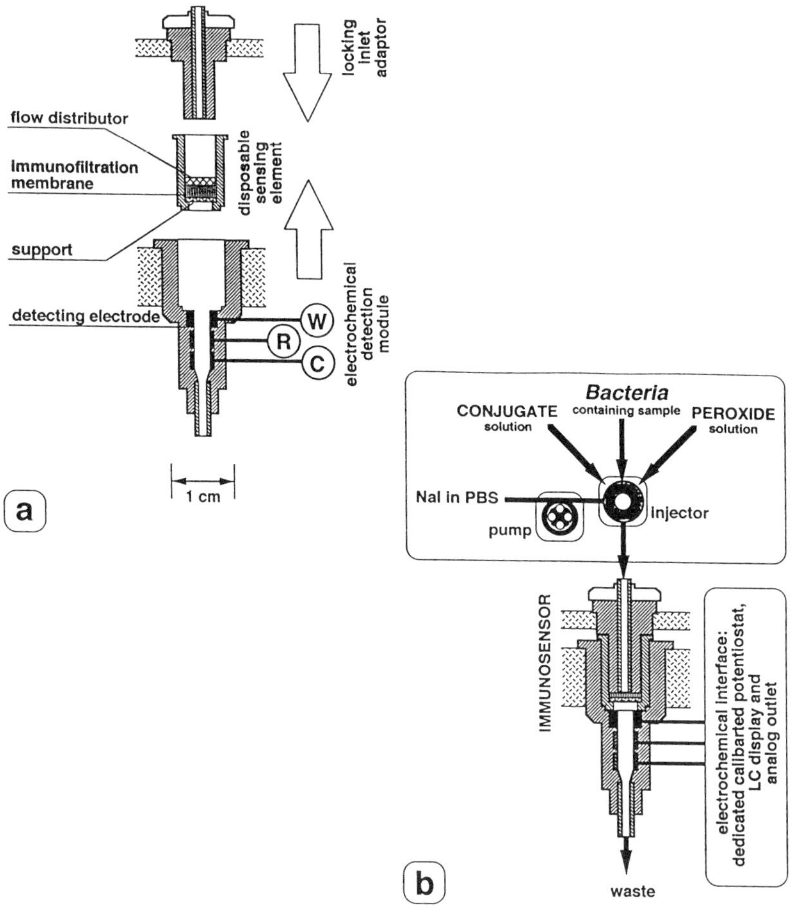

Figure 1. Immunosensor design (a) and schematic diagram of the flow-injection immunofiltration assay system (b).

This immunoassay system was engineered into a laboratory prototype device which includes all system components. The prototype is a compact (20 x 8 x 25 cm), self-standing device and has the potential for complete automation allowing its use in laboratory settings as well as in field conditions. In this work the prototype was used in a bench-top experimental set-up where the bacteria containing *sample* is fed into the *immunoassay system* device and the output is recorded on a PC-based *data acquisition system*.

Experimental Methods

Affinity purified antibodies to *E. coli* or *Salmonella* (KPL, Inc., Gaithersburg, MD) were immobilized on the surface of Nylon membranes (courtesy of Pall Corporation, NY) as follows. The Nylon membranes were cut into 1 cm discs and placed into separate wells of a polystyrene plate. Each membrane was incubated (while shaking) for 1 hour in 1 mL of 10% (w/v) carbodiimide solution (pH 5.0). The membranes were washed 3 times (3 minutes each) in 20 mM phosphate buffer solution (pH 5.6). The membranes were transferred to a new polystyrene plate and left 3 minutes to dry. Then, 20 µl stock solution (1 mg/ml) of anti-*E. coli* or anti-*Salmonella* antibodies were dropped onto each membrane and left to dry. Following that, each membrane was incubated for 2 hours with 2 ml of 0.5% BSA solution in 20 mM phosphate buffer, pH 5.6. The membranes were then washed 3 times as described above, incorporated into the disposable immunofiltration column and stored at 4 °C until further use. Mortalized bacteria were used in this study for safety reasons and to ensure a static (non-proliferating) sample population for quantitative purposes. Reference quantification of mortalized bacteria samples were performed by hemacytometer counts and used to prepare stock bacterial suspensions in phosphate-buffered saline.

The immuofiltration assay experiment is started by placing the disposable sensing element into the holder and the assembly is then manually raised upwards using the jack until the adaptor is inserted and fixed firmly in the immunocolumn to provide air-tight locking of the immunosensor. An analyte sample is then injected into the analyte supplier. The different solutions are pumped through the immunosensor and the flow controller is manually adjusted to select the desired solution. The working potential is applied to the immunosensor during the final washing procedure. Once the output signal is recorded, the working potential and the pump are switched off. The holder/sensing element assembly is then lowered down and the sensing element is removed and disposed of.

A "sandwich scheme" of immunoassay was employed (see Fig. 2), and the following steps were performed at a fixed flow rate of 0.1 ml/min. A defined volume of *bacteria* cells (suspension) was loaded into the injection loop (injection volume 0.1-1.0 ml) and then injected, by the pump, into the inlet of the immunosensor and flushing with a 20 mM phosphate buffer (pH 5.8), containing 0.15 M NaCl and 1 mM NaI for a duration of 3 minutes. Then, the same volume (0.1 - 1.0 ml) of HRP-conjugate solution (0.5-12 µg/ml) was loaded into the injection loop and injected into the inlet of the immunosensor. This was followed by flowing the washing buffer for a duration of 5 minutes. The last step was the electrochemical measurement of the reduction currents of iodine, a product of the enzymatically oxidized iodide-ions (see Fig. 2). The activity of the peroxidase label on the membrane surface was measured using an amperometric technique with iodide-ions as a mediator (see Fig. 2). A polarization potential of 0.0 V vs. Ag/AgCl was applied between the working and reference electrodes using an in-house built potentiostat (a component of the device electronics). The background signal was allowed to establish its steady-state value for 1 min. At this point, 1 ml of a 3 mM solution of H_2O_2 was loaded into the injection loop (injection volume 0.1-1.0 ml) and then injected into the inlet of the immunosensor. The analytical signal of the enzymatic reaction was estimated as the initial rate of current-change (the initial slope of the current transient expressed in nA/s).

It was found that a duration of 6 minutes for the electrochemical reaction was sufficient to produce an amperometric signal distinguishable for different bacterial concentrations. Hence the analytical signal was estimated as the slope of the current transient during the first 6 minutes. Control experiments were performed when the first incubation stage contained no *bacteria* cells in the sample in order to determine the background response. All experiments were carried out at room temperature (approximately 23°C).

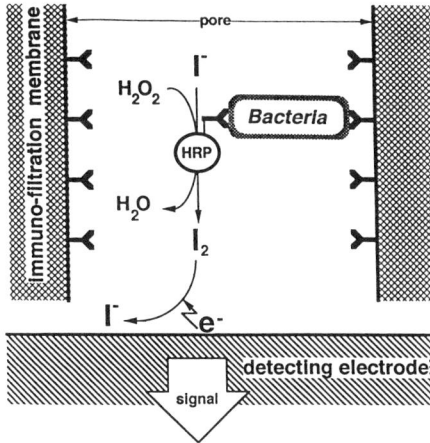

Figure 2. "Sandwich" scheme of the immunoassay conducted on the immunofiltration membrane after capturing the bacterial cells and labeling them with the HRP-conjugated antibody.

A conventional "sandwich" scheme ELISA procedure was performed as a validation experiment with standard 96-well polystyrene plates (Corning Co.). In order to modify each individual well with the primary antibody, 100 µl of a 1.0 mg/mL solution of antibody against total *E. coli* or *Salmonella* diluted in 0.1 M phosphate buffer solution, pH 7.2 (PBS) was passively adsorbed at 37° C for 1 hour. The wells were rinsed three times (3 minutes each) with 0.1 M phosphate buffer solution (pH 7.2) containing 0.5 M NaCl and 0.1 % Tween 20 (washing buffer). Aliquots of 100 µl of *E. coli* or *Salmonella* cells suspension were incubated in each well at room temperature for 2 hours and the wells were then rinsed three times (3 min. each) with washing buffer. Following that, 100 µl of the conjugate solution (1:1000 dilution) was added to each well and incubated for 1 hour. The wells were again rinsed as previously described. For the detection stage, 100 µl of the enzyme substrate solution containing 12 µl of 0.66 mg/mL of OPD in 0.1 M citric acid-phosphate buffer (pH 5.0), was mixed with 5 µl of 30 % (v/v) H_2O_2. The reagents were incubated for 15 min. The enzymatic reaction was stopped by adding 150 µl of 1 M H_2SO_4 to each well. The results of ELISA were measured by an spectrophotometric ELISA-reader at a wavelength of 492 nm.

Immunosensor Detection and Enumeration of Bacteria

Parameters affecting the performance of the immunofiltration assay system such as the type of immunofiltration membranes, membrane pore size, conjugate concentration, non-specific binding of conjugate molecules, the contact time between reagents and the surface of the antibody modified membrane, flow rate and sample volume were studied and optimized (27). We found that Tween 20 and BSA markedly reduce nonspecific adsorption of the enzyme conjugate to the Nylon membrane, but did not fully eliminate it in the stationary (no-flow) mode. In case of direct filtration of a buffer solution through porous membrane (the flow-through immunofiltration system) a much lower value of the background signal is obtained. It may be hypothesized that the main reason of the non-specific signal is accumulation of conjugate molecules in the pores of the membrane which is easily removed by buffer flow through he membrane in the flow-through format of immunoassay.

Figures 3, 4 and 5 present the calibration curves obtained with the electrochemical immunosensor operating under optimal conditions for different bacterial target analytes: E. coli O157:H7 (Fig. 3), total E. coli   (Fig. 4) and Salmonella (Fig. 5). This Figures compare the analytical performance of the flow-injection immunofiltration assay system (curves a) with those of a conventional "sandwich" ELISA assay performed with the same immunoreagents in a micro-titer plate with spectrophotometric detection of the enzyme label (curves b). In all three individual cases the advantage of the immunosensor assay in terms of sensitivity and low limits of detection is obvious. The working range for total E. coli   and Salmonella detection is from 50 to 200 cells/ml (for E. coli O157:H7 the upper limit is somewhat higher up to 600 cells/ml) and the overall analysis time does not exceed 40 min. The low detection limit of the amperometric flow-injection immunoassay system is 3 orders of magnitude lower than that of the standard ELISA assay (which takes several hours to develop). A number of factors contributed to this amplification. First, immunofiltration accelerates the diffusion-controlled rate of immunological, and enzymatic reactions to within the surface of the sensing element of immunosensor. This procedure overrides the effects of diffusion limitations. Secondly, microporous membranes offer 100 to 1000 times more available surface area for immobilization of antibodies than solid wells currently used in ELISA. The bacteria may be concentrated on the membrane surface from a large sample volume. We also found that cell adhesion to the antibody modified membrane surface was significant, and it increased with the amount of cells in the incubation sample. At the same time, the degree of adhesion of bacteria to the unmodified membrane surface was low. Capturing of the bacterial cells by the immunofiltration membrane here were confirmed by direct examination by SEM of the cell adhesion of E. coli O157:H7 to the antibody modified porous Nylon membranes surface (1.2 μm pore size). The bacterial cells were observed on the membrane surface after immunofiltration of the sample through the membrane.

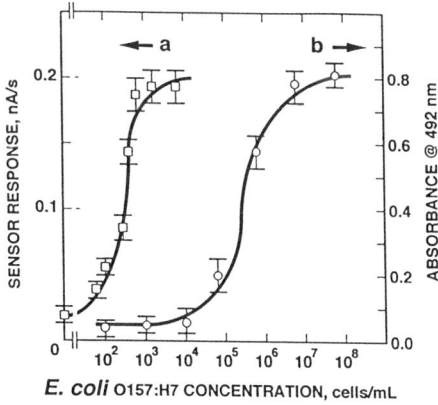

Figure 3. Calibration curves for E.coli O157:H7 enzyme immunoassay using the immunofiltration assay system with amperometric detection (a) and an enzyme-linked immunosorbent assay (ELISA) (b) with spectrophotometric detection.

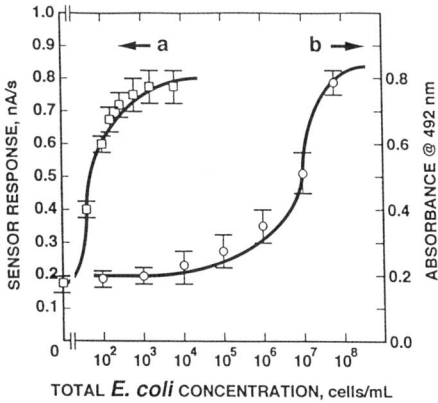

Figure 4. Calibration curves for total E.coli enzyme immunoassay using the immunofiltration assay system with amperometric detection (a) and an enzyme-linked immunosorbent assay (ELISA) (b) with spectrophotometric detection.

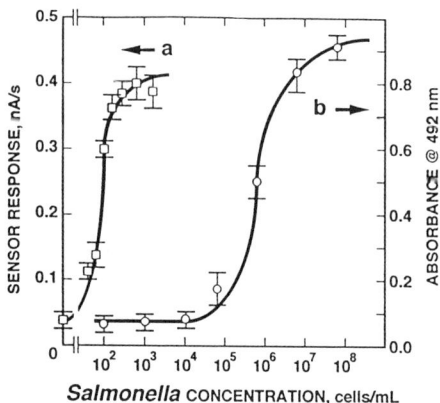

Figure 5. Calibration curves for Salmonella enzyme immunoassay using the immunofiltration assay system with amperometric detection (a) and an enzyme-linked immunosorbent assay (ELISA) (b) with spectrophotometric detection.

The preparation technique of the disposable sensing elements used in this flow-injection amperometric immunofiltration system demonstrated high reproducibility of the analytical parameters. The variation between the responses to a standard concentration of mortalized cells (100 cell/ml) obtained with membranes prepared from one and the same batch was 16 %. It should be noted that the data points in Figures 4, 5 and 6 (curves *a*) are obtained as an average of 4 independent measurements, each performed by an individual sensing element. The error bars in Figures 4,5 & 6 represent the standard deviation obtained from these 4 independent measurements.

Conclusion

Portable, fast and highly sensitive amperometric flow-injection immunofiltration biosensor systems have an excellent potential for application in food quality control, medical diagnostics, environmental monitoring, defense and other industries. The amperometric immunofiltration assay system permits the detection of bacteria cells with a lower detection limit of 50 cells/ml (as demonstrated for *E. coli* and *Salmonella*). The system is also able to distinguish between bacterial species and individual bacterial strains (as illustrated by *E. coli* O157:H7 assay) and to provide direct enumeration within the dynamic range of the sensor response from 50 to 600 or 1000 cells/ml. This technique is simple, facilitates the concentration of bacteria from relatively large sample volume, offers high signal-to-background ratio and the complete immunoassay is carried out in 40 min.

Acknowledgments

This research was supported in part by a grant from the DoE/Waste-Management Education and Research Consortium of New Mexico. Authors express also their gratitude to BioDetect, Inc. (Albuquerque, NM) for partial financial support and for providing the prototype immunoassay device for this experiments.

References

1. M.P.Doyle, L.R. Beuchat, T.J. Montville (eds). *Food microbiology: fundamentals and frontiers, ASM Press*, Washington DC, **1997**, Ch. 3, p. 127.
2. G.W. Beran, H.P. Shoeman, K.F. Anderson, *Dairy Food Environ. Sci.* **1991**, *11*, 189.
3. R.M. Atlas, *Critical Rev. Microb.*, **1998**, *24*, 157.
4. M. Dando (ed), *Biological warfare in the 21st century.* Macmillan, London, 1994, p. 258.
5. B. Swaminathan, P. Feng, *Ann. Rev. Microbiol.*, **1994**, *48*, 401.
6. USDA, *News Release No.0272.97*, Washington DC, Aug.12, 1997.
7. A.M. McNamara, *Bull. N.Y. Acad. Med.*, **1998**, *75*, 503.
8. L. Slutsker, S.F. Altekruse, D.L. Swerdlow, *Inf. Dis. Clin. N. Am.*, **1998**,*12*, 199.
9. M.E. Potter, S. Gonzalez-Ayala, N. Silarug, in Doyle, M.P., Beuchat, L.R., Montville, T.J. (eds.): *Food Microbiology: fundamentals and frontiers*, ASM Press, Washington, DC., **1997**, p. 376.
10. J. Meng J., M.P.Doyle, *Bull. Inst. Pasteur,* **1998**, *96*, 151.
11. P. Feng, J. Food Protection, **1992**, *55*, 927.
12. M. Tietjen, D.Y.C. Fung, *Crit. Rev. Microb.*, **1995**, *21*, 53.
13. K. Helrich, in: *Official Methods of Analysis of the Association of Official Analytical Chemists, Vol. 2*, Ch.17., AOAC, Inc., Arlington, VA, 15 ed., **1990**, p. 425.
14. B. Mirhabibollahi, J.L. Brooks, R.G. Krool, *Appl. Microb. Biotech.* **1990**, *34*, 242.
15. N. Nakamura, A. Shigematsu, T. Matsunaga, *Biosens. Bioelectronics,* **1991**, *6,* 575.
16. J.L. Brooks, B. Mirhabibollahi, R.G. Krool,*J. Appl. Bacteriology.* **1992**, *73*, 189.
17. H.J. Kim, H.P. Bennetto, M.A. Halablab, *Biotechnol. Tech.*, **1995**, *9*, 389.
18. J. Rishpon, D. Ivnitski, *Biosens. Bioelectronics*, **1997**, *12*, 195.
19. J.D. Brewster, R.S. Mazenko, *J. Immunol. Methods*, **1998**, *211*, 1.
20. J.D. Brewster, A.G. Gehring, R.S. Mazenko, L.J. Vanhouten, C.J. Crawford, *Anal. Chem.*, **1996**, *68*, 4153.
21. A.G. Gehring, C.G. Crawford, R.S. Mazenko, L.J. Van Houten, J.D. Brewster, *J. Immunol. Methods,* **1996**, *195*, 15.
22. C.R. Clark, K.K. Hines, A.K. Mallia, *Biotechnol. Tech.,* **1993**, *7,* 461.
23. S.M. Paffard, R.J. Miles, C.R. Clark, R.G. Price,*J. Immunol. Methods,* **1996**, *192*, 133.

24. P. Bouvrette, J.H.T. Luong, Int. J. Food Microbiol., **1995**, *27*, 129.
25. I. Abdel-Hamid, D. Ivnitski, P. Atanasov, E. Wilkins, *Anal. Chim. Acta,* **1999**, in press
26. I. Abdel-Hamid, D. Ivnitski, P. Atanasov, E. Wilkins, *Biosens. Bioelectronics,* **1999**, *14,* 309.
27. I. Abdel-Hamid, P. Atanasov, A.L. Ghindilis, E. Wilkins, *Sens. Actuators, B,* **1998**, *49,* 202.
28. A.L. Ghindilis, R. Krishnan, P. Atanasov, E. Wilkins, *Biosens. Bioelectronics,* **1997**, *12,* 415.

Chapter 18

Spot Assay for Rapid Detection of Blood Glucose

Ralph Ballerstadt and Jerome S. Schultz

University of Pittsburgh, Center for Biotechnology and Bioengineering,
300 Technology Drive, Pittsburgh, PA 15219
(phone: (412) 383–9700, fax: (412) 383–9710, email: jssbio@pitt.edu)

A quantitative receptor-polymer spot assay is presented which allows one to determine analyte concentrations by measuring the size of circular, colored spots on a cellulose-coated sheet. This biosensor method for glucose analysis is demonstrated by using concentrated dextran/Concanavalin A (ConA) dispersions. These dispersions were shown to possess a very high viscosity in the absence of glucose due to the crosslinking between terminal glucose residues in dextran and the carbohydrate receptor sites of ConA. However upon the addition of glucose, the viscosity is strongly decreased due to analyte-induced cleavage of intermolecular dextran/ConA affinity bonds. The spot assay allows one to measure glucose in full blood and blood plasma with a respective accuracy of ± 2mM and ± 1 mM between 5 and 20 mM. The excellent storage stability of ConA/dextran dispersion promotes this economic biosensing assay for its use in third world countries.

Introduction

Chemical and biological sensors gain more and more acceptance in the society as powerful detection methods for environmental and pharmacological monitoring due to their high specificity and the large variety of naturally existing detector molecules. Despite the ongoing development to establish self-contained automatic sensing devices, there is still a need for "manual" biosensing assays that provide rapid detection of blood components such as glucose or other pharmacologically relevant substances. For instance, the palm-sized blood glucose-testing devices (e.g., SureStep from Lifescan, or SpotPen from Medisense/Abbott) provide an easy and fairly convenient system for diabetics, informing them immediately (within a few minutes) of the actual blood glucose level. The principle of those devices is based on a specific reaction of an enzyme with glucose, resulting in reaction products which can be measured amperometrically or optically. Such palm-sized devices are popular and affordable for patients in developed countries, but still too expensive for the majority of people living in developing countries in Africa, Asia, or Latin America. Simpler assay kits at affordable costs may be desired in developing countries which would doubtlessly help to raise the health standards in those areas. Such an assay should be based on an easy-to-read parameter, such as the change of color or spot dimension, which could be easily assessed visually by comparing it with a reference chart. In this paper the basic principle and performance of a simple and fast spot test for quantification for blood glucose will be discussed.

Background And Principle Of Spot Test

The spot test described herein is based on the Viscometric Affinity Assay (VAA) described earlier [1, 2]. The principle of the VAA is based on the fact that the mixture of the glucose-specific lectin Concanavalin A (ConA, [3,4]) with a highly concentrated dispersion of 1,000 kDa dextran (wt 1-10%) yields a tremendous rise of the resulting viscosity (up to 20 times). Such viscous dispersions can adopt gel-like properties as discussed in detail by Ehwald et al. [5]. This viscous behavior is due to extensive intermolecular affinity crosslinking of dextran molecules by ConA (see Figure 1a).

a) Association of Con A and Dextran = High Viscosity Sol

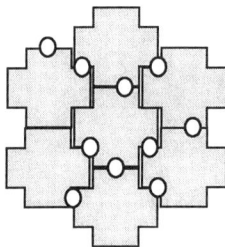

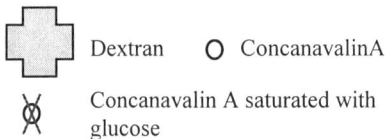

b) Dissociation of Con A and Dextran = Low Viscosity Sol

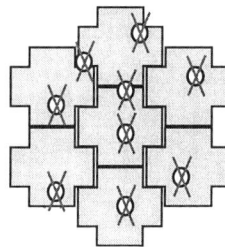

Dextran   O ConcanavalinA

Concanavalin A saturated with glucose

Figure 1. Schematics of mechanism of the Viscometric Affinity Assay (VAA).

However the introduction of the competing analyte glucose into this dispersion leads to a strong yet reversible viscosity decrease approaching that of water-like fluids if the amount of glucose added is at saturating concentrations (Figure 1b). The decrease in viscosity can be explained by the competitive glucose-induced dissociation between ConA and terminal glucose residues of dextran. Note that although the viscosity shows a drastic drop in viscosity, the network structure within the dispersion remains virtually unchanged regardless of the presence or absence of glucose. The decrease in viscosity strongly correlates with the increase in glucose concentration from 0 to 20 mM. The VAA has been investigated for its application in various biosensor schemes for glucose detection [1,6,7], but the challenge lies in the difficulty inherent when to continuously measure viscosity in a sub-microliter volume. However it is comparatively easier to design disposable one-time methods for blood glucose detection by miniaturizing known methodologies of viscometry for example measuring flow velocity in microcapillaries [1,2].

In this paper a novel variation of viscosity measurement is proposed: a spot test for glucose detection. When drops of ConA/dextran dispersion are placed on a highly wettable surface, we will show that the size of the obtained spots will be increased at high glucose concentration due to diminished viscosity of the dispersion. In order to attain an optimized spot size change, other factors in addition to sample viscosity have to be taken into account such as water absorbancy of the surface, humidity, and temperature. The competition of all these factors will affect the final spot size (Figure 2).

Experimental

The glucose-sensitive dispersions were prepared by slowly mixing 0.5 ml of a phosphate-buffered solution (8 mM, pH 7.3, 0.9 % NaCl, 0.05 % NaN$_3$) of ConA (22mg/ml) to 0.5 ml of a solution containing Rhodamine-labeled dextran (2,000 kDa, 12 mg/ml). The colored dextran makes the spot easily visible for size determination.

For measuring glucose concentrations in full blood and blood plasma, human blood (10 ml) was stabilized by addition of citric acid (0.5 %). The blood was centrifuged and dialyzed overnight against 4 l of saline PBS. Full blood and blood plasma samples were spiked with glucose (dissolved in PBS) in the volume ratio of 20 to 1. Aliquots of the spiked blood samples were then gently mixed with the ConA/dextran dispersion in volume ratio of 1 to 3.

Spot development was facilitated by using chromatographic paper consisting of a 160 micron-thick coating of microcrystalline cellulose layered on a flexible plastic base (Eastman Kodak Company, Rochester, N.Y., USA). The sample solutions (10 μl) containing various amounts of glucose were slowly aspirated into a pipette tip (10 μl) connected to a micro-pipette. While the pipette was held above the paper surface without touching it, the dispersion was slowly pushed out of the pipette tip forming a hanging drop that was carefully placed on the surface material. The temperature and humidity were 20 to 25°C and 60 to 70 % respectively during the experiment. After the drop was allowed to expand over the surfaces, the dried paper was scanned. The

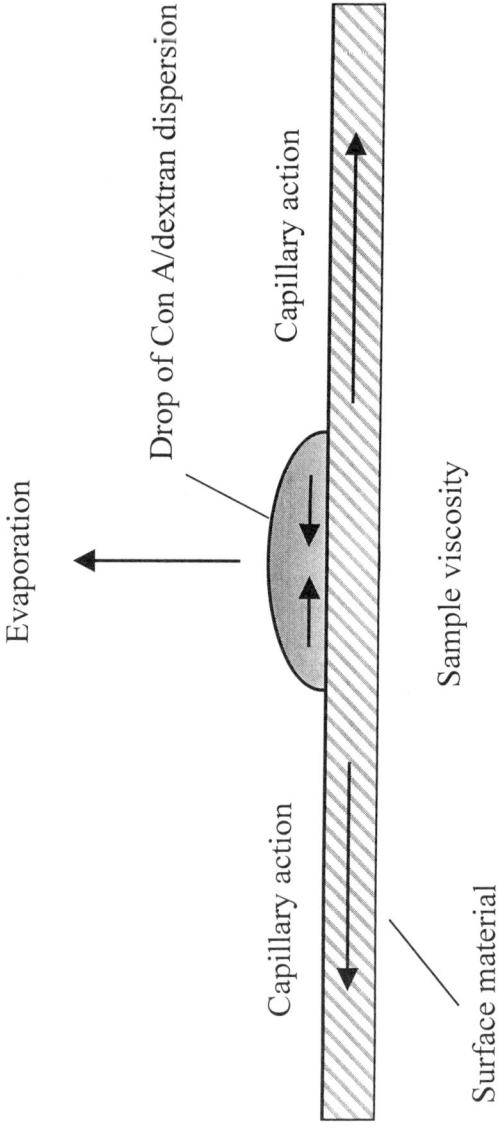

Figure 2. Factors affecting the spot size dimensions.

diameter of the scanned spots (n=3) was measured by means of a ruler designed by computer software (Microsoft Powerpoint). The averaged spot size diameter was plotted vs. glucose concentration.

Results And Discussion

To yield a high absolute change of spot diameter from 0 to 20 mM glucose, the ConA/dextran concentration ratio and the surface properties were varied. The latter were accomplished by employing microcrystalline cellulose powder layered on a thin plastic base. The surface of this chromatographic material forms an extremely homogenous texture with a high water absorbancy. While the surface texture ensures circular spots, the high water absorbancy translates into strong capillary forces, the opposite force to the sample viscosity. By keeping the dextran concentration low at wt 1.2 % and the ConA concentration high at wt 2.2 %, the glucose-induced spot size changed between 0 and 40 mM of glucose was significant and well measurable (see Figure 3).

The basis for the viscosity-dependent spot diameter lies in the competition between spot expansion and evaporation of water from the wetted surface. Experiments have demonstrated (not shown) that at medium to high temperatures (20 to 30°C) and low humidity (30 to 60 %), a large spot size change is obtained, while the opposite conditions result in a lesser change.

The specificity of ConA towards various types of sugars is relatively broad. The experiment described in the following confirms this. Solutions spiked with various amounts of glucose, mannose, and galactose were mixed with the test dispersion and the spot size was plotted versus their concentrations. Note that here the developed and dried spots were copied, cut and weighed. In Figure 4, the graph for each type of monosaccharide shows different slopes due to different affinity constants towards ConA. For mannose, the maximal spot size was reached already at 10 to 20 mM, while for glucose the graph shows an almost linear slope up to 20 mM before it starts to slowly level off. In contrast, galactose produced only a marginal increase in spot size at relatively high concentrations (30 mM). For blood sugar detection, the low binding specificity of ConA does not pose a major problem since glucose occurs in blood at high abundance, while mannose, although present in blood as well, is found at ten times lower concentrations.

To test the spot assay on real-world samples, human full blood and blood plasma, which were dialyzed prior to this, were spiked with glucose. The spiked blood samples were mixed with ConA/dextran dispersions and the diameter of the developed spots was measured. For both blood plasma and full blood, an almost linear graph from 5 to 20 mM glucose were obtained (see Figure 5). In this concentration range, the standard error was around 5 % translating in a test accuracy of ± 1 mM for blood plasma. Note the shallower slope for full blood which diminishes the precision of the assay to ± 2 mM glucose. This phenomenon can be attributed to the higher background viscosity, which was due to the viscosity contribution of red blood cells to the one of the glucose-sensing dispersion. Thus separation of red blood cells from

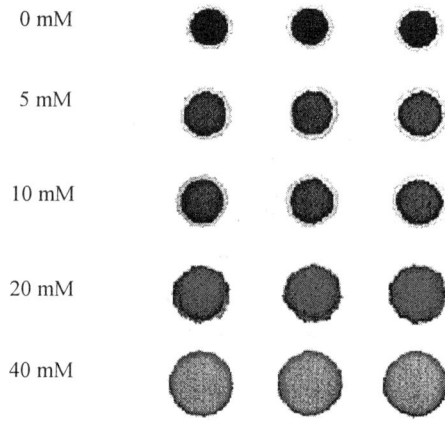

Figure 3. Spot size of ConA/dextran dispersions spiked with various concentrations of glucose

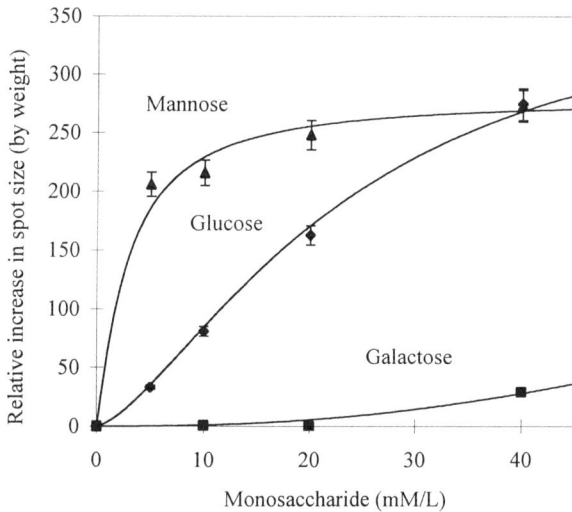

Figure 4. Plots of sugar concentration versus weight of cut spots using various types of monosaccharides. Note that the spot weight correlates with spot diameter.

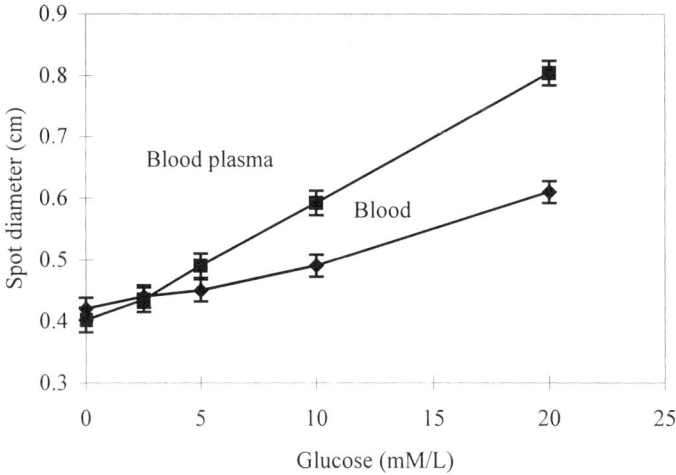

Figure 5. Plot of glucose concentration versus spot diameter using human blood and blood plasma spiked with glucose.

blood prior to the test is highly recommended to generate a significant spot size response.

To be able to measure low blood glucose concentrations between 0 and 5 mM with high accuracy, the glucose sensitivity of the spot assay has to be improved. The concentration of the dextran were already minimized to the possibly lowest concentration (wt 1.2 %) to induce a measurable viscosity change without compromising the solution stability with regard to precipitation. The law of mass actions governs that the lower the amount of fixed glucose residues (i.e., analyte-analog), the lower the amount of free glucose needed to dissociate ConA/dextran bonds. Thus this will require reducing the number of glucose residues by chemical modification of appropriate polymers. Under this condition, the concentration of dextran could be increased up to around 5 %, resulting in a high absolute viscosity change in the presence of low glucose concentrations.

As it was already mentioned above, environmental factors such as humidity and temperature have an effect on the spot size. By conducting a two-point calibration between 5 and 20 mM glucose, the spot size assay can be adjusted to the actual environmental conditions which can be quite extreme in tropical countries.

The long-term stability of the ConA/dextran dispersion is remarkable. Studies have shown [1] that the viscosity drop of the sensing dispersion upon reaction with glucose is extremely stable at 37°C over a 3-month period of time. This feature is very beneficial for the proposed application in countries with warm climates where cooling devices are often not available.

A spot test for blood glucose determination was described which excels by its simplicity in handling and results evaluation. The *See-and-Know* approach of this low tech, but fairly robust biosensing assay makes it attractive for applications in third world countries. Also the assay has the potential to be adapted to other analytes involved in diseases prominent in those countries. A patent for this technology has been applied for [8].

References

1. Ballerstadt, R.; Ehwald *Biosens.Bioelectron.* **1994**, *9*, 557-567.
2. Ehwald, R. German Patent 19501159A1, 1996.
3. Sumner, J. B.; Howell, S.F. *J.Bacteriology* **1936**, *32*, 227-237.
4. Goldstein, I.J., Hollermann, C.E.; Smith, E.E. *Biochemistry* **1965**, *4*, 876-883.
5. Beyer, U.; Ehwald, R.; Fleischer, L.-G. *Biotechnol. Prog.* **1997**, *3 (6)*, 722 –726.
6. Ballerstadt, R.; Schultz, J.S. *Sensors&Actuators* **1998**, *B 46*, 50-55.
7. Ehwald, R.; Ballerstädt, R.; Dautzenberg, H. *Anal. Biochem.* **1996**, *234*, 1-8.
8. Ballerstadt, R; Schultz, J.S., US-Patent (pending).

Nucleic Acid-Based Sensors

Chapter 19

A Critical Review of Nucleic Acid Biosensor and Chip-Based Oligonucleotide Array Technologies

Paul A. E. Piunno, Dalia Hanafi-Bagby, Lisa Henke, and Ulrich J. Krull[1]

Chemical Sensors Group, Department of Chemistry, University of Toronto at Mississauga, 3359 Mississauga Road North, Mississauga, Ontario L5L 1C6, Canada

The development of biosensors for sensitive and selective nucleic acid analysis in a timely fashion is a most desirous goal sought after by environmental and health care comunities. This article provides a brief review of the concepts and advantages of genetic testing methods followed by a survey of some of the more significant contributions to the development of biosensors and chip-based oligonucleotide array technologies for nucleic acid analysis. A critical analysis of the state of the art of this technology is then given followed by an exploration of possible future directions of this rapidly expanding field.

Introduction

Detection of Pathogenic Microorganisms

The detection and identification of microorganisms is a problem common to many areas of human and veterinary health. For example, the detection of pathogenic species such as *Salmonella typhimurium*, *Listeria monocytogenes*, and *Escherichia coli* 0157, which are causative agents of major food borne epidemics, is a great concern within the food industry with respect to the quality and safety of the food supply. In other areas of human and veterinary health care, detection and identification of infectious diseases caused by pathogenic microorganisms and viruses

[1]Corresponding author.

is a first step in diagnosis and treatment. For example, it is estimated that 10-15 million medical office visits per year in North America are for the detection and treatment of three major pathogens - *Chlamydia ssp.*, *Trichomonas vaginalis* and *Gradenerella vaginitis*. Infections by these organisms annually affect 3.75 million, 0.75 million and 1.5 million patients, respectively. Recently, there have been an overwhelming number of endeavors aimed at the development of tools for the detection of microorganisms and the screening of genetic mutations. The demand for such diagnostic methods in various industries stems from a need for techniques that are rapid, sensitive, inexpensive, simple and reliable.

Detection processes based on culturing require analysis times which are too lengthy for effective monitoring and timely intervention to prevent the spread of biohazardous agents or to treat disease. In order to shorten the time required to detect and identify pathogenic bacteria, rapid tests such as enzyme immunoassays (EIA) have been developed. An immunoassay is an analytical technique which uses antibodies or antibody related reagents for the selective determination of sample components, and have been widely reviewed [1 - 4]. Although immunoassay techniques can be very sensitive and effective, there are practical drawbacks that have restricted the use of these methods and many laboratories are still plagued by the necessity to utilize less technical procedures, like plating/culturing techniques. Such drawbacks include the need for skilled personnel, specialized equipment lengthy preparation and analysis times, and the large quantities of costly reagents that are required to do such analyses [5].

In addition, newer methodologies are hindered from entering the commercial marketplace as many are lacking the required recognition from regulatory bodies such as the International Organization for Standardization (IOS) and the Association of Official Analytical Chemists (AOAC), a process which can take several years [6]. In cases where novel diagnostic systems are used for screening, negative results are usually considered definitive. Positive results, however, normally require confirmation using conventional standards, which sometimes require a time of weeks [7]. For example, a sample found to contain *Mycobacterium tuberculosis*, the causative agent of tuberculosis, when identified by an EIA-type test is then submitted for confirmatory testing which requires an average of 34 days for proper identification plus an additional two to three weeks in order to determine drug sensitivity [8]. Furthermore, the *Mycobacterium* genus contains more than 25 species, all of which must be identified in order to eliminate the possibility of tuberculosis and to assess proper treatment. Such scenarios constitute a primary impetus for the development of optimized techniques in the clinical laboratory.

Nucleic Acids

Assays based on nucleic acid detection are now becoming commonplace in the diagnostic market scene. For example, DNA and RNA based diagnostic kits for detection of viruses and pathogenic organisms are commercially available from companies such as Gen-Probe [9], Chiron [10], and Digene [11]. Nucleic acids are responsible for the transmission and expression of genetic information required for both self-replication and cellular function [12-14]. Deoxyribonucleic acid, DNA, is the master template of genetic information and is located in the nucleus of eukaryotic cells. Prokaryotes, lacking a distinct cell nucleus, have two types of DNA: a large, circular DNA molecule in a region termed the nucleoid; and smaller, circular fragments known as plasmid DNA. Viruses are small entities of protein, lipid, and nucleic acid materials that differ from living organisms in that they do not contain the materials necessary for reproduction and proliferation, but rather rely on the host cell for multiplication following infection. Viruses have small genomes consisting of either single- or double-stranded ribo- or deoxyribonucleic acids. The DNA molecule is composed of adjoining nucleotide units that are composed of a sugar, a nitrogenous heterocyclic base and one phosphate group. DNA can be viewed as a linear information-bearing molecule with a data coding scheme based on a 4-letter alphabet. The predominant nitrogenous bases, which may be considered as the letters of the genetic alphabet, are the purines, adenine (A) and guanine (G), and the pyrimidines, cytosine (C), thymine (T).

While DNA functions as a repository for genetic information, the mediators for the processes governed by that information are the ribonucleic acids, RNA, located primarily in the cell cytoplasm. RNA is found to exist as long, single-stranded molecules that are also referred to as ribonucleotides. The composition of RNA molecules is similar to that of DNA in that they are both composed of a sugar-phosphate backbone functionalised by purine or pyrimidine nucleobases. Ribonucleotides differ physically from deoxyribonucleotides in that a uracil base replaces the thymine base found in deoxyribonucleotides and the pentose sugar of RNA is ribose (which differs from deoxyribose by the presence of a hydroxyl moiety at the 2' position). Ribonucleotides are generally much shorter than deoxyribonucleotide molecules, are more abundant by a factor of at least 10^4 in eukaryotic cells, and are found to participate in a variety of different functions. There exist three major types of RNA in all living cells, namely messenger RNA (mRNA), ribosomal RNA (rRNA), and transfer RNA (tRNA). This classification is based on the fact that each of the three kinds of RNA has a specific biological function, a characteristic molecular weight, and nucleotide sequence [15]. rRNA is complexed with proteins to provide the physical makeup of the ribosomes. Ribosomes exist as granular particles in the cytoplasm and are associated with protein biosynthesis. Messenger ribonucleic acid (mRNA) acts as a transient intermediary molecule that transmits the protein-coding instructions from the genes. A complementary mRNA strand is produced (a process called transcription) from the master DNA template for the information within a gene to be expressed. This mRNA is moved from the nucleus, nucleoid or plasmid out into the cellular cytoplasm, where it serves as the template for protein synthesis in the ribosomes. These protein-synthesizing engines within the cell then translate the mRNA codons (three successive nucleotides which code for a particular amino acid residue in the expressed protein) into a string of amino acids that will constitute the protein molecule for which it codes.

Towards Genetic Testing

Two major driving forces have been and continue to expedite the entry of nucleic acid diagnostics, namely, the Human Genome Project and nucleic acid amplification techniques, such as the polymerase chain reaction (PCR). The Human Genome Project, initiated in 1990, aims to elucidate all of the approximately 100,000 human genes by the year 2005 [16]. High-resolution physical sequence mapping of the entire human genome will ultimately provide information for the detection, treatment and possible prevention of diseases which manifest as a result of acquired genetic deficiencies. While individual gene mutations associated with nearly 1000 diseases have been identified [17], progress in this area has been described as hindered by the current methods used to detect genetic variations [18]. As a result, efforts have focussed on the development of rapid, cost-efficient, large-scale automated sequencing techniques in order to realize the full potential of gene sequence analysis, which have concurrently been applied to elucidation of sequence information from simpler non-human genomes. From this endeavor, high-resolution physical maps of the genomes of more simple organisms (*Escherichia coli* and *Saccharomyces cerevisiae*) have been completed with others (*Drosophila melanogaster*, *Caenorhabditis elegans*, and *Mus musculus*) well on their way. Of particular interest, studies on *M. musculus* (white mice, which have a genome 95% similar to the human genome) have already lead to the elucidation of morphological markers for oncogenes and provided many insights into the human biology of cancer.

Nucleic acid amplification techniques, such as PCR, have assisted the incipience of nucleic acid diagnostic methods. PCR is an enzymatically driven reaction that replicates specific DNA regions, potentially yielding millions of copies of a particular nucleic acid sequence in a time span of a few hours. This has made the *in-vitro* amplification of specific sequences from a portion of DNA or RNA possible, and detection of very low numbers of microorganisms has been demonstrated [19,20]. In brief, the process begins by setting the physical boundaries of the nucleotide sequence that is to be amplified. Oligonucleotide primers (one for each strand of the double-stranded parent nucleic acid) that are complementary to the flanking portions at the 3'-ends of the sequence to be amplified are mixed with the target DNA strands. Heating the sequence above the duplex melting-temperature (T_m) denatures the double-stranded structure of the target DNA to yield single-strands. Typically, a temperature above 90°C is sufficient for this purpose. The system is then cooled in the presence of the primers and nucleotide triphosphate monomers (that exist in a large excess relative to the parent nucleic acid) to allow for sequence-specific annealing to complementary nucleotides on the target DNA. The temperature is then raised to *ca.* 72 °C to allow the thermostable *Taq* DNA polymerase enzyme (from the hot springs organism *Thermus aquaticus*) to extend the annealed primers by assembling and connecting complementary nucleotide monomers. This process yields two new double-stranded DNA sequences that are identical to the original, and may be repeated to continue the enrichment of the DNA sequence as each new DNA sequence produced may then serve as a template for further amplification. Products accumulate at a rate of 2^n, where n equals the cycle number, so that enrichment factors on the order of 10^7 (from 30 cycles of amplification) may be realized in a time span of hours [21].

PCR-assisted hybridization assays are not extensively used in the clinical diagnostic laboratory for the detection of microorganisms due to the susceptibility of PCR to contamination. The presence of co-amplified contaminants tends to manifest as false-positive assay results. Contamination sources include products carried over from previous PCR, use of positive controls, and contamination during the collection and handling of samples [22]. As such, special rooms or areas for sample preparation and analysis are required to prevent contamination. Although the PCR technique is a remarkable tool for sample enrichment, quantitative measurement of amplified PCR products is difficult and PCR results alone (*e.g.* agarose gel electrophoresis of the amplified products) are not sufficient for conclusive results and must be substantiated by secondary methods of analysis. In general, PCR needs to be automated in conjunction with a rapid hybridization assay method for acceptance as a practical diagnostic tool.

Advantages of Genetic Testing

With knowledge of the genetic make-up of organisms and virulent factors and the ability to amplify small quantities of nucleic acids has come the development of detection schemes based on nucleic acid hybridization. This new convention of diagnostic testing, yet in its infancy, will offer many advantages over the traditional methods of clinical testing that are currently employed [23]. To begin, the omnipresence (with few exceptions) of the entire gene sets within the cells of an organism may greatly simplify sampling issues with respect to genetic disease testing in multi-cellular organisms. Non-invasive sampling methods may be exploited in such cases, thereby reducing the risk of infection or other complications resulting from invasive procedures. Testing for infection, disease or mutations will be possible shortly after identification and sequence analysis of marker genes has been done. This may ultimately lead to the ability for the clinician to test for every genetic disorder and pathogenic organism. Also, given that the basic assay format for nucleic acid hybridization will most likely be similar and done *via* a standardized protocol, development and optimization of new tests should proceed with unprecedented speed.

Another advantage offered by nucleic acid diagnostics rests in the stability of nucleic acids with respect to their proteinaceous counterparts. DNA samples can be extremely long lived ($>10^6$ years) under the appropriate storage conditions. The success of the nucleic acid hybridization assay is based on the ability of the single-stranded nucleic acids to selectively form a duplex with complementary strands by formation of a double-stranded secondary structure (β-helix). Proteins, however, have an activity that is dependent on tertiary structure, which is known to often rapidly denature once the protein is removed from the host organism. The long term physical stability of genetic materials provides the ability to re-test a sample at a much later date. Alternative targets and newly identified disorders may then be tested without having to obtain a new sample from the patient (provided that a disease brought on by a recent infection is not sought). In the case where a nucleic acid sample supply is small, enzymatic amplification techniques such as PCR may be used to provide the desired quantity of nucleic acid. Redundant testing may be greatly reduced by employing genetic testing strategies as this new testing motif is capable of immediately discriminating different types of disease and, for example, drug resistance in the case of diagnosing pathogenic infections. Hence, the need for screening procedures followed by in-depth testing to determine the exact disease type for proper treatment would be obviated. This would lead to reductions in the expense of laboratory testing and allow for more timely treatment, which may be invaluable to the patients in critical care situations (*vide supra* for a comparison with EIA screening of tuberculosis).

In addition to detecting manifested disease and mutations, genetic testing also offers the ability to detect predisposition to genetic diseases. For example, genetic testing may be done on a sample of DNA from a patient for early detection of inherited diseases such as cystic fibrosis, sickle-cell anemia, Tay-Sachs disease, fragile-X syndrome and myotonic dystrophy. The same sample may also be used for additional testing for predisposition to certain forms of cancer, certain mental illnesses and other complex diseases that are currently not well understood. In this way the patient's wellbeing may be improved by more rigorous supervision of the disease progress, and by permitting the opportunity for early intervention to arrest or slow the progression of the disease. These advantages, in addition to the elegance, simplicity and chemical stability of the DNA molecule itself, make the use of nucleic acids extremely attractive and well suited for use as a biorecognition tool.

Nucleic Acids as Biorecognition Elements

One of the keys to the success of nucleic acid detection strategies lies in the isolation of suitable nucleic acid probes for target identification. Nucleic acid probes typically consist of short sequences (10 to 30 units) of single-stranded nucleotides that are capable of hybridizing with a target nucleotide sequence that is unique to a particular species or genus of organisms [24]. Nucleic acid probes are normally composed of DNA due to the greater stability of the deoxyribonucleotide backbone against hydrolysis relative to that of ribonucleotides. The selectivity of the probe is not only a function of the base sequence, but also the conditions under which the probe is used. This allows for detection selectivity to range from genera to individual organisms. To illustrate, any two human beings (with the exception of identical twins) differ by *ca.* 3×10^6 nucleotides, or 0.1%, of their gene-coding DNA [16]. A probe for a particular individual may then be constructed and used for identification of that individual, for example, in a forensic application, or, by altering the probe sequence, length and/or stringency conditions, the selectivity of the assay may be broadened to include family members, and perhaps even further to include all human beings.

The length of nucleic acid probes that have been used for nucleic acid analysis has ranged from 10 to 10 000 nucleotides. The most commonly developed probes range in length between 10 to 30 nucleotides [24-27]. To identify an organism or virus based on nucleic acid hybridization techniques; a nucleic acid probe of a certain minimum number of nucleotides is required in order to guarantee statistical uniqueness and specificity. This number depends on the size of the genome, the proportion of A·T to G·C base pairs, and nearest-neighbor frequency [28]. To illustrate, the human genome is estimated to consist of a minimum of 100 000 genes which are of an average length of 3000 nucleotides [16]. Based on a random nucleotide distribution, the minimum number of nucleotides, n, in a nucleic acid probe required to be complementary to a single statistically unique sequence within the total number of coding nucleotides, CN, will be given by $4^n = CN$. For the case of the human genome, CN is equal to 3×10^8 and a minimum probe length of 14 nucleotides would then be required to identify a unique sequence within the coding nucleotides, provided that the nucleobase distribution was indeed random. However, when the subtleties of the genome under consideration are taken into account, this number is found to increase slightly. For example, in eukaryotic organisms, the proportion of A·T to G·C base pairs is typically *ca.* 3:2 and the sequence $^{5'}CG^{3'}$ (and not $^{5'}GC^{3'}$) is underrepresented by a factor of five with respect to that based on a random distribution [28]. A slightly larger minimum probe length of 17 nucleotides is then determined for identification of a unique sequence in the human genome [28,29]. Minimum probe lengths for *E. coli* (4.5×10^6 base pair genome) and HIV proviral DNA of 12 and 7 nucleotides, respectively, have been determined in this way.

The use of short nucleic acid probes (at or close to the minimum length for statistical uniqueness) is most advantageous for use as a biorecognition element. The primary reason for this stems from the nature of the assay. Given that identification is based on hybridization as opposed to a direct reading of the genomic sequence, only hybrids between the probe and the desired target sequence are desired. With increasing probe length comes the increased probability of formation of a stable hybrid containing a mismatch. This is the result of the energetic penalty associated with de-stacking caused by incorporation of one mismatched nucleotide pair being of reduced consequence relative to the ensemble free energy change for duplex formation with increased duplex length. Long probes also display slower hybridization kinetics and increase the time required to conduct an assay. For example, probes of several hundred nucleotides in length generally require hybridization times on the order of several hours under standard conditions employed for hybridization. However, hybridization of short oligonucleotides probes (10 – 20 nucleotides in length) under similar conditions of ionic strength, temperature and pH would occur on a time-scale of minutes. Another advantage offered by the use of short nucleic acid probes is the ease and speed with which they can be produced. Oligonucleotides can routinely be synthesized in the laboratory in high yield and short time (1-2 hours) by use of an automated solid-phase DNA synthesizer and phosphoramidite synthons. By the same token, they may be purchased from oligonucleotide synthesis companies at reasonable costs (typically $1 per nucleotide at the 0.2 μmol scale) and with fast delivery times (next day service).

Oligonucleotide Arrays

Recent efforts in nucleic acid hybridization strategies have led to the development of ordered, high-density oligonucleotide microarrays, which can serve as tools for nucleic acid analysis. The underlying principle with regard to the function of oligonucleotide arrays is sequencing by hybridization (SBH). SBH was first developed and patented in the early 1990's by Radoje Drmanac and Radomir Crkvenjakov [30], currently with Hyseq Inc. The initial assay was based on functionalising each well of a 96-well microtiter plate with a unique, short oligonucleotide probe. Identification of the nucleotide sequence of an analyte strand was done by analysis of the hybridization pattern observed in the 96-wells subsequent to treatment with the analyte. Since the development of this approach, many industrial and research consortiums have endeavored to advance this technology so that high throughput DNA sequencing on genosensor chips functionalised with 2-dimensional oligonucleotide probe arrays may one day be realized. Dr. Stephen P. Fodor and co-workers at Affymetrix Inc. [31] may be identified as one of the leaders in this rapidly expanding venture. The Affymetrix group has exploited photolithographic technology, the same as that employed to produce integrated circuits used in the computer and microelectronics industries, to fabricate glass chips (GeneChips®) functionalised with a two-dimensional array of oligonucleotides.

Instead of using ultraviolet light projected through a series of masks to remove photo-resist coatings and affect etching of the silicon chip, as done in the microelectronics industry to create electric circuits, the Affymetrix technology utilizes masks and electromagnetic radiation to remove protecting groups during solid-phase oligonucleotide synthesis. In the approach taken by Affymetrix, protecting groups that can be removed following irradiation with light of a suitable wavelength are placed at the sites for oligonucleotide extension (*i.e.* where the next nucleotide building block is to be added to the growing oligonucleotide strand). Masks are then placed over the glass slide onto which the oligonucleotide strands are being grown so that only some sites are irradiated during the deprotection step of the synthesis cycle so as to limit where nucleotide coupling is done. This is repeated several times using different masks so that a multi-element array with a full set of 4^n oligonucleotides of length n (nucleotides) may be assembled in $4n$ cycles, where each element of the array is functionalised with a unique nucleic acid sequence.

Initially, GeneChips® functionalised with 65,536 unique 8-nuleotide-long sequences were created. More recently, chips have been produced that were functionalised with 136,528 unique oligonucleotides, each contained within a 35 μm^2 area (total chip size estimated to be *ca.* 13 cm^2) with strand coverage densities of approximately $10^6 \cdot cm^{-2}$ [32]. The stepwise efficiency of oligonucleotide synthesis has been reported to be in the range of 92-94% [33]. Detection of hybridization of fluorescently tagged target strands to the immobilized probe sequences of the array is achieved using laser confocal fluorescence scanning microscopy. The collected images are then analyzed using software that allows multiplexing and internal redundancy for control. The controlled microfabrication of these devices eliminates the need for large quantities of reagents and labor-intensive preparations often associated with conventional hybridization assays. These oligonucleotide "chips" may find applications in all areas of nucleic acid analysis, with a strong application focus in the area of sequencing by hybridization and large-scale screening of genetic mutations. The multiplexing capability of these devices can assist in the diagnosis of diseases such as cystic fibrosis, which is known to be caused by approximately 200 different mutations [34]. Note that the multiplexed microarray concept is not limited to nucleic acid hybridization but can be modified for other biorecognition processes.

Researchers at Nanogen Inc. have reported an oligonucleotide chip device that relies on electric fields for the selective control of hybridization [35]. Fabrication using conventional microelectronic technology yielded a $1cm^2$ device with 25 central $80\mu m$ diameter electrodes in a $1mm^2$ test area. The electrode pads consisted of 500nm thick Pt deposited on a 20nm thick Cr adhesion layer above a thermally oxidized silicon substrate that had been sputter coated with Aluminum. The surface of the Pt electrodes was coated with low-stress silicon nitride onto which a *ca*. $1\mu m$ thick permeation layer of streptavidin-doped agarose was then formed *via* a spin-casting method. This layer served to provide a means of ssDNA attachment, in addition to preventing electrochemical DNA damage while maintaining ion movement. Each electrode in the array was connected positive, negative or neutral with respect to the power supply. Experiments were operated in a constant current mode to maintain a constant electric field in the bulk solution above the permeation layer. Biotinylated single-stranded DNA (ssDNA) probes were then immobilized to the agarose layer by applying a 100nA current to the buffer-equilibrated chip. Attachment was facilitated by binding between the biotin moiety on the probe strand and the streptavidin in the agarose layer, a reaction that is characterized by a binding constant on the order of $10^{15}M^{-1}$ [36]. Studies employing ssDNA bis-functionalised with biotin and the fluorescent probe Bodipy Texas Red were done and showed that immobilization occurred only at those sites (to within the detection limit of the experiment) which were connected positive. The reaction was deemed irreversible after no discernable loss of material was observed following several washes and reversal of the electric field, as confirmed by confocal fluorescence microscopy employing a cooled CCD camera for image analysis. Hybridization investigations were carried out and revealed that hybrids formed 25 times faster at positive electrical potential sites than at neutral sites indicating that kinetic limitations of passive, diffusion-limited hybridization on solid supports may be overcome. The device was demonstrated to be capable of distinguishing single-base mismatches as revealed by a decrease in fluorescence signal of approximately 90% after 50 seconds of application of an alternating field of negative bias ($0.6\mu A$, 0.1s on, 0.2s off, repeated for 150 cycles) to the sites containing the mismatch hybrid. A decrease in fluorescence intensity of 30% for those sites that contained fully complementary sequences was observed under similar treatment. The current applied during denaturation studies for single base pair mismatch discrimination was found to depend on the energetics of the duplex system, where higher current magnitudes were required for systems of oligonucleotides of greater length and G+C content. Measurement of the fluorescence from sequentially layered $10-20\mu m$ thin film solutions containing known concentrations of fluorescently labeled oligonucleotide over an unpatterned Pt surface was done to provide calibration of the signal intensity to quantity of labeled oligonucleotide. Full response times and regeneration times of the surface bound probes were not reported. As flagged by the authors, though this process is novel, a more detailed analysis of the physical properties of the system needs to be done. This endeavor was undertaken and reported in a subsequent publication by researchers at Nanogen [37] and is worthy of detailed review as many of these issues must be realized and contemplated for all sensing systems based on immobilized polyanionic nucleic acid layers for biorecognition, especially with regard to electrochemical methods.

In general, once nucleic acid sequences of known composition and length are selected, the energetics underlying hybrid formation resulting from the annealing of two complementary nucleic acids is governed by three independent variables, namely, temperature, pH and ionic strength. Alteration of any one of these parameters will affect hybrid stability and hence stringency. The concept of electric field assisted hybridization introduces another variable affecting hybrid stability, namely, electric field strength, which now adds a new dimension of complexity to the system as it is interdependent with the three aforementioned key variables.

One of the immediate consequences of passing a current through an electrolyte solution is the onset of joule heating, which will primarily be a function of the conductivity of the solution and the magnitude and duration of the current passed between electrodes. In order to assess the magnitude of temperature change experienced within the permeation layer above the Pt electrode and onto which the probe DNA was attached, the Nanogen group incorporated temperature-sensitive encapsulated liquid crystals into the permeation layer. In all cases, the temperature was found not to exceed 37°C. This indicates that joule heating was not of sufficient magnitude (or at least sufficiently conducted away *via* the permeation layer and underlying metal circuits) to noticeably affect the formation of hybrids [35].

It has been well established that an ionic strength gradient will accumulate over time under conditions of applied potential that causes the flow of electrical current of constant magnitude in electrolyte solutions [38]. For the case under consideration here, in which an electrolyte solution commonly employed for passive hybridization (*e.g.* 5× SSC: 0.75M NaCl, 75mM sodium citrate, pH 7.0) containing a low concentration (*e.g.* nanomolar) of an oligonucleotide is biased under an electrical field, two key trends are observed with respect to the resultant distribution of the oligonucleotides. As electrical double-layer formation around the electrodes accumulates and new ionic species are generated with the onset of Faradaic current flow, a decrease in the rate of electrophoretic DNA accumulation above the active electrode will be observed owing to the effective decrease in the transport number of the oligonucleotide. With the increased accumulation of DNA at the electrode then comes a concomitant increase in the rate of Fickian diffusion of DNA away from the electrode into bulk solution as a result of the amassed concentration gradient. This may ultimately lead difficulties in terms of calibration as the amount of DNA above the active anode may vary non-linearly with solution concentration.

Another factor that must be considered is the resultant alteration in pH in the vicinity of the electrode caused by generation of protons at the anode by hydrolysis of water. In all experiments done for active accumulation of oligonucleotides at test sites on the permeation layer, the magnitude of the applied potential (*ca.* 3 V) was sufficient for water hydrolysis. This lead to formation of an acidically biased pH gradient that extended from the Pt electrode surface, through the permeation layer, and out into bulk solution. Studies of pH above the test sites were done using a micro-pH electrode and showed that a dramatic decrease in pH was observed in non-buffered solutions, even at low current levels. For example, a decrease in pH from 7 to 3 was observed in *ca.* 20 seconds for an unbuffered solution of 50mM KCl under application of a constant current of 200nA. A drop from neutrality to a pH of *ca.*1 was observed for the same system when a 500nA current was drawn [37]. In addition to the obvious consequence of non-neutral pH levels on hybrid stability, an additional detrimental effect was observed. As the pH was decreased from neutrality and the pI of the oligonucleotide was approached, the net charge of the oligonucleotide decreased, hence, leading to a decrease in oligonucleotide transport to the test area. This would act to enhance the aforementioned effect of decreased oligonucleotide mobility with the duration of applied current when non-buffered supporting electrolytes are used.

In order to overcome these limitations associated with high ionic strength aqueous buffers, the Nanogen group then looked to the use of low conductivity buffers. The resulting study included low conductivity (<100μSiemens/cm) aqueous buffers based on the species: cysteine, γ-amino-butaric acid (GABA), β-alanine and glycine, which form zwitterions at near neutral pH, and D- & L-histidine, carnosine, 1-methylhistidine, 3-methylhistidine, pyridine, collidine and imidazole. In all cases accumulation of oligonucleotides was observed to improve dramatically in the presence of low ionic strength buffers. Direct proportionality between the initial rate of accumulation and the inverse of the solution conductance was observed. However, for investigations of rate of accumulation versus applied current, deviations from linearity with current densities above *ca.* 500nA were observed. This trend was similarly observed for all buffer species investigated where the total amount of accumulated oligonucleotide was observed to plateau at high current densities. This effect is most likely associated with more rapid increase in concentration of ionic constituents or ionic breakdown products of the buffer under the higher voltage and current conditions [37].

Investigations of hybridization under low ionic strength conditions revealed no hybrid formation under passive conditions. The zwitterionic buffers, GABA, β-alanine, and glycine also failed to provide suitable conditions to support stable hybrid formation even under electric field mediated conditions. It was speculated that the low ionic strength buffers did not supply sufficient ionic strength to provide adequate charge-screening of the phosphodiester backbones of the complementary oligonucleotides so as to permit annealing and stable hybrid formation. To further examine this effect, experiments were then done using peptide nucleic acids (PNA) immobilized on the permeation layer. The PNA backbone is uncharged and studies of chimeric hybrids of PNA and DNA have shown the formation of these complexes to be largely independent of ionic strength while maintaining similar affinity characteristics [39]. PNA:DNA chimeric hybrids were observed to form , thereby demonstrating the necessity for and ionic strength dependence of oligodeoxyribonucleotide annealing. Other buffers investigated that were based on an imidazole-like parent structure did support hybridization between DNA:DNA oligonucleotides under electric field mediated conditions.

Buffers that most successfully supported electric field mediated hybridization were those that were non-conducting yet had titrable substituents with pKa values at or near neutrality. Further investigations of buffer species consisting of imidazole-like parent compounds revealed that formation of protonated species that effectively buffered the pH occurred with the onset of proton production at the anode. These protonated species also provided ionic compounds that offered charge-screening of the anionic phosphodiester backbone of nucleic acids to provide conditions suitable to support hybridization while offering the ability to allow for electroactive accumulation of DNA. The use of low conductivity buffer species, such as histidine, in comparison to conventional highly-conductive sodium chloride based buffers revealed that *ca.* 40% less hybrid formation was sustained with the histidine buffer system. Therefore, a buffer system that offers more effective shielding would be required for optimization of the electric field mediated system. Given that a combination of factors including concentration, maintenance of the pH near neutrality and the generation of a cationic species suitable for shielding the nucleotide phosphodiester backbones all contribute to electronically mediated hybridization of DNA:DNA complexes, much work is required for system optimization and full characterization of the nucleic acid systems so that quantitative results may be determined.

Biosensors

The other major strategies employed for the detection of nucleic acids make use of the biosensor concept. A biosensor is a device that consists of a biologically active material connected to a transducer that converts a selective biochemical reaction into a measurable analytical signal [40, 41]. The advantages ideally offered by biosensors over other forms of analysis include ease of use (by non-expert personnel), low cost, ease of fabrication, small size, ruggedness, facile interfacing with computers, low detection limits, high sensitivity, high selectivity, rapid response, and reusability of the devices. Biosensors have been used to selectively detect cells, viruses, other biologically significant materials, biochemical reactions and immunological reactions by using detection strategies that involve immobilization of enzymes, antibodies or other selective proteins onto solid substrates such as quartz and fused silica (for piezoelectric and optical sensors) or metal (for electrochemical and surface plasmon resonance sensors) [42, 43]. However, such sensors are not widely available from commercial sources due to problems associated with the long-term stability of the selective recognition elements when immobilized onto solid surfaces [44, 45]. An alternative approach that may potentially be used to create biosensors with long-term chemical stability takes advantage of the stability of DNA. With the recent advent of DNA probe technology (*vide supra*), a number of selective oligomers which interact with the DNA of important biological species, have been identified [46-49]. These have been used to provide a new type of biorecognition element which is highly selective, stable, and can be easily synthesized in the laboratory [50-52]. The three transducer platforms encountered most often in nucleic acid biosensor technology are piezoelectric, optical and electrochemical.

Electrochemical Biosensors

Nucleic acid biosensors using electrochemical detection schemes have been reported [53-57]. These biosensors normally involve immobilization of single-stranded DNA (ssDNA) at an electrode surface and hybridization is measured by monitoring changes in the redox properties of an electroactive indicator that is physically associated with the nucleic acid material.

Mikkelsen and co-workers [53,58,59] have investigated voltammetric detection of DNA hybridization on the surface of oxidized glassy carbon and carbon paste electrodes. This was accomplished by immobilizing polynucleotide probe sequences which were rich in guanosine or terminated with a 96-nucleotide-long poly(dG) sequence to the oxidized electrode surfaces *via* carbodiimide mediated amide formation between the exocyclic amine moieties of the guanosine residues and the activated carboxylate functionalities present on the electrode surface. Detection of target-probe hybridization was achieved by cyclic voltammetry in conjunction with the redox active metallo-intercalators tris(2,2'-bipyridyl)cobalt(III) perchlorate $(Co(bpy)_3(ClO_4)_3)$ and tris(1,10-phenanthroline)cobalt(III) perchlorate $(Co(phen)_3(ClO_4)_3)$. This was generally achieved *in-situ* by treating the nucleic acid functionalised electrode with 60µl of a target nucleic acid in 5mM Tris buffer (pH 7.0) containing 20mM NaCl and 60µM of the redox agent for various times at room temperature. In cases where high ionic strength hybridization buffers (*i.e.* containing 0.5M NaCl) were employed, the sensor was removed from the hybridization buffer prior to electrochemical analysis, rinsed and placed in the buffer described above for electrochemical transduction. The mechanism of transduction was determined to be based on enhanced pre-concentration of the redox active intercalant with increased concentrations of double-stranded DNA at the electrode surface due to selective partitioning of the intercalant into double-stranded DNA. This conclusion was based on investigation of peak current versus scan rate and peak separation versus scan rate, which suggested intercalant migration to the electrode surface occurred in a diffusion limited manner and that electron transfer occurred in a heterogeneous transport mechanism, the kinetics of which decreased with increasing double-stranded DNA on the electrode surface, respectively.

Investigations based on carbon paste electrodes (modified with octadecylamine or stearic acid to provide amine or carboxylic acid functionalities) were quickly abandoned owing to problems with day-to-day reproducibility and precision. Regeneration methods used to denature the double-stranded DNA on the nucleic acid functionalised glassy-carbon electrode surfaces were found to destroy the carbon paste electrodes. Regeneration of sensors based on glassy carbon electrodes was reported to be effectively achieved by treatment with hot (100°C) distilled water for 10 minutes and that no significant peak current deterioration was observed to occur after 10 such cycles. Cyclic voltammetry data recorded from a glassy carbon electrode functionalised with oligo(dT) $_{20}$(dG)$_{98}$ and hybridized with complement, regenerated with hot water, and successively re-hybridized with the same target nucleic acid, revealed a *ca.* 15% reduction in cathodic peak intensity for the subsequent re-analysis with respect to that originally observed [59].

A dichotomy with regard to device sensitivity and response time was reported by the Mikkelsen group [53]. Hybridization reactions done at high ionic strength (0.5M NaCl) yielded a detection limit three orders of magnitude lower with respect to those done in low ionic strength (20mM NaCl) buffers. Under the low ionic strength conditions, however, the response time of the sensor when treated with 10μg/ml complement was found to be *ca.* 60 minutes whereas 10 minutes was determined to be sufficient time for generation of maximum peak currents under high ionic strength conditions. Thus, for optimal results, the luxury of *in-situ* hybridization and transduction must be sacrificed for a three-step method of analysis involving separate hybridization and transduction steps with an interstitial washing procedure. A detection limit for poly(dA)$_{4000}$ on a poly(dT)$_{4000}$ electrode was calculated to be 10^7 molecules (based on reported data [53]) so that the detection limits for a hybrid probe of more realistic size (*i.e.* 15-40 nucleotides) would then be *ca.* 10^9 molecules assuming the amount of signal provided by the electroactive intercalator was linearly dependent on oligonucleotide length.

An alternative approach, employed by Wang *et al.* [55,56,60-68], uses the technique of chronopotentiometric stripping analysis (PSA), a method normally used for trace-metal analysis. Preparation of electrodes and subsequent hybridization analysis followed an eight-step procedure of electrode pre-treatment, adsorptive accumulation of the oligonucleotide probe, hybridization, accumulation of electroactive intercalant, followed by chronopotentiometric transduction with interstitial washings. A carbon based electrode (either a carbon paste electrode or a screen-printed thick film carbon electrode) was pretreated by application of a potential of +1.8V (note: all potentials are reported relative to a Ag/AgCl reference electrode) for 1 minute in a stirred acetate buffer solution (0.2M, pH 5.0). This was found to enhance subsequent adsorptive accumulation of probe oligonucleotides owing to increased surface roughness and greater hydrophilicity of the electrode surface [64]. Application of a positive potential (+0.5V) to the electrode following introduction of a similarly buffered solution containing the oligonucleotide probe was then done for 2 minutes in order to adsorb the nucleic acid onto the electrode. The electrodes are then subjected to a brief (10s) washing procedure in a phosphate buffered saline solution (pH 7.0, the ionic strength of which depended on the probe and target species under investigation). Hybridization of adsorbed probe with complementary targets was done by dipping the washed electrode into a stirred buffer solution (pH 7.0, 0.02M phosphate, 0 – 0.75M NaCl) containing the target oligonucleotide at a potential of +0.5V for times which ranged from one to thirty minutes. Another brief (10s) washing procedure was then done by rinsing the oligonucleotide functionalised electrode with a solution of 0.02M Tris-HCl (pH 7.0). Binding of the electroactive intercalant, $Co(phen)_3^{3+}$, was then done by treatment of the electrode held under a potential of +0.5V for one minute with a solution of 0.02M Tris-HCl (pH 7.0) containing a desired concentration of the intercalant for the probe and target system under investigation. A subsequent rinse of the electrode with Tris-HCl buffer for 10 seconds was then done prior to transduction. The quantity of accumulated intercalant in the adsorbed dsDNA layer was then quantitatively measured by chronopotentiometric stripping analysis. This was done by setting the initial working electrode potential to +0.5V, followed by recording the time dependent alteration in potential (30 kHz sampling frequency) while maintaining a constant negative current (4 - 6µA). Wang *et al.* have shown chronopotentiometric analysis methods to be advantageous over the use of more traditional voltammetric methods. The plot of the derivative of time with respect to potential, $\partial t/\partial E$, against potential provides a peak-shaped response with a well-defined baseline. When directly compared to cyclic voltammograms, the chronopotentiometric signal offered a lower background response and greater sensitivity. In order to guarantee selective signal transduction rather than false positive signals from subsequent adsorption of more single-stranded material, the guanine oxidation signal of the functionalised electrode was used to ensure full-capacity loading of the electrode surface with the probe oligonucleotide. Electrode saturation was found to be complete in an average time of 2 minutes when oligodeoxyribonucleotide probes were accumulated under a potential of +0.5V from solution of 5-7 µg·ml^{-1}.

The selectivity of the carbon electrode system developed by Wang *et al.* [64], was shown to be relatively poor in investigations of oligodeoxyribonucleotides. Signal magnitudes of 20-25% of those observed for fully complementary oligonucleotides were generated when the system was challenged with sequences containing one- or two- base mismatches. Dramatic improvements in selectivity were demonstrated when PNA oligomers were used as biorecognition elements. In a collaborative study done by Wang, Nielsen and co-workers [69] it was shown that for one particular system of capture probes consisting of PNA or DNA pentadecanucleotides, a signal intensity of 90% that observed for a fully complementary oligonucleotide target was observed when the device was challenged with a one base-pair mismatched target sequence for the DNA·DNA system, while only 19% interference was observed for the chimeric PNA·DNA system. In general, it was shown that the use of PNA as a biorecognition element yielded selectivities that were better by factors ranging from 2 to 7 with respect to biorecognition elements composed of identically sequenced oligodeoxyribonucleotides. However, true specificity was not realized as interference levels of as high as 21% were observed for single base mismatches, even for the chimeric system. Analysis for small quantities of single base mutations in a sample rich in wild-type nucleic acid would be futile unless drastic improvements are realized.

Investigations to elucidate the formation of higher-order nucleic acid structures were also done by Wang *et al.* [68]. Triple-helical nucleic acid formation in solution, in particular $C^+ \cdot GC$ and $G \cdot GC$, followed by immobilization of the nucleic acid complex onto electrode surfaces of mercury and carbon paste was elucidated by reductions in the magnitude of the guanine oxidation peak and disappearance of a binding signal from the electroactive intercalant, $Co(phen)_3^{3+}$. Experiments involving formation of these triple-stranded complexes about one previously adsorbed strand, either poly(C) or poly(G), were also attempted. The results for the second study proved to be extremely complex wherein definitive triplex formation could not be discerned. The authors suggested that a variety of not well understood and perhaps unique polymorphic structures may have assembled on the electrode surface. Preliminary investigations were done to identify the effects of the degree of structure of the immobilized strand and the resultant influence on the observed signal. The results of the study indicated that more optimal results may be realized when using electrode-adsorbed capture oligonucleotides that *a priori* exhibit minimal amounts of structure. Single-stranded sequences rich in T, C and U that do not display traits of self-complementarity have then been identified as being preferred for use as capture probes on electrodes. For example, sensitivity was increased by a factor of two when poly(U) was used as a capture probe as opposed to poly(A), which is known to be more structured. More detailed investigations are underway in order to determine methods which may permit the *in-situ* formation and subsequent detection of triple-helical complexes.

Wang *et al.* have shown other types of analysis to be possible [64,70-72]. For example, detection of small molecules such as pollutants (*e.g.* polyaromatic hydrocarbons, hydrazine) or drugs (*e.g.* phenothiazine tranquilizers) which bind to DNA and radiation-induced DNA damage may be determined. This may be achieved either by direct chronopotentiometric transduction in the case where the target compound is electroactive, or *via* a competitive binding format against an electroactive intercalant for the case where the target compound is not. Alterations of the guanine oxidation signature as a consequence of binding may also be exploited for transduction.

A novel approach for the electrochemical elucidation of nucleic acid hybridization was undertaken by H. Thorpe *et al.* [73]. The strategy employed monitoring the onset of a catalytic current between guanine moieties of the target strand and a redox active metallo-intercalator with a slightly greater oxidation potential than that of guanine. The approach began with immobilization of oligonucleotide probes onto a Cyclopore™ poly(ethylene terephthalate) track etched membrane (0.4µm pore size). The polymeric membrane was first activated by treatment with a solution of $KMnO_4$ in 1.2N sulfuric acid to produce carboxylate functionalities on the polymer surface. The carboxylate moieties were subsequently activated by treatment with a water-soluble carbodiimide followed by washing and coupling of oligonucleotide probes to the membrane by formation of amide-type bonds between the exocyclic amine moieties of the nucleobases and the activated carboxylate functionalities on the membrane. The membrane was then mounted onto an indium tin oxide working electrode which was placed into an electrochemical cell that included a Ag/AgCl reference electrode and a platinum counter electrode. Measurements were done by cyclic voltammetric or chronoamperometric methods of analysis, which were used to determine the oxidation current of the metallo-intercalator $Ru(bpy)_3^{2+}$ (bpy = 2,2'-bipyridine) in the presence of the immobilized nucleic acid and conjugate complexes thereof. Unlike the redox reporters chosen by Mikkelsen and Wang, which undergo redox exchange at potentials well below that of the nucleobases (in particular, below a maximum of 1.06V at which guanine is known to be oxidized [74]), the concept for the case of Thorp *et al.* was to have the metallo-intercalator carry electrons from guanosine in the nucleic acid complex to the electrode in a mechanism suggested by the following two equations [73]:

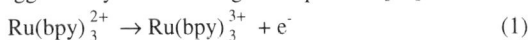

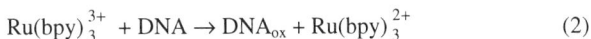

$$Ru(bpy)_3^{2+} \rightarrow Ru(bpy)_3^{3+} + e^-$$ (1)

$$Ru(bpy)_3^{3+} + DNA \rightarrow DNA_{ox} + Ru(bpy)_3^{2+}$$ (2)

By choosing to immobilize probe oligonucleotide strands composed of A, C, and T nucleotides or those containing inosine as a replacement for guanosine, the catalytic current enhancement brought on by interactions of the oxidized metallo-intercalator with guanosine residues in the target nucleic acid could then be used to signal binding of the target sequence. The sensor was functionalised with a 21-nucleotide probe sequence based on the *ras* oncogene and challenged with complement and non-complement (14-base mismatch) oligonucleotides. This was done by placing the nucleic acid functionalised polymer disk into a 200μl solution of hybridization buffer (0.8M NaCl, 50mM sodium phosphate) containing 1nmol of the target nucleic acid, heating the system to 50°C for 1 hour and then permitting the system to slowly cool to room temperature. Prior to electrochemical analysis, the polymer was removed from the hybridization buffer and washed twice with 20mM sodium phosphate and then once with water. After affixing the polymer to an indium tin oxide working electrode, the polymer film was permitted to equilibrate in a buffer solution of 50mM phosphate (pH 5.0) containing 100μM $Ru(bpy)_3^{2+}$. Cyclic voltammetric measurements revealed the sensor was selective for the complementary sequence, however, the non-complementary nucleic acid provided a significant level of interference (23%), owing to non-specific adsorption of target oligonucleotides to the polymer support. No current above background levels were reported for the oligonucleotide modified electrodes in the absence of $Ru(bpy)_3^{2+}$ before or after hybridization, indicating that the polymeric matrix in addition to the immobilization strategy was sufficient to prevent direct electron transfer from the immobilized DNA to the electrode. A small background current ($> 10^{-5}$ A) was observed for electrodes functionalised with probe oligonucleotides containing inosine in the presence of $Ru(bpy)_3^{2+}$. This result was not unexpected given that inosine mononucleotides were demonstrated to have an appreciable electron-transfer rate to $Ru(bpy)_3^{3+}$ (97 $M^{-1} \cdot s^{-1}$), although this is three orders of magnitude less than that of guanosine nucleotides (6.4×10^5 $M^{-1} \cdot s^{-1}$) under similar conditions. Subsequent analysis of any given sample is not possible with this detection scheme owing to the destructive nature of the assay where all the guanosine moieties of the target are oxidized as part of the signal transduction process. The authors did not report attempts to regenerate ssDNA at the polymer membranes for subsequent analysis. Control experiments employing the use of ^{32}P-labeled oligonucleotides were done to characterize the quantity of oligonucleotide immobilized onto the polymeric films and the hybridization efficiency of the immobilized nucleic acids with complementary targets. These investigations revealed an average loading of 8.2 ± 0.8pmol of oligonucleotide per polymer disk (8mm diameter) and a hybrid capture efficiency of $5.2 \pm 0.5\%$ relative to the quantity of immobilized nucleic acid when challenged with an excess of radio-labeled complementary oligonucleotide. Similar to the findings of Mikkelsen, Thorp et al. found more efficient electron transfer was observed when solvents of lower ionic strength were used. Studies were also done in which the sensor was challenged with samples of crude PCR product containing complementary and non-complementary amplified sequences. Clear signals were observed for the amplified products with interference levels of not more than 19% for the oligonucleotide systems studied. Negative interference was observed in one experiment, indicating possible occlusion of the electrode surface by one or more of the components in the crude PCR mixture, hence causing a reduction in the availability of the immobilized nucleobases for binding or redox exchange with the electroactive mediator.

An interesting approach to hybrid recognition has been reported by Korri-Youssoufi et al. [75] that involved molecular wiring of oligonucleotides onto electrodes via a polypyrrole tether. Operation of the device was not based on the use of intercalant reporter molecules, but rather monitoring the alterations in the potential at which oxidation (doping) of the polypyrrole tether occurred subsequent to formation of double-stranded oligonucleotide at the terminus of the molecular wire. A reduction in the number of degrees of freedom of the polypyrrole tether occurs upon oxidation where a more rigid, planar quinoid structure is formed. The potential at which this occurs in the absence of conjugated oligonucleotide was at +0.1V/SCE for the system of Korri-Youssoufi et al. A linear dependence between peak current and scan rate demonstrated that the polypyrrole film was in intimate contact with the underlying electrode with no diffusive contribution to the signal. With the introduction of a single-stranded quadradecanucleotide on the terminus of the electro-polymerized polypyrrole, the doping potential was observed to shift to −0.2V/SCE. Upon hybridization with increased quantities of complementary sequences, further shifts in the oxidation potential with a concomitant decrease in the peak area were observed due to the greater energetic penalty imparted on the electroactive tether as the size of the complex bound at the tether terminus increased. By configuring the system to monitor peak current at constant potential, a non-linear calibration curve was produced which yielded a maximum sensitivity of 1µA/nmol complement strand and a detection limit of 10^{12} molecules. The sensitivity of this technique was found to be dependent on the size of the target sequence, thereby making calibration difficult. Another limitation faced by this transduction strategy lies in the non-selectivity to the nucleic acid structure that is formed. For example, formation of triplex-structures could be followed by this methodology, but it is possible that triplex-formation may not be discriminated from duplex-structures containing mismatch-induced bulges which would more readily occur under the lower temperature, higher ionic strength conditions required for triplex-formation.

The use of ferrocene modified oligonucleotides have been reported by Ihara et al. [76] in a sandwich-type sensing scheme based on the annealing of a capture probe and a ferrocene conjugated reporter oligonucleotide to a target nucleic acid. The capture-oligonucleotide consisted of a hexadecanucleotide that was synthetically created to include five successive phosphorothioate internucleotide linkages at the 5'-

terminus. This provided a means of attachment of the capture probe to a polished gold electrode surface *via* well-established thiol-gold interactions. The ferrocenyl oligonucleotide was created by attachment of an activated ester of ferrocene-carboxylic acid to the terminus of an amino-terminated oligonucleotide. Functionalisation of the electrode with the capture-oligonucleotide was verified by cyclic voltammetric measurements in the presence of a ferrocyanide/ferricyanide redox couple. Decreases in the peak current were observed for the functionalised electrode relative to that of a gold electrode in the presence of a similarly sequenced oligonucleotide lacking internucleotide phosphorothioate moieties, signifying surface modification of the electrode. The stability of the immobilized capture-oligonucleotide was tested by heating the electrode to 80°C for 30min in water. No change was detected in the voltammetric traces and the unusually high stability that was observed was attributed to the multi-point anchoring *via* the five phosphorothioate moieties per strand. Hybridization experiments were done by incubating the oligonucleotide-modified electrode in a 50μM solution of an equimolar mixture of the ferrocenyl-modified reporter probe and the target nucleic acid in a buffered aqueous solution (0.1M KCl, 5mM Tris-HCl, pH 8.0) for 24 hours at 5°C. Following incubation, the electrode was rinsed with buffer and differential pulse voltammetric measurements were made in aqueous buffers. Good stringency for this system was reported where a single-base-mismatch control provided an interference level of only 7.5% with respect to the signal for the fully matched target sequence. The authors did not report sensitivity and detection limits. The authors observed that there existed a sensitivity of the signal to the temperature of the rinse solution that was used to wash the electrode surfaces prior to electrochemical analysis. A 20°C wash solution was observed to cause a decrease of 75% in the magnitude of the signal from hybridization with the fully complementary material while signals from the control oligonucleotides (including the one base mismatch) were unaffected. This lead the authors to speculate that the signals observed from the control samples, and perhaps the observed temperature dependence, could provide a means to determine contributions of non-specific adsorption to the total signal.

A hybrid of the two previously described concepts of Korri-Youssoufi [75] and Ihara [76] is under development at Clinical Micro Sensors (CMS). This company is founded and lead by Thomas Meade and Jon Kayyem , and has reported development of a prototype hand-held DNA sensor [77]. The CMS strategy involves the use of molecular wiring in conjunction with the principle that double-stranded DNA supports long-range electron transfer with significantly higher efficiency than that of single-stranded DNA or double-stranded DNA complexes containing mismatches [78-80]. Though the rate of electron transfer, and hence the efficiency of electron transfer, is dependent on the length of the double-stranded oligonucleotide, appreciable conduction can be observed over 3 to 100 base pairs. The observations of Meade *et al.* [81] have suggested that a distance of 15 base paired nucleotides is optimal.

The basic method begins with immobilization of a capture oligonucleotide *via* a phenylacetylene tri-mer onto a gold electrode. The molecular wire is connected to the capture oligonucleotide *via* a 2'-amino functionality at the 3'-terminus of the oligonucleotide and subsequent linkage to the electrode is done by use of a terminal thiol group on the phenylacetylene conductor. The remaining exposed portions of the gold electrode surface are then insulated from bulk solution by treatment of the modified electrode with alkane-thiol of similar length to the molecular wire. A protective layer is then self assembled onto the electrode surface so that effectively only the capture oligonucleotide is exposed to solution. One disclosed method of operation [81] involves treating the modified electrode with a solution containing a non-complementary strand containing a covalently attached ferrocynium redox reporter and a target strand which contains regions complementary to the capture strand and reporter strand. Hybridization of the three oligonucleotides in a sandwich complex, analogous to that reported in the method of Ihara, then closes the circuit between the ferrocynium redox center and the gold electrode *via* the phenylacetylene molecular wire and the bioconductive double-stranded DNA. Electrochemical analysis for the presence of molecularly wired ferrocynium can then be done with attomole detection limits. Kayyem has recently reported that measurements of target DNA have successfully been made in "dirty environments, including blood" [77].

Acoustic Wave Devices

Another commonly pursued strategy in the development of nucleic acid biosensors makes use of piezoelectric, or acoustic wave transduction [82-86]. More specifically, these sensors employ AT-cut quartz crystals operating in thickness shear mode (TSM) for transduction of hybridization events. In contrast to most electrochemical and optical detection schemes, TSM devices do not require the use of labels. Often misnamed as "quartz-crystal microbalances", these biosensors simultaneously provide information on the interfacial microviscosity and charge distribution upon hybridization, in addition to surface mass changes. Thompson and co-workers [82] have described a TSM biosensor for the detection and study of DNA hybridization. The sensor employs a gold-plated 9 MHz AT-cut quartz crystal coated with a layer of palladium oxide. Single-stranded DNA or RNA (approximately 4000 base pairs in length) and ^{32}P-labelled as a means for internal calibration, was immobilized onto PdO surfaces *via* adsorption. Studies were done in air to determine the extent of surface coverage on the PdO layer. The results showed that coverage densities were comparable to values obtained when the same experiments were performed on conventional nylon filter supports equal in size to the TSM sensor surface. Hybridization events at the sensor surface were subsequently analyzed by acoustic network analysis, which permits measurement of the series resonant frequency, in addition to several parameters of the equivalent sensor circuit such as minimum and maximum impedance and maximum phase angle. Alterations in the series resonant frequency that were 18-fold greater than that predicted by mass changes alone were observed upon challenging the sensor with complementary nucleic acid of similar length to the immobilized probe nucleic acid. The use of an effective blocking agent, single-stranded salmon sperm DNA, prevented non-selective adsorption to the electrode surface so as to ensure the measured signal was generated by genuine hybridization events. In a subsequent and more detailed study, Thompson *et al.* [87] reported that the use of these blocking agents was not necessary as the TSM sensors in conjunction with network analysis was capable of direct detection of hybridization in the presence of non-specific adsorption. Since the transverse shear wave of the sensor and duplex length are both on the order of micrometers, the authors indicated that the large increase in the series resonant frequency was a result of alteration of the interfacial viscosity upon hybridization. While the system reported was referred to as unoptimized, the TSM sensor can detect hybridization in real time and has the potential to operate in an on-line configuration. The latest reported detection limits of this device are *ca.* 1×10^{10} molecules·cm^{-2} (20fmol·cm^{-2}) for both oligonucleotide hybridization and peptide binding [88]. Chemometric processing of the multi-dimensional data provided by the network analysis technique may soon provide great insight into the complex interfacial processes associated with immobilized nucleic acids.

Wang *et al.* [85] have reported the preparation of a nucleic acid sensor based on acoustic wave transduction and PNA for biorecognition. The PNA oligonucleotides were linked to gold plated electrodes on AT-cut quartz crystals with a fundamental frequency of 5MHz by a thiol terminated ethylene glycol linker on the terminus of the peptide nucleic acid backbone. The coverage of the gold electrode surface with PNA was found to be 7.4 pmol·mm^{-2} (total of 310 pmol) where consequent cyclic voltammetry experiments with ferrocyanide indicated 87% coverage of the electrode. Detection of hybridization was done *in-situ* at 25°C by monitoring alterations in the fundamental frequency of the crystal. Fast response times were observed (3-5 min) for full evolution of analytical signals. Sample to sample reproducibility was investigated based on six measurements on six separate crystals to provide a standard deviation in measurement of 18%. The device showed no discernible response when challenged with a pentadecadeoxyribonucleotide containing a centrally located single base mismatch, even at a concentration five times greater than that used to provide substantial signals when challenged with complementary deoxyribonucleic acid.

Optical Biosensors Based on Surface Plasmon Resonance

The phenomenon of surface plasmon resonance (SPR) has been exploited for use as a transduction motif for immunoassays and more recently for nucleic acid biosensors. The technique is based on the generation of plasmons (charge density oscillations) in a metal film deposited at the interface between a dense but optically transparent medium (usually a glass prism), and a liquid [89]. In the case where a thin (*ca.* 50 nm) metal film is used, plasmons are restricted to propagating back and forth on the surface of the metal, resulting in what is termed a surface plasmon [90]. Impinging light of the proper wavelength, phase velocity, and incidence angle onto the metal film may induce surface plasmons.

Light of proper polarization incident on the metal film at the resonance angle, θ_R, ideally provides no reflected radiation. This phenomenon is the result of the photon energy being coupled into surface plasmons. θ_R is sensitive to the dielectric of the outer medium. As the evanescent field associated with the plasmon (normal to the metal film) extends only a few hundred nanometers from the metal surface, the resonance condition is affected by variations in the dielectric constant (and thus refractive index) in regions near to the metal surface. As such, the surface plasmon resonance technique is most suitable for the study of thin films and interfacial binding events. Detection of binding events is done by monitoring alterations in interfacial dielectric resulting from mass loading at the interface between the solution phase and a metal surface capable of sustaining a surface plasmon [91]. These alterations are most commonly reported as changes of the intensity of the reflected beam, where the angle if incidence is fixed for the entire experiment. Alternatively, data may be reported with respect to the exact resonance angle, *i.e.* relative reflectivity *versus* incidence angle.

In essence, this technique allows for the detection of surface binding interactions in real time without the use of labels. Immunosensors that rely on these principles are commercially available from Biacore AB (formerly Pharmacia Biosensor). These are based on the popular BIACORE® 3000, BIACORE® 2000 and BIACORE®probe instruments in conjunction with disposable sensor chips, which may be purchased with affinity capture chemistries in place (*e.g.* SA5 Sensor Chips with streptavidin bearing surfaces for capture of biotinylated compounds) [92]. These devices incorporate micro-fluidic systems for delivery of reagents required for analysis and are capable of measuring affinities and kinetics of antibody-antigen interactions in addition to analyte concentrations, which can be determined to the picomolar range [93].

A number of research groups have used these commercially available instruments to investigate the kinetics of interfacial nucleic acid hybridization, formation of triple-stranded complexes, develop assays for selective detection of PCR amplified nucleotides and investigate the use of PNA capture probes to enhance selectivity [94-102]. This is generally accomplished by treatment of the commercially available SA5 sensor chips with biotinylated capture oligonucleotides prepared by standard solid-phase reagents and assembly techniques. Immobilization of the capture strand is followed in real time by monitoring alterations in the intensity of the reflected beam at a fixed incidence angle, commonly measured in Reflectance Units (RU, arbitrary value), so as to provide an estimate of the degree of coverage of the sensor surface. Following a brief washing procedure, subsequent exposure of the DNA functionalised sensor to complementary nucleic acid sequences provides further alteration in the measured reflectivity from the sensor chip owing to the change in the dielectric value of the interfacial layer. Triple-helical detection has been reported by Neidle *et al.*[101] wherein subsequent exposure of the sensor chip with immobilized duplex oligonucleotide to a third oligonucleotide capable of Hoogsteen binding with the immobilized duplex provided further decreases in reflectivity owing to triple-strand formation.

Of particular interest is the work done by Corn, Smith and co-workers [96,97], who have identified the limitations of the SPR strategy and have initiated investigations of methods to enhance signals from hybridization. One of the ultimate goals of this research team is to enhance the sensitivity of this technique so that SPR interrogation of oligonucleotide arrays can realize comparable sensitivities to those methods employing fluorescence detection. Detailed investigations have been reported involving secondary methods (*e.g.* Polarization Modulation - Fourier Transform Infrared Reflection Absorption Spectroscopy) to evaluate nucleic acid immobilization and the resulting hybridization efficiency of the immobilized oligonucleotides. The use of a long-chain (>35 atoms) substrate linker terminated with a thymidylic acid pentadecanucleotide to further space the oligonucleotide probe sequence from the substrate surface was found to provide a coverage of *ca.* 50% of a monolayer which yielded a hybridization efficiency of 60±20%. Methods investigated to overcome the current minimum detection limit of 10^{11} immobilized molecules·cm^{-2} included the use of biotinylated target sequences onto which a layer of streptavidin may be coupled in order to enhance the interfacial mass change associated with hybridization events. This provided a 4-fold increase in the hybridization signal over that observed with the unmodified target sequence.

Fluorimetric Fiber Optic Biosensors

Optical transduction techniques are also being used for development of nucleic acid biosensors. The high sensitivity of fluorescence techniques has recently been highlighted by Weinfeld *et al.* [103] who have demonstrated zeptomole detection limits for the detection of radiation induced DNA damage. Most efforts toward fluorimetric optical biosensor development incorporate optical fibers in the sensor design. The use of optical fibers in biosensing configurations is considered attractive due in part to their multiplex capability, since radiation of different wavelengths can propagate simultaneously in different directions along the fiber, in addition to their small size and potential use in applications of remote and continuous sensing [104]. Fluorimetric detection of nucleic acid hybridization where the nucleic acids are immobilized onto optical substrates is the most commonly employed strategy.

Perhaps the first such reported undertaking was that of Squirrell *et al.* [105]. In this work, single-stranded nucleic acid sequences ranging in length from 16-mer oligonucleotides to 204-base oligomers functionalised with an aminohexyl linker at the 5' terminus were covalently attached to optical fibers functionalised with 3-aminopropyl triethoxysilane *via* a gluteraldehyde linkage. Investigations of nucleic acid hybridization were done by monitoring fluorescence intensity in an intrinsic mode configuration using complementary target strands that had been pre-labeled with a fluorescein moiety. This yielded a reusable assay system in which signal generation was observed to occur within minutes and nanomolar detection limits were achieved.

Abel and co-workers [106] at Ciba-Geigy Ltd. have reported an automated optical biosensor system. Their device utilizes 5'-biotinylated-16-mer oligonucleotide probes bound to an optical fiber functionalised with avidin to detect complementary oligonucleotides pre-labeled with fluorescein moieties in a total internal reflection fluorescence (TIRF) evanescent wave motif similar to that of Squirrell. Immobilization of nucleic acid probes onto the optical fiber substrate was achieved by functionalisation of the surface with (3-aminopropyl)triethoxysilane (APTES) or mercaptomethyldimethylethoxysilane (MDS). Onto the short alkylsilane layer was coupled a layer of biotin (by treatment of the aminosilane functionalised fibers with NHS-LC-biotin or by treatment of thiolsilane functionalised fibers with biotinylated BSA). The biotinylated fibers were then treated with avidin or streptavidin, followed by coupling of 5'-biotinylated oligonucleotide capture probes. Each assay consisted of a 3 minute pre-equilibration, 15 minute hybridization time, 10 minute washing procedure, followed by a 5 minute regeneration cycle (chemical or thermal). A chemical denaturation scheme (50% aqueous urea) was observed to be the preferred method for sensor regeneration, as exposure of the optical sensor to temperatures exceeding 52°C caused irreversible damage due to denaturation of the avidin used for immobilization. Assays for the detection of complementary target strands pre-labeled with fluorescein showed a working range of almost five decades and a detection limit of 24fmol (1.4×10^{10} molecules). In order to detect nucleic acids not pre-labeled with fluorescein, Abel et al. employed a competitive binding assay. Detection of the unlabelled analyte began with pre-treatment of the sensor with fluorescein labeled "tracer-DNA". The experiments involved monitoring of the decreases of the fluorescence intensity from the sensor upon exposure to and subsequent displacement of the tracer-DNA by complementary nucleic acid. The dose-response curves reported by Abel et al. show a detection limit of 132 pmol (7.95×10^{13} molecules) with a working range of three decades for this detection strategy.

Duveneck et al. [107], also from Ciba-Geigy, have investigated the use of planar waveguides as a platform for an optical nucleic acid biosensor. The device consisted of a single-mode planar waveguide composed of a tantalum pentoxide film of ca. 100nm thickness deposited on a glass substrate. The waveguide material was chosen so as to provide a high refractive index ($n = 2.2$) so as to maximize evanescent field strength in the biorecognition layer immobilized above the waveguide. Coupling of excitation radiation into the waveguide was accomplished by use of a grating coupler that was etched into the glass substrate onto which the waveguide was deposited. Detection was done in a "volume collection mode" wherein luminescence that was evanescently excited and isotropically emitted in the half-sphere outside of the waveguide was collected and quantitatively measured. Collection and guiding of the emitted light to a detector was achieved by use of a high numerical aperture lens and intensity measurements were done by use of either a photodiode coupled to a high gain amplifier or a photomultiplier tube in combination with a photon-counting unit. Immobilization of capture oligonucleotides onto the waveguide surface was done by a direct immobilization scheme wherein the waveguide surface was functionalised with glycidyloxypropyltrimethoxy silane followed by extension of the substrate linker with hexaethylene glycol and assembly of the capture sequence by automated solid-phase oligonucleotide synthesis. Hybridization experiments were done in-situ and revealed full analytical signal evolution in a time of ca. six minutes with a detection limit of 100 attomoles (6×10^7 molecules) for target strands labeled with red cyanine-type fluorophores. Experiments were done in which one sensor was used for 20 consecutive hybridization assays, with regeneration being achieved by treatment with 50% urea. After the first two cycles, no discernable change in sensitivity was observed with each complete hybridization/regeneration cycle requiring ca. 40 minutes using an unoptimized fluidics system and flow cell geometry.

The most sensitive optical biosensor device reported to date is that of Bier *et al.* [108]. The intrinsic mode fiber optic sensor provided a detection limit of 3.2 attomoles (1.9 × 10⁶ molecules) of non-labeled fully complementary nucleic acid target sequence in the presence of a double-strand selective fluorochrome with full regenerability for over 60 cycles of application without discernible loss of sensitivity. Capture oligonucleotides were immobilized onto the optical fiber substrate by either an avidin-biotinylated oligomer approach or by covalent attachment of the capture oligomer by carbonyldiimidazole activation. The avidin-biotin method involved adsorption of avidin onto the surfaces of fused silica optical fibers, which was subsequently cross-linked *via* treatment with gluteraldehyde. 5'-Biotinylated oligonucleotides were then coupled onto the immobilized avidin layer. In the covalent coupling strategy, clean optical fibers were incubated with carbonyldiimidazole overnight in dry acetone, washed and then treated with nucleic acid probes activated with 1-ethyl-3-(3-dimethylaminopropyl)-carbodiimide. Following a washing procedure, the fibers were used immediately or stored for up to three months at -18°C. Hybridization experiments were done using a protocol involving exposure of the sensor situated in a continuous flow apparatus (100µl·min⁻¹ flow rate, 100µl volume exposed to the fiber-optic sensor to the target nucleic acid in hybridization buffer for 60-180s, followed by a 120-180s wash with buffer, 45-60s treatment with a solution of fluorochrome (either Pico-Green or YOYO-1), 30s wash with buffer (at 2.0ml·min⁻¹), and finally immersion in buffer for 210-255s (steady state for data collection)). Experiments done involving the use of biotin-avidin immobilized oligonucleotides provided a hybridization efficiency of 55%, however, the sensor could not be regenerated owing to the high stability of the double-stranded oligonucleotide fluorochrome complex. Chemical denaturation (*e.g.* treatment with NaOH) did not regenerate the sensor. Thermal regeneration of the sensor at high temperature (90°C) could denature the nucleic acid fluorochrome complex (T_m = 88°C for Pico-Green/dsDNA, T_m > 90°C for YOYO-1/dsDNA, T_m = 45°C for dsDNA alone) and regenerate the sensor, however, the avidin-biotin linkage of the oligonucleotide to the waveguide substrate would also be compromised. This lead Bier *et al.* to abandon the avidin-biotin immobilization protocol for covalent attachment methods, where complete regeneration was possible using thermal denaturation *via* a stepwise heating to 90°C following hybridization and staining. Single-base mismatch discrimination was also investigated. It was demonstrated by Bier *et al.* that under appropriate conditions of temperature, pH and ionic strength, a high degree of selectivity for detection of a central single-base (T-T) mismatch could be achieved for a tridecanucleotide.

Efforts in our research laboratory have also been directed toward the development of fiber-optic nucleic acid biosensors [109-112]. In general, we have chosen to pursue a sensing scheme based on covalent immobilization of oligonucleotides *via* long-chain polyether substrate linker molecules. Following substrate functionalisation with these linker molecules terminated with dimethoxytrityl protecting groups, assembly of oligonucleotide capture probes is done by automated solid-phase oligonucleotide synthesis. We have switched to the use of polyether substrate linkers as these types of tethers (in contrast to alkyl chains) have been shown to reduce non-selective binding of peptides or proteins that could be present as contaminants in real-world samples [113]. Hybridization of the immobilized capture oligonucleotides to target oligonucleotides was done by treating the optical sensor immersed in hybridization buffer with various quantities of complementary and non-complementary nucleic acids in a hybridization buffer (1.0M NaCl, 50mM phosphate, pH 7.0). Detection of hybrid formation was done by addition of ethidium bromide to the reaction vessel at various concentrations and times to affect *in-situ* staining of the immobilized hybrids with the intercalant fluorophore. Detection of stained hybrids was done by evanescent excitation of fluorophores along the length of the optical sensor followed by recovery and quantitative measurement of guided fluorescence by use of a photomultiplier tube. Using this strategy we have been able to detect as little as 10^{10} molecules of fully complementary hybrid molecules with linear calibration over a one decade concentration range in a non-optimized methodology.

Our research group has also investigated detection of triple-helix formation. Elucidation of triple-strand hybridization in Hoogsteen and reverse-Hoogsteen motifs between an immobilized single-stranded probe oligonucleotide and linear or branched complementary oligonucleotides was done by monitoring alterations in the fluorescence temperature coefficient of intercalated ethidium bromide. Ligand exclusion concomitantly occurred with the onset of triple-strand formation upon decreasing the temperature of the system below the T_m for triplex formation, causing the temperature coefficient of the fluorescence signal to switch from negative to positive. For most systems of triplex forming oligonucleotides, the inversion of the fluorescence temperature coefficient could be used to determine the melting temperature of the triple-stranded complex.

Our research work has lead to development of a reagentless sensor that incorporates the transduction element into the sensor, thereby fulfilling the true definition of a biosensor. The use of a fluorescent double-stranded nucleic acid binding ligand tethered onto the terminus of the immobilized strand provides a small quantity of background fluorescence when in the presence of single-stranded DNA and exposed to solution. This provides a means for internal calibration of the device *via* normalization of the background signal in terms of monitoring photobleaching and drift in the excitation source intensity and detector gain. A fluorescent reporter molecule that selectively binds to double-stranded nucleic acids and provides increased quantum efficiency when bound to the double-stranded target is desirous as a tethered transducer. Our initial investigations have lead to the creation of a tethered analogue of the ethidium fluorophore that was attached to the end of a 5'-amino functionalised oligonucleotide *via* a 19-carbon molecular tether. Such sensors have provided reagentless response times on the order of minutes.

Another interesting approach to fiber-optic sensor development has been reported by Walt and co-workers [114,115], who have created multiplexed extrinsic-mode fiber optic microarray biosensor. The sensor employed fiber optic bundles (2-3 feet in length) composed of many optical fibers each 200-350μm in diameter. The microarray was fabricated by immobilizing a different oligonucleotide probe sequence onto the distal end of each fiber. Individual fibers within the bundle were made reactive for oligonucleotide immobilization by immersing the bundle in a solution of monomeric acrylamide functionalised with succinimidyl ester residues. Ultraviolet radiation (350nm) was transmitted down the desired fiber to the distal terminus to polymerize acrylamide monomers, yielding an amine-reactive polymer matrix solely above the illuminated fiber. Subsequent oligonucleotide attachment was then done by incubating the terminus of the fiber bundle in a solution containing the desired 5'-amino-functionalised oligonucleotide capture probe. Following incubation with the capture oligonucleotide probe, any remaining reactive sites were capped by treatment with ethanolamine. Hybridization experiments were done by dipping the functionalised fiber tips directly into a solution containing fluorescently-tagged target strands. Fluorescence detection was achieved by coupling the proximal end of the fiber bundle into a cooled CCD camera and measuring the intensity of fluorescence emission from the distal tips. Such a microarray biosensor arrangement provides the potential for remote multiplexed oligonucleotide sequence analysis with reported detection limits of 1.3nM (in a 200µl reaction volume to provide an absolute detection limit of 1.5×10^{11} molecules) and with a response time of *ca.* 7 minutes. Full regeneration was achieved in less than 10s upon heating of the fiber bundle terminus to 65°C in hybridization buffer, and no discernable decrease in sensitivity was observed after three cycles of application. The unique aspect of this biosensor format is that in addition to intensity measurements, fluorescent images are also obtained by incorporating imaging fibers in the sensor design. Imaging fibers are simply bundled fibers that have their termini fully aligned at the distal and proximal ends. Thus each fiber behaves as a "pixel" to provide images with spatial resolutions on the order of micrometers [116].

Gerdt and Herr [117] describe detection of nucleic acid hybridization based on alterations in the quantity of light transmitted from one optical fiber in a coupled fiber system (similar to that of a Mach-Zehnder interferometer) to the second fiber of the waveguide system. The quantity of light transferred is a function of the refractive index of the media surrounding the waveguides. Refractive index alterations affect the penetration depth of the evanescent wave emitted from the first waveguide into which optical radiation is launched. This standing wave of electromagnetic radiation subsequently penetrates into (and thus transfers optical radiation to) the second waveguide. Therefore, the device is sensitive to refractive index alterations occurring within a volume surrounding the first waveguide with a thickness of *ca.* one wavelength of the light propagating within that waveguide. One of the arms of the waveguide may be functionalised with immobilized nucleic acid molecules which serves to provide selective binding moieties. The change in refractive index of the thin film of nucleic acids on the first waveguide caused by hybridization with target nucleic acid sequences alters the quantity of light transferred to the second waveguide, thereby providing a means of signal transduction. Hybridization events may then be identified based on changes in the output ratios of the two-waveguide arms in the coupled fiber system. One limitation of this technology lies in the fact that any alterations in refractive index near the surface of the waveguides will provide alterations in the output ratios of the two fibers. Therefore, non-specific binding events (such as protein adsorption) will provide false positive results.

Current Limitations and Future Directions

Perhaps the first substantive undertakings toward the creation of nucleic acid biosensors were independently reported in 1992 by Squirell and Mikkelsen [58,105]. Six years has since passed and the technology has moved to the point where perhaps most all of the common sensing methods and motifs have been investigated to some degree. However, for this technology to proceed into maturity where application in field as commercially available devices is realized, much work toward proper characterization and transcendence beyond the limitations of the current sensor embodiments is required. There are some commonalties in terms of the limitations of the technology and some limitations that are rather specific to each technique, as shall now be addressed.

Strategies for Immobilization of Biorecognition Elements

Although extremely easy to carry out, immobilization techniques based on adsorption of the capture oligonucleotide strand face many complications that limit the applicability of this approach. Subsequent desorption of the capture oligonucleotide may occur, especially under the conditions used for hybridization and after binding with complementary sequences, leading to decreased sensitivity with time. Though the magnitude of this deleterious effect may be minimized by maintaining a positive potential at the interface, the issue of selectivity is then raised. Given that an unknown proportion of the nucleobase moieties may be unavailable for complementary binding, sites that can accommodate mismatches without providing a destabilizing energy penalty then exist.

The next easiest method for immobilization of oligonucleotides involves the use of the avidin-biotin affinity pair. 5'-biotinylated oligonucleotides may be readily prepared by automated synthesis using commercially available biotin synthons and standard cyanoethylphosphoramidite chemistries, and then can be linked to substrates functionalised with avidin. Though the binding affinity for this pair is perhaps the highest of all non-covalent interactions, $K_d = 10^{-15}$ M, the proteinaceous avidin moiety is susceptible to long-term and thermally induced denaturation, as evidenced by a number of researchers who have used this strategy and attempted thermal sensor regeneration. In addition, avidin functionalisation of the interaction surface leads to the formation of a proteinaceous interface that may offer many sites for non-selective adsorption by other organic components in a sample.

Covalent immobilization methods are the most difficult to employ, but this disadvantage is offset by the control afforded over packing density and strand orientation. The use of covalent attachment of oligomers has been observed to provide a very stable means of oligonucleotide attachment and, in conjunction with substrate linker molecules of sufficient length (>25 atoms) have demonstrated fast kinetics of hybridization where analytical signal generation was observed to occur in minutes as opposed to hours. In order to identify the surface derivatisation conditions which provide the optimal sensor response characteristics such as response time, hybridization efficiency, regenerability, and selectivity, characterization of immobilized oligonucleotide films by a secondary method is required. Such characterization of surface structure appears to be lacking in much of what has been reported in the recent literature.

Pre-fabricated oligonucleotide capture probes may be deposited in a single step onto an activated substrate. The deposition of pre-fabricated oligonucleotide probes is advantageous with respect to the integrity of the immobilized capture probe as chromatographic purification of the oligonucleotide prior to deposition may be done, ensuring that all of the recognition elements are of the desired sequence and length. Control over surface coverage is limited though, given the steric and electrostatic factors associated with these long polyanionic molecules. Nucleotide by nucleotide assembly using small monomeric and electrically neutral synthons provides the highest degree of control of strand orientation and surface packing density. However, at best each coupling reaction is only 99.5% efficient. For the case where a twenty-nucleotide capture probe is assembled on a solid substrate, only 90% of the assembled strands would be of full length. Note that sequences onto which successful nucleotide coupling did not occur can be quantitatively capped by acylating reagents to prevent further extension and prevent the creation of probes of undesired sequence. This limitation of sequence integrity may however be overcome if transduction moieties are affixed to the terminus of only the full-length oligonucleotides, as for example, in the case where a tethered intercalant fluorophore is coupled to the terminus of the full-length immobilized strand.

Interfacial Polyanionic Films

Perhaps one of the most overlooked characteristics of nucleic acid biosensors which needs to be addressed by the genosensor community is that of the interfacial properties of nucleic acid films. An understanding about how nucleic acids that are immobilized as a densely packed polyanionic monolayer film in a high ionic strength media behave with respect to their counterparts in bulk solution is of significance to device response characteristics. Consider, for example, the case of a bare metal electrode immersed in an aqueous solution. Organization of water molecules about the metal film creates a region of aligned dipoles that imparts a potential at that surface. In order to maintain electroneutrality, an electrical double-layer is formed around the immersed electrode, all in the absence of applied potential. A film of polyanionic molecules deposited at an interface will dramatically influence the ionic strength in an interfacial region due to formation of an electrical double-layer structure. In one of our earlier reported experiments [110], we observed the T_m of double-stranded oligonucleotides at and interface to be greater by 2.4°C than that of the counterpart system in bulk solution in a high ionic strength buffer (1.0 M NaCl, 50mM phosphate, pH 7.0). This difference may be greatly exaggerated for systems in which a lower ionic strength hybridization buffer is employed and perhaps be used to preferentially affect interfacial hybridization in competitive binding scenarios.

Effects on the interfacial pH must also be contemplated. A preponderance of hydronium ion will be found in the double-layer region at negatively charged interfaces; the activity being a function of surface potential and interfacial ionic strength. Therefore, formation of a pH gradient from low values, which are dependent on interfacial electrostatics, to that of bulk solution pH will be observed to correlate with distance from the interface. Formation of hybrids at the interface could lead to alterations in the amount of charge accumulated at the interaction surface. This will result in dynamic rearrangements of the double-layer structure, effective ionic strength, and local pH, and may ultimately provide undesired ramifications with respect to control of hybridization efficiency and stringency. Electrochemical methods in particular will be the most difficult to characterize owing to the interdependence of electrode potential and local electrical double-layer structure, and other processes must be considered, e.g. local joule heating processes with the onset of Faradaic current flow in the case where redox reporters are used or electro-accumulation is employed.

Signal origins

SPR, acoustic wave and interferometric sensors provide the advantage of operating in a label free motif so that external reagent treatment to label the interaction pair is not required. This allows for these technologies to investigate a plethora of analytes and binding interactions, such as DNA/DNA duplex and triplex formation, DNA/protein and DNA/drug interactions. As a consequence of the capability to detect ensemble alterations in interfacial structure, interference from non-specific binding interactions is also embraced by these methods. Similar to the use of avidin films for oligonucleotide capture, the use of peptide nucleic acid recognition elements to boost the selectivity of the chemistry to permit single-base mismatch discrimination may also provide a surface that permits a multitude of non-specific binding interactions. In order to avoid the problem of interferent binding leading to false-positive results, especially when keeping the goal of analyzing real-world samples (e.g. blood) in mind, a transduction strategy that is sensitive to the structure of the binding pair is preferred. This limitation may be overcome by using reporter molecules that associate selectively with the binding pair.

By associating a selective transduction element with the biorecognition element, the device may function without the need for external reagent treatment and obviate the need to collect and dispose of hazardous waste. Such a technology readily lends itself to automated and in-line analysis and precludes the need for skilled technicians to partake in the analysis procedure or disposal of waste (provided the sample itself is not biohazardous).

Another series of advantages is provided by the incorporation of a selective reporter, and is related to the concept of internal calibration. The associated selective reporter may provide a means to determine the quantity of immobilized nucleic acid on the interaction surface. The reporter in the presence of single-stranded nucleic acid may provide a unique baseline signal to which all signals can be referenced, hence providing meaningful analytical data and perhaps insight into the physical properties of the interaction surface. Also, the useful lifetime of the device can be determined from alterations in the background signal from the reporter molecules over time. Therefore, by including a selective reporter, an internal reference marker and diagnostic tool for the device status is included as an integral part of the biosensor.

Concluding Remarks

It has been estimated [118,119] that by the year 2000, most individuals may have access to a limited genetic profile based on genetic testing of 20 to 50 genes. By the year 2010 these profiles may be extended to 5000 genes pending the completion of the human genome endeavor. By the year 2030, everyone may have their genome sequenced and take home a CD-ROM to study their genetic profile on a personal computer. This is a tall order to fill with regard to the advancement of genetic testing, one in which biosensor technology may play an integral part. More detailed characterizations of the device technology must be done and the chemistry of the system better understood in order to provide devices capable of functioning in real-world applications (*e.g.* point-of-care, in-field and in-stream continuous environmental monitoring). Ultimately this will require careful control over fabrication of the biorecognition interface and more detailed characterization of the thermodynamics of the binding events. The development of small, portable devices, capable of internal calibration and appropriate signal conditioning with little effort required in terms of sample pre-treatment or other involvement by a skilled technician may one day be realized with such knowledge in hand.

References

1. Hage, D.S. *Anal. Chem.* **1993**, *65*, 420R-424R.
2. Braun, T.; Klein, A.; Zsindely, S.; Lyon, W.S. *Trends Anal. Chem.* **1992**, *11*, 5-7.
3. Ngo, T.T. *Cur. Opin. Biotech.* **1991**, *2*, 102-9.
4. Engvall, E. In *Biomedical Applications of Immobilized Enzymes and Proteins*; Chang, T.M.S., Ed.; Vol. 2, Plenum Press, New York, **1977**.
5. Bright, F.V.; Betts, T.A.; Litwiler, K.S. *Anal. Chem.* **1990**, *62*, 1065-1069.
6. Tietjen M.; Fung, D.Y.C. *Crit. Rev. Microbiol.* **1995**, *21*, 53-58.
7. Feng, P.J. *Food Prot.* **1992**, *55*, 927-931.
8. Shinnick, T.M.; Good, R.C.; *Clin. Infect. Dis.* **1995**, *21*, 291-297.
9. Gen-Probe Inc., 10210 Genetic Center Drive, San Diego, California 92121-4362 U.S.A.
10. Chiron Corporation, 4560 Horton Street, Emeryville, California 94608 U.S.A.
11. Digene Corporation, 2301-B Broadbirch Drive, Silver Spring, Maryland 20904 U.S.A.
12. Rawn, J.D.; *Biochemistry*, Neil Patterson Publishers, Burlington, NC, **1989**.

288

13. McGilvery, R.W. *Biochemical Concepts*, W.B Saunders Company, Philadelphia, PA, **1975**.
14. Keeton, W.T.; Gould, J.L. *Biological Science*, W.W. Norton and Company, New York, NY, **1986**.
15. Lehninger, A.L. *Principles of Biochemistry*, Worth Publishers Inc., New York, NY, **1982**, pp. 794-795.
16. Cantor, C.; Spengler, S. "Primer on Molecular Genetics", taken from U.S. Department of Energy *Human Genome 1991-1992 Program Report*. Available: [Online]: http://www.ornl.gov/TechResources/Human_Genome/publicat/primer/intro.html, June, **1992**.
17. Galjaard, H. *Path. Biol.*, **1997**, *45*, 250-256.
18. Arlinghaus, H.F.; Kwoka, M.N. *Anal. Chem.* **1997**, *69*, 3747-3753.
19. Rossen, L.; Holmstroem, K.; Olsen, J.E.; Rasmussen, O.F. *Int. J. Food Micro.*, **1991**, *14*, 145-151.
20. Notermans, S.; Wernars, K.; Soentoro, P.S.; Dufrenne, J.; Jansen, W. *Int. J. Food Micro.* **1991**, *13*, 31-40.
21. Reynolds, R.; Sensabaugh, G. *Anal. Chem.* **1991**, *63*, 2-15.
22. Yap, E.P.H.; Lo, Y.M.O.; Fleming, A.; McGee, J.O.D. In *PCR Technology: Current Innovations*, Griffin, H.G.; Griffin, A.M., Eds., CRC Press, Boca Raton, FL, **1994**.
23. Beese, K. "Human Genome Analysis, Genetic Screening and Gene Therapy: New Biomedical Services in the Light of Modern Techniques, Growing Insights and Ethical/Social Limits", IPTS Draft Discussion Paper 3, February 15, **1996**, available online at: http://www.jrc.es/~beese/screening.htm.
24. Wolcott, M.J. *Clin. Microbiol. Rev.* **1992**, *5*, 370-386.
25. Tenover, F.C. *Clin. Microbiol. Rev.* **1988**, *1*, 82-101.
26. Hilborne, L.H.; Grody, W.W. *Lab. Med.* **1991**, *22*, 849-856.
27. Rubin, F.A.; Kopecko, D.J. In *Nucleic Acid and Monoclonal Antibody Probes*, Swaminathan, B.; Prakash, G., Eds., Marcel Dekker Inc., New York, NY, **1989**, pp 185-219.
28. Thuong, N.T.; Hélène, C. *Angew. Chem. Int. Ed. Eng.* **1993**, *32*, 666-690.
29. Hélène, C.; Toulme, J.J. In *Oligodeoxynucleotides: Antisense Inhibitors of Gene Expression*; Cohen, J.S., Ed., McMillan Press, London, **1989**, pp. 137-172.
30. Drmanac, R.T.; Crkvenjakov, R.B. *Method of Sequencing by Hybridization of Oligonucleotide Probes*; United States Pat. No. 5695940, June 5, **1995**.
31. Affymetrix Inc., 3380 Central Expressway, Santa Clara, CA, 95051, USA.
32. Chee, M.; Yang, R.; Hubbell, E.; Berno, A.; Huang, X.C.; Stern, D.; Winkler, J.; Lockhart, D.J.; Morris, M.S.; Fodor, S.P.A. *Science*, **1996**, *274*, 610-614.
33. McGall, G.H.; Barone, A.D.; Diggelmann, M.; Fodor, S.P.A.; Gentalen, E.; Ngo, N. *J. Am. Chem. Soc.* **1997**, *119*, 5081.
34. Sheldon, E.L.; Briggs, J.; Bryan, R.; Cronin, M.; Oval, M.; McGall, G.; Gentalen, E.; Miyada, C.G.; Masino, R.; Modlin, D.; Pease, A.; Solas, D.; Fodor, S.P.A. *Clin. Chem.* **1993**, *39*, 718.
35. Sosnowski, R.G.; Tu, E.; Butler, W.F.; O'Connell, J.P.; Heller, M.J. *Proc. Natl. Acad. Sci. USA.* **1997**, *94*, 1119-1123.

36. Wilchek, M.; Bayer, E.A. *Anal. Biochem.* **1988**, *171*, 1.
37. Edman, C.F.; Raymond, D.E.; Wu, D.J.; Tu, E.; Sosnowski, R.G.; Butler, W.F.; Nerenberg, M.; Heller, M.J. *Nuc. Acids Res.* **1997**, *25*, 4907-4914.
38. Newman, J.S. *Electrochemical Systems*, Prentice Hall, Englewood Cliffs, NJ, 1991, Chapter 11.
39. Egholm, M.; Buchardt, O.; Christiensen, L.; Behrens, C.; Freier, S.M.; Driver, D.A.; Berg, R.H.; Kim, S.K.; Norden, B.; Nielsen, P.E. *Nature*, **1993**, *365*, 566-568.
40. Thompson, M.; Krull, U.J. *Trends Anal. Chem.* **1984**, *3*, 173-178.
41. Guilbault, G.G. *Cur. Opin. Biotech.*, **1991**, *2*, 3-8.
42. Andrade *et al*. *Biosensor Technology: Fundamentals and Applications*; Buck, R.P.; Hatfield, W.E.; Umana, M.; Bowden, E.F.; Eds., Marcel Dekker Inc., New York, NY, 1990, p. 219.
43. Wise, D.L. *Bioinstrumentation: Research, Developments and Applications*; Butterworth Publishers, Stoneham, MA, 1990.
44. Kallury, K.; Lee, W.E.; Thompson, M. *Anal. Chem.* **1992**, *64*, 1062-1068.
45. Krull, U.J.; Brown, R.S.; Vandenberg, E.T.; Heckl, W.M. *J. Elec. Micros. Tech.* **1991**, *18*, 212-222.
46. Symons, R.H. *Nucleic Acid Probes*, CRC Press, Boca Raton, FL, **1989**.
47. Bock, L.C.; Griffin, L.C.; Latham, J.A.; Vermaas, E.H.; Toole, J.J. *Nature*, **1992**, *355*, 564-568.
48. Tay, F.; Liu, Y.B.; Flynn, M.J.; Slots, J. *Oral Micro. Immun.* **1992**, *7*, 344-351.
49. Sherman, M.I.; Bertelsen, A.H.; Cook, A.F. *Bioorg. Med. Chem. Lett.* **1993**, *3*, 469-473.
50. Letsinger, R.L.; Lunsford, W.B. *J. Am. Chem. Soc.* **1976**, *98*, 3655-3660.
51. Beaucage, S.L.; Caruthers, M.H. *Tet. Lett.* **1981**, *22*, 1859-1862.
52. Alvarado-Urbina, G.; Sathe, G.M.; Liu, W-C.; Gillen, M.F.; Duck, P.D.; Bender, R.; Ogilvie, K.K *Science*, **1981**, *214*, 270-276.
53. Millan, K.M.; Saraullo, A.; Mikkelsen, S.R. *Anal. Chem.* **1994**, *66*, 2943-2948.
54. Hashimoto, K.; Ito, K.; Ishimori, Y. *Anal. Chem.* **1994**, *66*, 3830.
55. Wang, J.; Cai, X.; Gustavo, R.; Shiraishi, H.; Farias, P.A.M.; Dontha, N. *Anal. Chem.* **1996**, *68*, 2629-2634.
56. Wang, J.; Cai, X.; Jonsson, C. *Anal. Chem.* **1995**, *67*, 4065-4071.
57. Mishima, Y.; Motonaka, J.; Ikeda, S. *Anal. Chim. Acta* **1997**, *345*, 45-52.
58. Millan, K.M.; Spurmanis, A.J.; Mikkelson, S.R. *Electroanal.* **1992**, *4*, 929-932.
59. Millan, K.M.; Mikkelson, S.R. *Anal. Chem.* **1993**, *65*, 2317-2323.
60. Wang, J.; Cai, X.; Tian, B.; Shiraishi, H. *Analyst.* **1996**, *121*, 965-970.
61. Wang, J.; Rivas, G.; Cai, X.; Dontha, N.; Shiraishi, H.; Luo, D.; Valera, F.S. *Anal. Chim. Acta*, **1997**, *337*, 41-48.
62. Wang, J.; Rivas, G.; Ozsoz, M.; Grant, D.H.; Cai, X.; Parrado, C. *Anal. Chem.* **1997**, *69*, 1457-1460.
63. Wang, J.; Rivas, G.; Cai, X.; Parrado, C.; Chicharro, M.; Grant, D.; Ozsoz, M. *Electrochem. Soc. Proc.*, **1997**, *19*, 727-740.
64. Wang, J.; Cai, X.; Rivas, G.; Shiraishi, H.; Dontha, N. *Biosens. Bioelectron.* **1997**, *12*, 587-599.

65. Wang, J.; Rivas, G.; Cai, X.; Chicharro, M.; Parrado, C.; Dontha, N.; Begleiter, A.; Mowat, M.; Palecek, E.; Nielsen, P.E. *Anal. Chim. Acta.* **1997**, *344*, 111-118.

66. Wang, J.; Rivas, G.; Cai, X. *Electroanal.* **1997**, *9*, 395-398.

67. Wang, J.; Rivas, G.; Parrado, C.; Cai, X.; Flair, M.N. *Talanta* **1997**, *44*, 2003-2010.

68. Cai, X.; Rivera, G.; Shirashi, H.; Farias, P.; Wang, J.; Tomschik, M.; Jelen, F.; Palecek, E. *Anal. Chim. Acta* **1997**, *344*, 65-76.

69. Wang, J.; Palecek, E.; Nielsen, P.E.; Rivas, G.; Cai, X.; Shiraishi, H.; Dontha, N.; Luo, D.; Farias, P.A.M. *J. Am. Chem. Soc.* **1996**, *118*, 7667-7670.

70. Wang, J.; Chicharro, M.; Rivas, G.; Cai, X.; Dontha, N.; Farias, P.A.M.; Shiraishi, H. *Anal. Chem.* **1996**, *68*, 2251-2254.

71. Wang, J.; Rivas, G.; Cai, X.; Palecek, E.; Nielsen, P.; Shiraishi, H.; Dontha, N.; Luo, D.; Parrado, C.; Chicharro, M.; Farias, P.A.M.; Valera, F.S.; Grant, D.H.; Ozsoz, M.; Flair, M.N. *Anal. Chim. Acta* **1997**, *347*, 1-8.

72. Wang, J.; Rivas, G.; Ozsoz, M.; Grant, D.H.; Cai, X.; Parrado, C. *Anal. Chem.* **1997**, *69*, 1457-1460.

73. M.E. Napier, C.R. Loomis, M.F. Sistare, J. Kim, A.E. Eckhardt and H.H. Thorp, *Bioconjugate Chemistry*, **1997**, *8*, 906-913.

74. Steenken, S.; Jovanovic, S.V. *J. Am. Chem. Soc.* **1997**, *119*, 617-618.

75. Kori-Youssoufi, H.; Garnier, F.; Srivastava, P.; Godillot, P.; Yassar, A. *J. Am. Chem. Soc.* **1997**, *119*, 7388-7389.

76. Ihara, T.; Nakayama, M.; Murata, M.; Nakano, K.; Meada, M. *Chem. Commun.* **1997**, 1609-1610.

77. Wilson, E.K. *Chem. Eng. News*, **1998**, *5*, 47-49.

78. Murphy, C.J.; Arkin, M.R.; Ghatlia, N.D.; Bossmann, S.; Turro, N.J.; Barton, J.K. *Science*, **1993**, *262*, 1025-1029.

79. Murphy, C.J.; Arkin, M.R.; Ghatlia, N.D.; Bossmann, S.; Turro, N.J.; Barton, J.K. *Proc. Natl. Acad. Sci. USA* **1994**, *91*, 5315-5319.

80. Meade, T.J.; Kayyem, J.F. *Angew. Chem. Int. Ed. Eng.* **1995**, *34*, 352-354.

81. Meade, T.J.; Kayyem, J.F.; Fraser, S.E. U.S. Patent 5,780,234, 1998.

82. Su, H.; Kallury, K.M.R.; Thompson, M. *Anal. Chem.* **1994**, *66*, 769-774.

83. Ito, K.; Hashimoto, K.; Ishimori, Y. *Anal. Chim. Acta* **1996**, *327*, 29-35.

84. Okahata, Y.; Kawase, M.; Miikura, K.; Ohtake, F.; Furuswa, H.; Ebara, Y. *Anal. Chem.* **1998**, *70*, 1288-1296.

85. Wang, J.; Neilsen, P.E.; Jiang, M.; Cai, X.; Fernandes, J.R.; Grant, D.H.; Ozsoz, M.; Begliter, A.; Mowat, M. *Anal. Chem.* **1997**,*69*, 5200-5202.

86. Caruso, F.; Rodda, E.; Furlong, D.N.; Niikura, K.; Okahata, Y. *Anal. Chem.* **1997**, *69*, 2043-2049.

87. Su, H.; Chong, S.; Thompson, M. *Biosens. Bioelectron.* **1997**, *12*, 161-173.

88. Prof. Michael Thompson, Department of Chemistry, University of Toronto, 80 St. George Street, Toronto, Ontario, Canada, M5S 1A1, personal communication.

89. Kovacs, G. In *Electromagnetic Surface Modes*; Boardman, A.D.; Ed.; **1982**; Vol. 4.

90. Raether, H. *Surface Plasmons on Smooth and Rough Surfaces and on Gratings*; Springer-Verlag, Berlin, **1988**.

91. Granzow, R.; Reed, R. *Biotech.* **1992**, *10*, 390-406.

92. Biacore AB; Rapsgatan 7; S-754 50; Uppsala, Sweden.

93. Griffiths, D.; Hall, G. *Trends Biotech.* **1993**, *11*, 122-128.

94. Bianchi, N.; Rutigliano, C.; Tomassetti, M.; Feriotto, G.; Zorato, F.; Gambari, R. *Clin. Diag. Virol.* **1997**, *8*, 199-208.
95. Rutigliano, C.; Bianchi, N.; Tomassetti, M.; Pippo, L.; Mischiati, C.; Feriotto, G.; Gambari, R. *Int. J. Oncol.* **1998**, *12*, 337-343.
96. Thiel, A.J.; Frutos, A.G.; Jordan, C.E.; Corn, R.M.; Smith, L.M. *Anal. Chem.* **1997**, *69*, 4948-4956.
97. Jordan, C.E.; Frutos, A.G.; Thiel, A.J.; Corn, R.M. *Anal. Chem.* **1997**, *69*, 4939-4947.
98. Kruchinin, A.A.; Vlasov, Y.G. *Sens. Actutors B*, **1996**, *30*, 77-80.
99. Jensen, K.K.; Orum, H.; Nielsen, P.E.; Nordén, B. *Biochem.* **1997**, *36*, 5072-5077.
100. Persson, B.; Stenhag, K.; Nilsson, P.; Larsson, A.; Uhlén, M.; Nygren, P. *Anal. Biochem.* **1997**, *246*, 34-44.
101. Bates, P.; Dosanjh, H.S.; Kumar, S.; Jenkins, T.C.; Laughton, C.A.; Neidle, S. *Nuc. Acids Res.* **1995**, *23*, 3627-3632.
102. Gotoh, M.; Hasegawa, Y.; Shinohara, Y.; Shimizu, M.; Tosu, M. *DNA Res.* **1995**, *2*, 285-293.
103. Le, X.C.; Xing, J.Z.; Lee, J.; Leadon, S.A.; Weinfeld, M. *Science*, **1998**, *280*, 1066-1069.
104. Thompson, R.B.; Ligler, F.S. In *Biosensors with Fiberoptics*; Wise, D.L.; Wingard, L.B.; Eds.; Humana Press, Clifton, NJ, **1991**.
105. Graham, C.R.; Leslie, D.; Squirrell, D.J. *Biosens. Bioelectron.* **1992**, *7*, 487-493.
106. Abel, A.P.; Weller, M.G.; Duveneck, G.L.; Ehrat, M.; Widmer, H.M. *Anal. Chem.* **1996**, *68*, 2905-2912.
107. Duveneck, G.L.; Pawlak, M.; Neuschäfer, D.; Bär, E.; Budach, W.; Pieles, U.; Ehrat, M. *Sens. Actuators B*, **1997**, *38-39*, 88-95.
108. Kleinjung, F.; Bier, F.F.; Warsinke, A.; Scheller, F.W. *Anal. Chim. Acta* **1997**, *350*, 51-58.
109. Piunno, P.A.E.; Krull, U.J.; Hudson, R.H.E.; Damha, M.J.; Cohen, H. *Anal. Chim. Acta*, **1994**, *288*, 205-214.
110. Piunno, P.A.E.; Krull, U.J.; Hudson, R.H.E.; Damha, M.J; Cohen, H. *Anal. Chem.* **1995**, *67*, 2635-2643.
111. Uddin, A.H.; Piunno, P.A.E.; Hudson, R.H.E.; Damha, M.J.; Krull, U.J. *Nuc. Acids Res.* **1997**, *25*, 4139-4146.
112. Krull, U.J.; Piunno, P.A.E.; Wust, C.; Li, A.; Gee, A.; Cohen, H. In *Biosensors for Direct Monitoring of Environmental Pollutants in Field*; Nikolelis, D.P.; Ed.; Kluwer Academic Publishers, Netherlands, **1998**, 67-77.
113. Lee, J.H.; Kopecek, J.; Andrade, J.D. *J. Biomed. Mat. Res.* **1989**, *23*, 351-368.
114. Ferguson, J.A.; Boles, T.C.; Adams, C.P.; Walt, D.R. *Nat. Biotech.* **1996**, *14*, 1681-1686.
115. Healey, B.G.; Matson, R.S.; Walt, D.R. *Anal. Biochem.* **1997**, *251*, 270-279.
116. Pantano, P.; Walt, D.R. *Anal. Chem.* **1995**, *67*, 481A-489A.
117. Gerdt, D.W.; Herr, J.C. U.S. Patent 5,494,798, 1996.
118. Fowler, G. *The Human Genome Project: Ethics in the Real World.* Presentation at the International Conference "Doing the Decent Things with Genes", Turku, August 9-11, **1995**.
119. Goodfellow, P.N. *Science*, **1995**, *267*, 1609.

Chapter 20

Molecular Beacons: A New Approach for Detecting *Salmonella* Species

Wilfred Chen, Grisselle Martinez, and Ashok Mulchandani

Department of Chemical and Environmental Engineering, University of California, Riverside, CA 92521

Molecular beacons are oligonucleotide probes that become fluorescent upon hybridization. We developed a new approach to detect the presence of *Salmonella* species using these fluorogenic reporter molecules and demonstrated their ability to discriminate between similar *E. coli* species in real-time PCR assays. A detection limit of 1 CFU per PCR reaction was obtained. The assays were carried out entirely in sealed PCR tubes, enabling fast and direct detection in semiautomated format.

Salmonella is an important food and water-borne pathogen associated with acute gastrointestinal illnesses around the world. The infective dose can be as low as 15-20 cells, depending on the age, health status of the host and differences between *Salmonella* strains. It is estimated that over 4 million cases of *Salmonella* infections and one thousand deaths occur in the United States (1).

Human infection usually occurs from consuming raw or undercooked foods. The largest Salmonellosis outbreak in the United States, involved 16,000 cases and was caused by milk products from a Chicago dairy in 1985. Although no outbreaks associated with drinking water have been reported, recent studies found *Salmonella* widely distributed in Southern California surface water (2). Currently, *Salmonella* is one of many waterborne pathogens regularly screened by water authorities.

Current detection methods involve primarily cell culturing in selective broth. Although a variety of selective media and protocols are available, most of these methods require a series of sequential enrichment and culturing steps. These are lengthy, labor-intensive procedures which require further confirmation analyses. In addition, false-negative results may be obtained when *Salmonella* is outgrown by other organisms.

The need for rapid and selective methods for detection of *Salmonella* has prompted the development of several rapid detection methods including new media

292

formulations or modifications, miniature biochemical tests, nucleic acid-based assays, antibody-based assays and automated systems (3). Amplification of nucleic acid using the polymerase chain reaction (PCR) or reverse transcription-PCR (RT-PCR) for the detection of mRNA is the current preferred method, which provides the best sensitivity. Typical detection methods of PCR products involve visual detection of an appropriately sized DNA band followed by specific hybridization with a labeled DNA probe, which could take up to 15 hrs and is approximately 50% of the total time requirement for detection. New improvements in both the sensitivity and speed of the final detection step, preferably real-time monitoring of PCR products within 1 or 2 hr, must be developed to realize this powerful PCR-based method for practical applications.

Molecular beacons are stem-loop shaped oligonucleotide probes labeled with a fluorophore at the 5' end and a quenching moiety at the 3' end. Minimal fluorescence is emitted when the beacons are self-hybridized due to the close proximity of the quencher to the fluorophore, which are held together by the stem structure (4). Hybridization of the loop to a complimentary DNA sequence forces the stem arms apart, separating the quencher from the fluorophore. This separation results in a significant increase in fluorescence (Figure 1). Molecular beacons incorporated into a PCR reaction mix become brightly fluorescent as they hybridize to complimentary target sequences amplified during the reaction. The fluorescent signal is proportional to the amount of amplicons synthesized. Detection of this signal provides a powerful tool for real-time monitoring of PCR amplification products. The interaction of molecular beacons with their targets is extraordinarily specific. No increase in fluorescence is observed even in the presence of a target strand that contains only a single nucleotide mismatch (5).

This study describes the development of a real-time PCR assay employing molecular beacons for detection of *Salmonella*. This assay combined the speed and specificity of PCR with the specificity and high sensitivity of molecular beacons. A 122 bp region of the *himA* gene was used as the target (6). The real-time PCR assay is highly specific and a detection limit of one single cell was achieved.

Experimental

Bacterial strains, media and culture conditions. *Salmonella typhimurium* LT2 and *Escherichia coli* W3110 were obtained from the American Type Culture Collection (ATCC). Seven *Salmonella* isolates were obtained from the County Sanitation Districts of Los Angeles County. These isolates were identified and labeled: 1) *Salmonella* OC1, 2) *Salmonella* Group C-Factor 7, 3) *Salmonella choleracious*, 4) *Salmonella enteriditis*, 5) *Salmonella heidelberg*, 6) *Salmonella newport* and 7) *Salmonella thompson*. All cultures were grown in Nutrient Broth (DIFCO Laboratories) overnight at 37°C. The cell count of overnight cultures was estimated by enumeration using the spread plate technique and expressed as the number of colony forming units (CFU).

DNA Extraction. 1.5 ml of each culture was harvested and resuspended in 567 μl of TE (10 mM Tirs and 1 mM EDTA) buffer. 30μl of 10% SDS and 3 μl of 20 mg/ml proteinase K were added to each sample and incubated in a water bath at

37°C for 1 hour. 100 µl of 5 M NaCl and 80µl of a 10% CTAB (hexadecyltrimethyl ammonium chloride) in 0.7 M NaCl solution were added to each sample, followed by incubation in a water bath at 65°C for 10 minutes. The solution was then extracted with 0.8 ml of chloroform/isoamyl alcohol (24:1). The aqueous supernatants were removed and mixed with 400 µl of isopropanol until DNA precipitates were visible. The precipitates were pelleted by centrifuging. The pellets were washed with 200 µl of ice-cold 70% ethanol. After removing the ethanol, the pellets were allowed to dry in a vacuum oven at 50°C for 5 minutes and resuspended in 100 µl of TE buffer. DNA recovery was confirmed by gel electrophoresis.

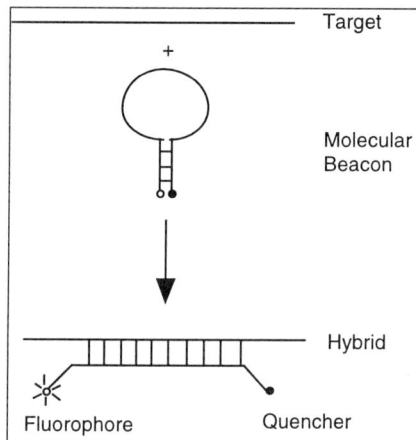

Figure 1. Principle of operation of molecular beacons.

Design of Molecular Beacons. The molecular beacon (BHMA1) 5'-FAM-CGCTATCC-GGGGCGTAACCCGTAGCG-3' DABCYL was synthesized by MIDLAND Certified Reagent Company. The target sequence was designed to be perfectly complementary to the *himA* gene of *Salmonella*, but contained two nucleotide mismatches relative to the same region in the *himA* gene of *Escherichia coli*. Beacons were resuspended in TE buffer, stored at -20°C and protected from light. 16 mM aliquots were prepared and used for subsequent studies.

Thermal Denaturation Profiles. Thermal denaturation profile studies were conducted to determine the optimum annealing temperature for the real-time PCR. The changes in fluorescence of a 50 µl solution containing 0.3µM of the beacon probe with or without 0.9uM of a perfectly complimentary single stranded oligonucleotide. The samples were place in a Perkin-Elmer ABI Prism 7700

Sequence Detector System and heated to 90°C for 5 minutes. The temperature was then reduced at 1°C per minute increments to 20°C. Data were recorded at each temperature interval. The optimal hybridization temperatures for each beacon were determined from these plots.

PCR Primers. Primers SHIMAF 5'-CGTGCTCTGGAAAACGGTGAG-3' and SHIMAR 5'-CGTGCTGTAATAGGAATATCTTCA-3' (Genosys) specific for a 122 bp fragment of the *him*A gene were previously reported by Bej et al. (6). To amplify the *him*A region of *E. coli*, a similar pair of primers EHIMA-Forwrad: 5'-CGCGCTCTGGAAAACGG-3'and EHIMA-Reverse: 5'-CGTGCTGTAATGGG-AATATC-3' were designed.

PCR conditions. The Perkin-Elmer ABI Prism 7700 sequence Detection System was used for real-time analyses. Thermal cycling conditions were specified as follows: initial melting at 98°C for 10 minutes, followed by 40 cycles of melting at 95°C for 1 minute, annealing at 57°C for 1 minute and extension at 72°C for 1 minute. These reactions ended with a final extension step at 72°C x 5 minutes. Fluorescent measurements were recorded during each annealing step. At the end of each PCR run, data were automatically analyzed by the system and amplification plots were obtained.

Results and Discussion

Thermal Denaturation Profiles. In the absence of a complimentary target, the molecular beacon expressed approximately 40% fluorescence (relative percentage) at 80°C (Figure 2), indicating that the stem structure had been denatured and the beacons had assumed a coiled conformation. As the temperature was decreased, stable hybridization of the stems was noted at 50°C, when the fluorescence had decreased to 0%. The melting temperature of the stem was determined to be 50-55°C.

Beacon-target hybrids also expressed 40% of total fluorescence at 80°C (Figure 2). As the temperature was decreased, a slow but steady increase in fluorescence was noted, indicating the re-naturation of the hybrids. The melting temperature of these hybrids was between 60-65°C. The optimal annealing temperature for the beacon, which allows a clear discrimination between randomly coiled beacons and beacon-target hybrids, was determined to be between 55-60°C. Initial studies in real-time PCR were conducted between this temperature range and the highest sensitivity was obtained at 57°C. This temperature was chosen as the annealing temperature for subsequent real-time PCR studies.

Real-Time PCR. Ten fold serial dilutions of template DNA were used in the real-time PCR assay. Figure 3 shows the resulting amplification plot. ΔRn is the normalized fluorescence of each sample. As expected, samples that contain a higher number of template molecules, produced higher signals. As few as one CFU was detected with the real-time assay. The critical cycle (Ct), the cycle at which a significant increase in fluorescence is first recorded, increased as the initial number of template molecules DNA decreased. This was expected, as samples containing low concentrations of template DNA would require more PCR cycles in order to replicate enough copies to produce a significant increase in fluorescence. Since the critical cycle is inversely proportional to the logarithm of the initial number of target molecules (7), these data can be used to formulate a standard quantification curve for the detection of *Salmonella*.

Target Specificity.   The specificity of the beacons was evaluated. Real-time PCR assays were conducted with *Salmonella* and *E. coli* as the initial templates. Although a 122 bp PCR fragment was amplified from both the *Escherichia coli* and *Salmonella* templates, only the *Salmonella* template elicited very strong fluorescence. This result clearly demonstrates the ability of molecular beacons to discriminate between very similar sequences. (Figure 4).

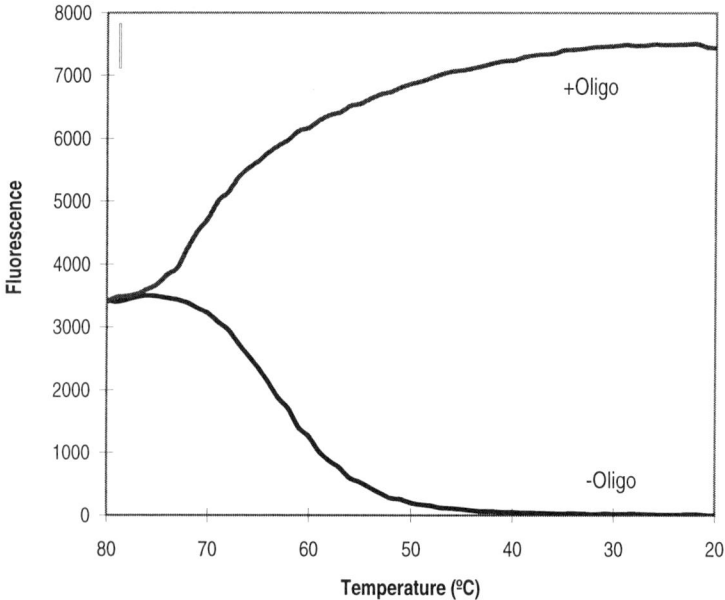

Figure 2. Thermal denaturation profile of molecular beacons.

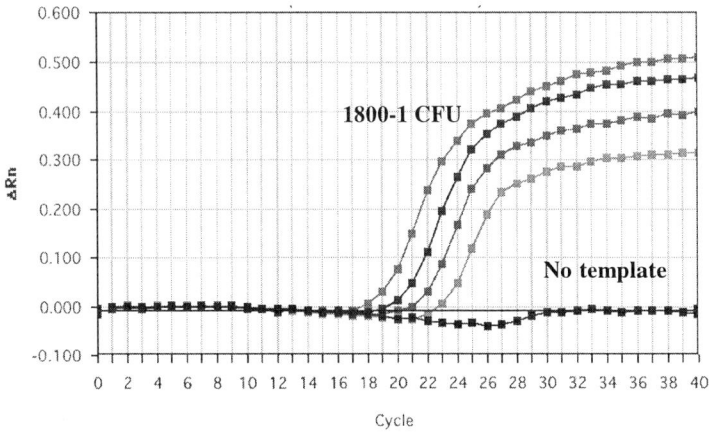

Figure 3. Real-time PCR assay. Initial template concentrations ranging from 1 to 1800 CFU.

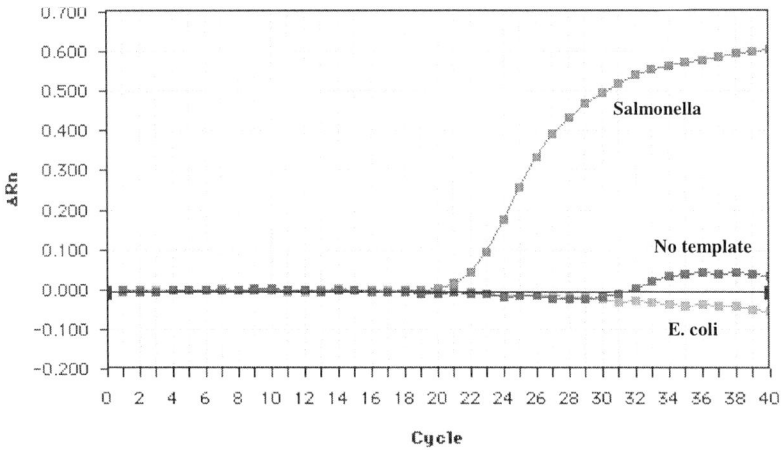

Figure 4. Real-time PCR amplification of *E. coli* and *Salmonella himA* genes. Beacon specific for *Salmonella* failed to detect the *himA* homologue in *Escherichia coli*.

Conclusion

Molecular beacon assays are simple and fast. Reagents are mixed in one step and reactions are carried in closed tubes, thus preventing contamination. Data are recorded during each cycle and results are automatically analyzed immediately after the reaction is completed, usually within 2-3 hours. Due to their high specificity and high sensitivity, molecular beacons can be effectively incorporated into real-time PCR assays and provide a quick and accurate method for detection of specific nucleic acid sequences in homogeneous solutions. We envision that the speed and sensitivity of bacterial pathogen detection based on PCR-assay method can be greatly enhanced with the application of molecular beacons.

Acknowledgments

This research was supported by the Water Environmental Research Foundation. We wish to thank Tammy Chen for her help on DNA sequencing and real-time PCR assays.

References

1. U. S. Food and Drug Administration. **1998**. Foodborne Pathogenic Organisms and Natural Toxins Handbook. Center for Food Safety and Applied Nutrition. http://vm.cfsan.fda.gov/list.html/
2. Moriginio, M. A.; Cornax, R. and Munoz, P.R. *Water Resource* **1990**, 24, 117
3. Feng, P. *J. of Food and Drug Analysis* **1993**, 1, 119.
4. Tyagi, S. and Kramer, F.R. *Nature Biotechnology* **1996**, 14, 303.
5. Tyagi, S.; Bratu, D. P. and Kramer, F. R. Multicolor molecular beacons for allele discrimination. *Nature Biotechnology* **1998**, 16, 49.
6. Bej, A. K.; Mahbubani, M. H.; Boyce, M. J. and Atlas, R. M. *Applied and Environmental Microbiology* **1993**, 60, 368.
7. Heid, C.A.; Stevens, J.; Livak, K.J. and Williams, P.M. *Genome Methods* **1996**, 6, 986.

Chapter 21

DNA-Based Biosensors: A Tool for Environmental Analysis

Michael Mecklenburg[1,2], Bengt Danielsson[2], Henrik Boije[1],
Ioana Surugiu[2], and Brigitta Rees[2]

[1]BT Biomedical Technology AB, Ole Römersväg 12,
S-223 70 Lund, Sweden
(Fax: +46 70 6119732)
[2]Pure and Applied Biochemistry, Lund University,
Box 124, 22 100 Lund, Sweden

The specific aim of this study was to determine the sensitivity and selectivity of our assay for a large number of metals ions in order to determine the applicability of broad range sensing for detecting this class of environmental pollutants. Our previous results, which indicated that the assay was capable of detecting metal ions, have been confirmed in the present study. The metals studied include: Ag, Al, Au, Cd, Ce, Co, Cr, Cu, Fe (II), Fe(III), Ga, Gd, Hg, In, La, Mg, Mn, Mo, Ni, Pt, Rb, Ru, Th, Ti, Y, Yb and Zn. The sensitivity varies from submicromolar for Ga, Ru and Th to essentially undetectable for Cd, Mo and Rb. The metal ion selectivity profile of this assay is to our knowledge unique. Analytical schemes based on broad range recognition, multivariate data analysis and nonlabel detection provide us with the tools required to address complex multicomponent interactions. At present however, the lack of appropriate broad range sensing elements limits the general application of these detection schemes in bioanalysis. In an effort to extend the utility of this approach, we have developed a DNA based broad range recognition scheme. Strategies for generating recognition diversity to create arrays of broad range sensing elements based on this assay are discussed.

Biosensor technology has at its finger tips an nearly unlimited resource, biodiversity. This scarcely tapped resource contains materials for modulating molecular recognition, solubility, stability, and orientation, as well as paradigms for constructing integrated sensing devices far more intricate than any man-made device. The challenge lies in mapping out strategies for mimicking complex biological sensing that will provide a foundation for developing artificial sensors with similar features. What then are these essential features and how can they be incorporated into artificial biological sensing devices?

Biologists have identified three key features including: 1) nonlabel detection, 2) broad range recognition and 3) neural network based data analysis (1). Nonlabel signal detection is achieved by hard wiring the signal transduction mechanism into the recognition element itself, i.e. binding results in conformational changes that set in motion a series of signaling events. In more advanced organisms, this allows the graded response patterns obtained from arrays of broad range recognition with overlapping selectivities to be evaluated using pattern based analysis, i.e. neural networks. This dramatically increases the efficiency by reducing the number of elements required to recognize a given set of molecules. Moreover, the scheme provides a mechanism for identifying 'new' compounds as well as built-in redundancy.

Biomimetic sensing devices based on the 'reverse engineering' of biological sensing paradigms were pioneered by Lundström and coworkers (2). The 'Electronic nose' used the variable chemical reactivity of patterned metallic surfaces to mimic the broad range recognition capability of natural sensing elements and multivariate data analysis to mimic neural networks. Numerous biomimetic schemes have since been developed based on a wide range of detection and recognition strategies (3), and are capable of detecting a variety of substances including solvent mixtures, fermentations, food stuffs, as well as diseases (4,5). Expansion of this strategy into traditional areas such as bioanalysis and diagnostics has been limited by the inability to create appropriate recognition diversity required for studying more complex biological interactions. And more specifically, to match the 'broadness' of the recognition range of a given array with the ability of the multivariate analysis software to evaluate the data set.

An alternative strategy would be to use the innate binding interactions of a given biomolecule as a sensing element. One approach has been to use fluorescent intercalating nucleic acid dyes as probes for studying DNA interactions. This strategy was used to probe nucleic acid interactions (6), as well as for detecting modified DNA (7). Subsequently, intercalator based biosensors were developed by the Weetall group using optical strategies (8), and by the Wang (9)and Palecek (10) groups using electrochemical approaches. The tight fit of the intercalators between the DNA basepairs provides a homogeneous local environment which if disturbed leads to a dramatic change in the characteristics of the intercalator. The optical and electrochemical schemes indirectly detect the interaction of compounds with nucleic acids via changes in the fluorescence of the intercalated dye or in the oxidation potential of the nucleotide bases, respectively. Intercalating fluorescent nucleic acid dyes exhibit greatly enhanced fluorescence upon binding to DNA that eliminates the need to separate unbound dye, thereby allowing a homogeneous assay format.

We have developed a similar optical detection scheme employing the long wavelength intercalating nucleic acid dye ToPro-3 (TP3) (11, 12). A drawback of homogeneous assays is interference from organic compounds that absorb and/or emit light that makes optical measurements below 600 nm very difficult (13). TP-3 has usually long excitation and emission maxima of 642 and 661 nm, respectively which dramatically reduces interference from organic contaminants. In addition, the dye has extremely low fluorescence in the unbound state that greatly decreases the noise which results in a corresponding increase in sensitivity (14).

Previously, we have shown that the assay scheme is capable of detecting known intercalators, a variety of environmental pollutants, organic solvents, heavy metal ions, carcinogens, natural products, as well as discriminate between structurally related acridine compounds (11, 12). Metal ions are known to interact strongly with the negatively charged phosphate backbone of DNA and were chosen early on as targets for investigation due to their importance in environmental analysis (15). On this basis, we expected that the metal ion standard curves would have similar profiles. On the contrary, initial studies indicated that the assay showed a significant degree of selectivity, suggesting that the DNA-TP3 complex was capable of differentiating metals. Here we report the analysis of an expanded series of metals ions using the TP3:DNA assay.

Materials and Methods

The chemicals were purchased from Merck/Schuchardt (Hannover, Germany) and were of *p.a.* grade, unless otherwise stated. The TP3 (a monomeric thiazole orange derivative) was obtained from Molecular Probes (Eugene, Oregon, USA). The metal ions and Salmon sperm DNA were purchased from Sigma Chemical company (St. Louis, Mo, USA). The microfuge tubes and pipette tips were lot tested and were obtained from Sarstedt (Nuerenbrecht, Germany). The glass capillaries KTW120-3 were bought from World Instruments (Sarasota, Fl., USA).

The fiber fluorometer was designed and constructed at the Royal Veterinary and Agricultural University, Copenhagen, Danmark by Dr L. Øgendal and Dr. A. Weber (Figure 2). The instrument has been described in detail elsewhere (11). Briefly, a Siemens (Munich, Germany) 645 nm light emitting laser (LED) diode was used for excitation and two Siemens SFH 450 silicon photodiodes were used to detect the fluorescence (reference/signal). A 605 +/- 35 nm bandpass filter was positioned in front of the laser and a 675 nm long pass filter was positioned in front to the detector. The entire unit can be powered either by a battery pack or an AC-DC power converter. The signal output was read using a standard voltmeter.

The DNA was prepared as follows: 10 mg of Salmon sperm DNA was dissolved in 10 ml Tris EDTA (TE) buffer (1mM EDTA and 10 mM Tris, pH 8) by gently shaking the mixture at room temperature for 4 days. The mixture was sonicated and the solution aliquoted and stored at -20°C. The TP3 stock solution (100 µM,) was aliquoted and stored at -20°C. Stock solutions of the test compounds were prepared in assay buffer.

The assay was performed as follows: the TP3-DNA solution (adequate for testing 8 samples in triplicate) was prepared by mixing in the following order 1163 µl Millipore

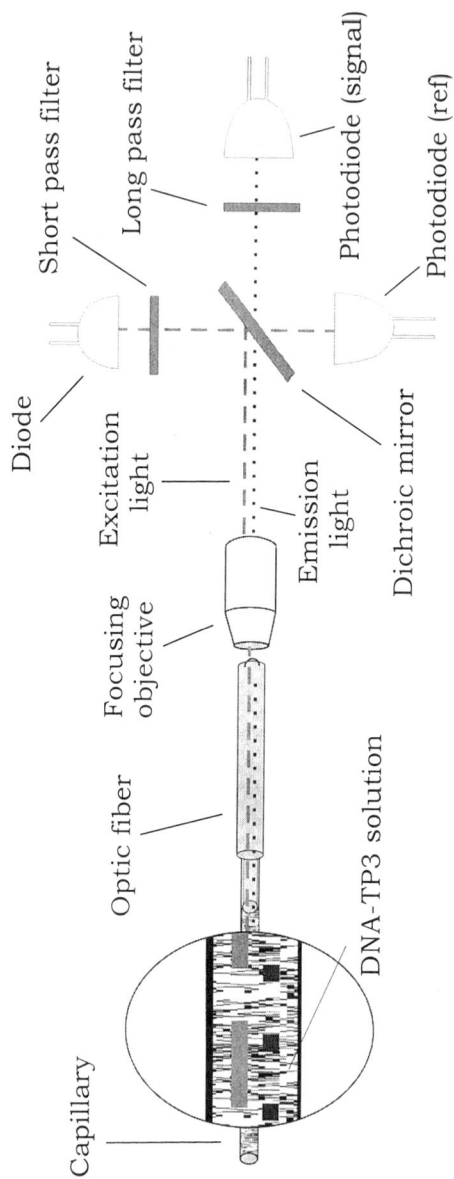

Figure 1. Overview of the capillary fluorometer and assay set-up.

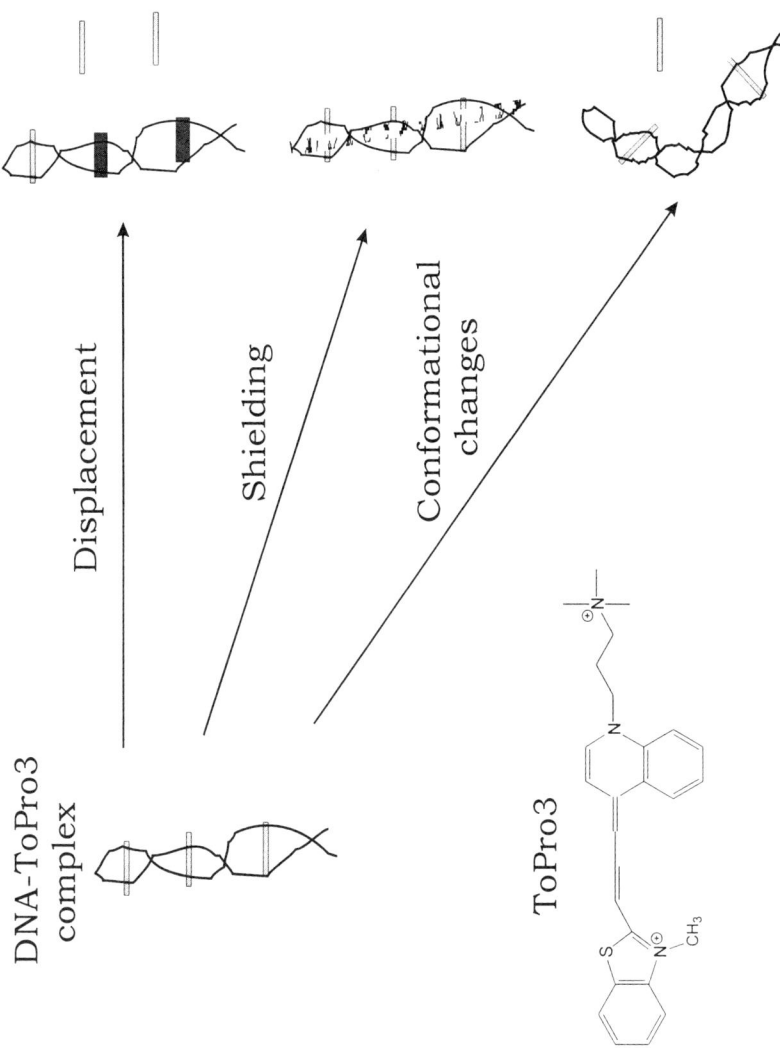

Figure 2. Schematic of the common DNA interaction modes and the chemical structure of fluorescent intercalating dye, TP3.

purified water, 150 µl Tris (1.0 M, pH 7.5), 3 µl EDTA (0.5 M, pH 8), 18 µl TP3 (100 µM) and 15 µl DNA (sheared, 250 µg/ml in TE buffer). The mixture was vortexed prior to and immediately after the addition of DNA. After 1 hr in the dark, 6 µl of the competitors were portioned out into microfuge tubes and then 54 µl of the equilibrated DNA-TP3 solution was added to each sample. After 60 min, the samples were sequentially transferred to capillaries tubes (50 µl) which were then slid onto the end of the fiber optic. All measurements were performed in triplicate.

Results

The studies described here were performed on a home-made fiber optic based fluorometer (Figure 1). The samples are loaded into capillary tubes, slid onto the end of the optic fiber, and the measurement taken. We and others have proposed three types of interactions that can effect the fluorescence of bound intercalators, displacement, shielding and structural modification (Figure 2). In reality, the interaction of a particular molecule with DNA does not rigidly fall into a specific category but rather is a combination of these interactions. Nevertheless exposure of the DNA-TP3 complex to a given test compound results in a highly reproducible shifts in the equilibrium of the complex as measured by changes in the fluorescent signal intensity.

The broad range character of the assay scheme requires numerous controls in order to eliminate artifactual results, such as quenching of the fluorescent signal or degradation of the TP3 or DNA. The routine controls that are performed include: solubility (turbidity), sample coloration, absorption spectra, fluorescent emission using 642 nm excitation and changes in the blue color of the TP3 solution with and without DNA. If any of these controls are positive, then the metal was not tested. Out of the 32 metals that were subjected to this screening procedure only Ba, Bi, Ir, Ta and Tl tested positively and were not analyzed.

Standard curves were prepared for the metals shown in Table 1. Each standard curve was performed three times with each point being assayed in triplicate. The metal ion concentration that resulted in a 50% reduction in fluorescent signal intensity (F_{50}) is given. The assay is capable of detecting Ga, Ru, and Th in the 500 nM range which corresponds to 25 pmoles. (Figure 3). These standard curves are linear over 3 to 4 logs and differ marked in shape from, for example, Ag and Au which show a rather sharp drop in signal intensity over a single log (Figure 4). Surprisingly some metals, such as Mo, Cd, and Rb, had essentially no effect on the TP3-DNA complex at concentration up to 5 mM, despite the fact that it is known that all cationic metal ions interact with the negatively charge phosphate backbone of DNA (Figure 5).

The majority of the metal ion standard curves showed a typical concentration dependent reduction in signal intensity with the exception of Al, Hg and Mg (Figure 6). The Al standard curves slowly decreases but then abruptly rises to then continue decreasing again. This would seem to indicate a concentration dependent shift in Al complexation. The Hg and Mg are the only metals that show an increase in signal

Table1. Overview of metal ions interaction analysis.

Element	Symbol	F_{50} (µM)
Silver	Ag	90
Aluminium	Al	20
Gold	Au	30
Cadmium	Cd	>10000
Cerous	Ce	80
Cobolt	Co	2000
Chromium	Cr	200
Copper	Cu	2000
Iron	Fe (III)	60
Iron	Fe (II)	900
Gallium	Ga	300
Gadolinium	Gd	300
Mercury	Hg	NA
Indium	In	90
Lanthanum	La	300
Magnesium	Mg	NA
Manganese	Mn	>10000
Molybdium	Mo	>10000
Nickel	Ni	600
Platinic	Pt	700
Rubidium	Rb	>10000
Ruthentnium	Ru	30
Thorium	Th	80
Titanium	Ti	>10000
Yttrium	Y	200
Ytterbium	Yb	900
Zinc	Zn	2000

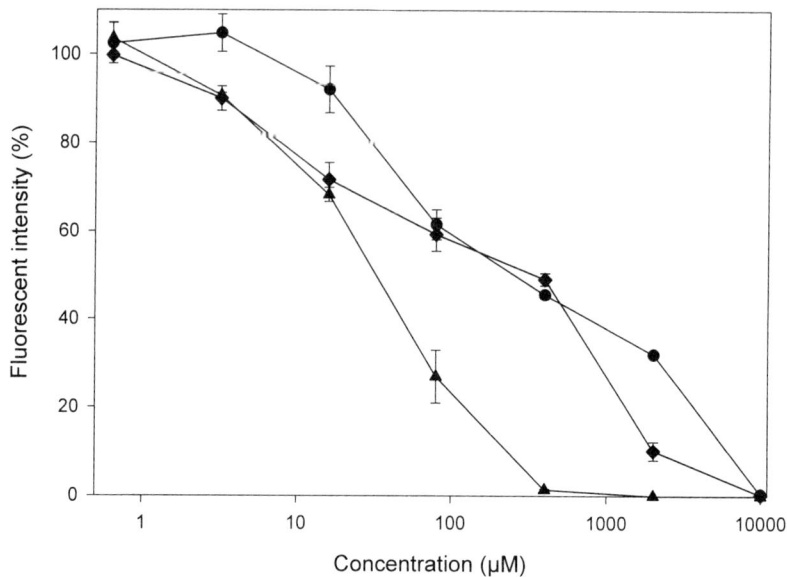

Figure 3. Standard curves for Ga (●), Ru (▲), and Th (◆).

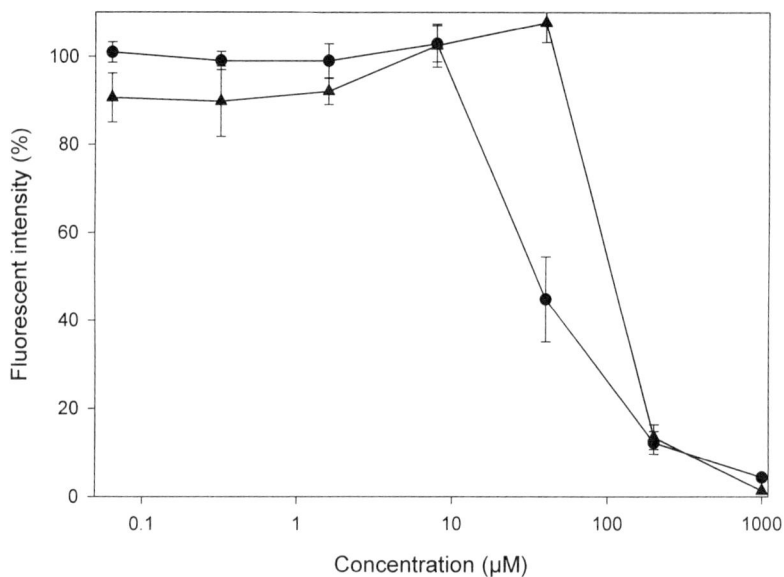

Figure 4. Standard curves for Ag (▲) and Au (●).

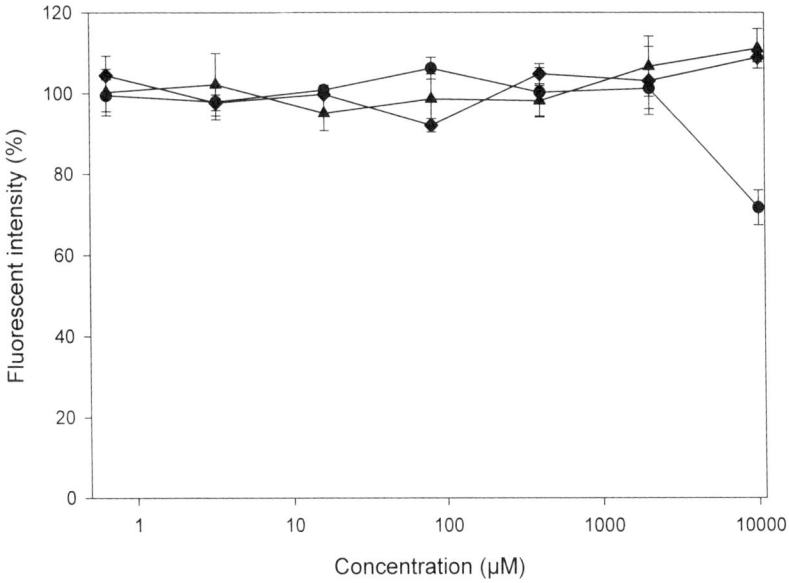

Figure 5. Standard curves for Cd (●), Mo (▲), and Rb (◆).

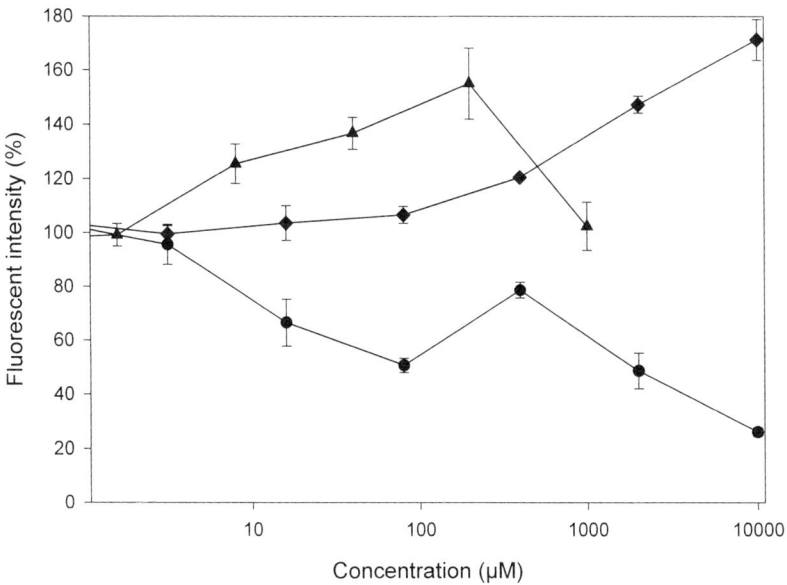

Figure 6. Standard curves for Al (●), Hg (▲) and Mg (◆).

intensity. The Hg standard curve shows a gradual rise beginning at 1 µM but then drops above 150 µM, while the Mg curve is flat until 150 µM after which it increases linearly. The gradual rise observed for Mg is especially interesting given that Mg is known to stabilize DNA. These standard curves have been repeated at least 5 times with essentially identical results.

Electronic configuration effects were investigated by assaying Fe (II) and Fe (III) (Figure 7). The 17-fold increase in sensitivity of Fe(III) over Fe(II) indicated that charge density plays an important role in these interactions. A comparative analysis of the metals was carried out. The charge density was calculated using the empirically determined atomic radii and plotted as a function of the fluorescent signal intensity obtained at 80 µM (Figure 8). Not surprisingly, a rough correlation between metal ion charge density and signal intensity was observed. In addition, there seems to be a clear division between the strong and weak binding metals.

Discussion

The specific aim of this study was to determine the sensitivity and selectivity of our assay for an expanded set of metals ions in order to determine the applicability of broad range sensing for detecting this class of environmental pollutants. A secondary aim was to ascertain if the assay could be used as a general approach for classifying nucleic acid - metal ion interactions. Our previous results, which indicated that the assay was capable of detecting metal ions, have been confirmed in the present study. The metal ion selectivity profile of this assay is to our knowledge unique.

First some general observations. A breakpoint in the standard curves often times occurs in the 140 µM region. This is significant since the assay solution contains the equivalent of 140 µM of basepairs and thus represents the saturation point of the phosphates in the backbone of the DNA helix. This can be most clearly seen in the Ag, Au, Hg, Mg and Ru standard curves. This is also the region in which the dip in the Al standard curve occurs. However, other metals such as Fe (II), Ga and Th do not follow this rule. While the reason for these effects is unclear, it provides qualitative information that can be useful in the initially classification of these compounds.

The highest sensitivity, submicromolar, was obtained for Ru and Th. The next most sensitive group of metals was Ga, Ag and Au. The strong interaction of Au is somewhat surprising given the fact that Au is so widely used in, for example dentistry, due to its high biocompatibility. Interestingly, Au complexes are also used in the treatment of a variety of diseases including arthritis (16). The mode of action of Au in arthritis therapy remains unclear (17). If Au does interact strongly with DNA as our results indicate, this will have major implications for the development of Au based drug design and treatment regimes.

Of the more than 500 compounds tested in this assay to date, only Hg and Mg have resulted in a rise in the fluorescent signal intensity. Magnesium is used ubiquitously by enzymes involved in the synthesis and modification of nucleic acids, and is known to stabilize DNA duplexes. Considering the fact that Mg is one of the few metals that

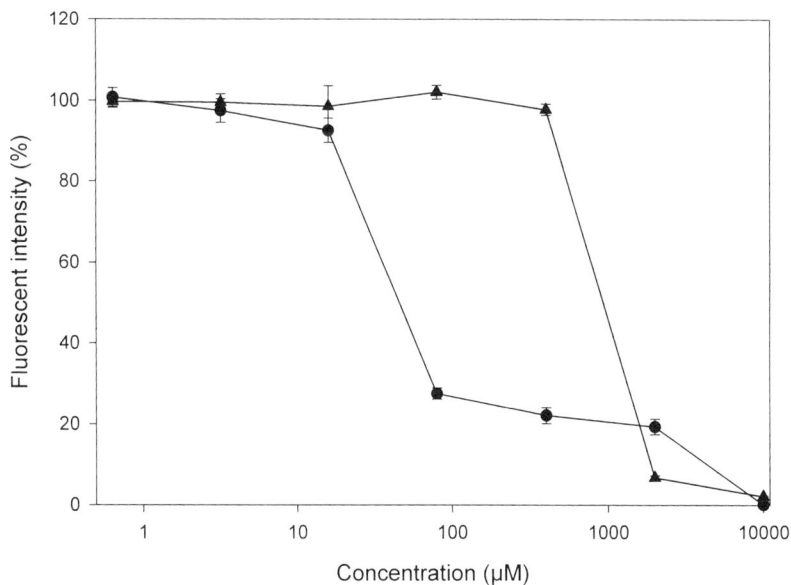

Figure 7. Standard curves for Fe(II) (▲) and Fe(III) (●).

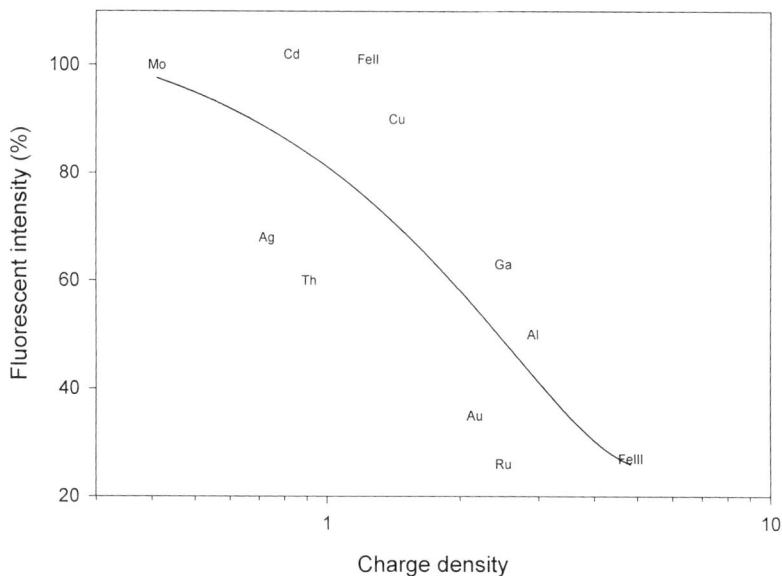

Figure 8. A representative selection of metals are plotted charge density vs signal intensity at 80 µM metal ion concentration.

stabilizes DNA, it is not surprising that Mg has a distinctive standard curve. Moreover, it supports the notion that the assay is indeed capable of detecting biological relevant interactions. Furthermore, the point at which the Mg curve begins its linear increase is near the 140 μM saturation point. Interestingly, the 140 μM break point is also near where the Hg curve begins its fall. Hg is one of the sulfhydryl-reactive metals that can effect a variety of biochemical and nutritional processes through their pro-oxidative effects (18). DNA is not implicated as a primary target in Hg toxicity. The simplest interpretation of our results would be that Hg, like Mg, stabilizes DNA

The environmentally important heavy metal Cd was undetectable in this assay, as were Mo and Rb, despite the fact that all cationic metal ions are known to interact with nucleic acids. Why then does this assay detect only a discrete subset of these interactions? The simplest explanation is that only specific structural changes in the DNA helix effect the affinity of the dye. Attempts to compare our results with the vast literature in this area has been complicated by the fact that the experiments are rarely carried out under the same conditions. We and others have shown that the stability of DNA metal ion complexes are extremely sensitive to minor changes in the assay condition, as well as contaminants which further complicates comparative analysis.

The simplicity and speed with which our assay can be performed makes it possible for us to analyze numerous compounds under essentially identical conditions. Access to large data sets makes it possible to carry out comparative studies (Figure 8). The correlation between charge density and binding affinity, while not surprising, does show the potential of the approach. Closer inspection of the data shows that the metals fall into two groups, low to medium binders (Mo, Cd, FeII, Cu Ga, Al and FeIII) and the high affinity binders (Ag, Th, Au and Ru).

Clearly, this assay is considerably less sensitive and selective than standard analytical techniques used to quantitate these metals. However, the assay is not designed to compete with established analytical chemistry but rather to reduce the number of parameters to be tested and to identify 'points of interest', which can then be scrutinized using traditional analytical methodology. Informationally rich assays, such as the one described here, are needed to provide the large, consistent, unbiased data sets required to prepare compound profiles or fingerprints. This could be used to initially characterize and possibly group the thousands of new compounds developed each year by simply comparing their interaction profiles with those of other known compounds. Similarly, metal profiles could be made for disease states, single compounds, bioactivity and bioavailability, complex mixtures metabolic and reduction break down products.

Expansion of the recognition diversity to create arrays of sensing elements would be essential. A number of alternatives are under investigation including the use of specific DNA sequences, addition of metal ions to the DNA-TP3 complex, and the use of RNA, i.e. RNA-TP3. These DNA sensing element arrays will be used in conjunction with other sensor arrays to extend the range and redundancy of these artificial sensing devices. To this end, we are currently working to develop a variety of broad range recognition strategies capable of sensing many classes of biomolecules.

The development of integrated arrays of broad range recognition elements (in combination with multivariate analysis) is an important stepping stone on the path towards creating artificial sensing devices.

Fabrication and multivariate data analysis schemes exist. Recognition strategies, on the other hand, are very limited despite considerable effort (19). Recent advancements are encouraging (20), but greater effort needs to be directed towards developing schemes for creating recognition diversity. Biosensor developers can assist in this process by providing strategies for mimicking biological sensing schemes for fabricating sensing devices and as tools to further our understanding of biological sensors.

Acknowledgements

We thank A. Weber for use of the fiber optic instrument. Funding for this project was obtained from NUTEK and TRF.

References

1. Murphy, C. *Olfaction and Taste XII: An International Symposium*; Annals of the New York Academy of Sciences; New York Academy of Sciences: New York, NY, 1999, Vol. 855.
2. Lundström, I; Erlandsson, R.; Fryleman, U.; Hedborg, E.; Spetx, A.; Sundgren, H.; Welin, S.; Winquist, F. *Nature*, **1991**, 352, 47.

3. Dickinson, T.; White, J.; Kauer, J.; Walt, D. *Nature*,**1996**, 382, 697.

4. Gopel, W.;Ziegler, C.; Breer, H.; Schild, D.; Apfelbach, R.; Joerges, J.; Malaka, R. *Biosens Bioelectron*, **1998**, 13, 479.
5. Ziegler, C.; Gopel, W.; Hammerle, H.; Hatt, H.; Jung, G.; Laxhuber, L.; Schmidt, H. L.; Schutz, S.; Vogtle, F.; Zell A. *Biosen Bioelectron* **1998**, 13, 539.
6. McCoubrey, A,; Latham, H.C.; Cook, P.R.; Rodger, A.; Lowe, G. *FEBS Letters*, **1996**, 380, 73.
7. Frey, T. *Cytometry*, **1994**, 7, 310.
8. Pandey, P.C.; Weetall, H.H. *Anal. Chem.* **1995**, 67, 787.
9. Wang, J.; Chicharro, M.; Rivas, G.; Cai, X.; Dontha, N.; Farias,P.; Shiraishi, H. *Anal. Chem.* **1996**, 68, 2251.
10. Palecek, E. *Electroanalysis* **1996**, 8, 7.
11. Mecklenburg, M.; Grauers, A.; Rees-Jonsson, B.; Weber, A.; Danielsson, B. *Anal. Chem Acta* **1997**, 347, 79.

12. Mecklenburg, M.; Danilsson, B. In *Biosensors for Direct Monitoring of Environmental Pollutants in Field*; Nikolelis, D.P., et al., Ed.; Kluwer Academic Publishers: Amsterdam, Netherlands, 1998; 87-95.

13. Song, R.; Li, K-P. *Applied Spectroscopy* **1993**, 47, 1604.

14. Milanovich, M.; Suh, M.; Janowiak, R.; Small, G.J.; Hayes, J.M. *J. of Phys. Chem.* **1996**, 100, 9181.

15. Sigel, A.; Sigel, H. *Probing of Nucleic acids by metal ion complexes of small molecules* in Metal ions in Biological Systems; Marcel Dekker: New York, NY, 1996; Vol. 33.

16. Danning, C.L.; Boumpas, D.T. *Clin. Exp. Rheumatol.* **1998**, 16, 595.

17. Merchant, B. *Biologicals* **1998**, 26, 49.

18. Nickle, R. *Drug Chem Toxicol,* **1999**, 22, 129.

19. Borkholder, D. A.; DeBusschere, B. D.; Kovacs, G. T. *Proceedings of the Solid-State Sensor and Actuator Workshop*, Hilton Head, SC, June 8 - 11, 1998, pp. 178 - 182.

20. Shi, H.; Tsai, W.B.; Garrison, M.D.; Ferrari, S.; Ratner, B.D. *Nature*, 1999, 398, 593-7.

INDEXES

Author Index

Subject Index

A

Acetone, experimental gas chromatography–electronic nose (GC–EN) data, 51*t*

Acoustic wave devices
signal origins, 286
thickness shear mode (TSM) biosensor for detection and study of DNA hybridization, 276–277
transduction and peptide nucleic acids (PNA) for biorecognition, 277

6-Acryloyl-2-dimethylaminonaphthalene (acrylodan). *See* Genetically engineered periplasmic binding proteins

Affinity biosensors
artificial immunosensors for detection of metal ions, 218, 220
bioassay definition, 212
characterization of sensing elements, 212
concentration for polychlorinated biphenyl (PCB) antigen, 214
conjugates in enzyme-linked immunosorbent assay (ELISA) format for determining metal ions, 218, 220
design of artificial interfaces with predefined functions, 209
design of templates for identification and detection of endocrine disrupting chemicals (EDCs), 210*f*
differentiating binding of interfering compounds, 214–215
electrochemical immobilization, 211
ELISA checkerboard titration of PCB coating antigen and PCB antibody, 213*f*
experimental instrumentation, 208–209
ideal screening system for EDCs, 208
PCB analogs, 209, 211
polymer synthesis using various pyrrole derivatives, 211–212
principle of metal detection using compounds of 2-pyridylazo (PAR)–

bioconjugate, 219*f*
production of templates and antibody, 209, 211
relative potency for PCB and structurally related compounds, 216*f*
relative potency of analyte and cross reactant, 214
relative potency of two EDCs, 215
response determining biological activity of analytes, 212
responses for cyanazine-hapten monolayer electrode to different antibody concentrations, 217*f*
sensor preparation and characterization, 211–212
separating components based on chemical structure, 208
testing synergistic effects of structurally similar analogs of PCBs under different concentrations of target compound, 214
triazine analogs and antibody, 211
triazine sensors, 215
two-site assay showing relevance of charge-to-size ratios in metal detection, 221*f*
UV/vis spectrometry showing metal ions binding to PAR conjugated to proteins, 219*f*

Aflatoxin B1
detection limits, 179, 181
stress fingerprints by treatment with aflatoxin B1 with and without metabolic activation, 180*f*
stress fingerprints with and without metabolic activation, 178–179

Air pollutants
environmental monitoring demands, 2
sulfur dioxide, 10
See also Sulfur-containing species

Alginic acid
coenzyme activity of Alg–$NADP^+$, 160–161

317